Lecture Notes in Computer Science 15471

Founding Editors

Gerhard Goos
Juris Hartmanis

Editorial Board Members

The series Lecture Notes in Computer Science (LNCS), including its subseries Lecture Notes in Artificial Intelligence (LNAI) and Lecture Notes in Bioinformatics (LNBI), has established itself as a medium for the publication of new developments in computer science and information technology research, teaching, and education.

LNCS enjoys close cooperation with the computer science R & D community, the series counts many renowned academics among its volume editors and paper authors, and collaborates with prestigious societies. Its mission is to serve this international community by providing an invaluable service, mainly focused on the publication of conference and workshop proceedings and postproceedings. LNCS commenced publication in 1973.

Martin Fränzle · Jürgen Niehaus · Bernd Westphal
Editors

Engineering Safe and Trustworthy Cyber Physical Systems

Essays Dedicated to Werner Damm
on the Occasion of His 71st Birthday

 Springer

Editors
Martin Fränzle
Universität Oldenburg
Oldenburg, Germany

Jürgen Niehaus
SafeTRANS
Oldenburg, Germany

Bernd Westphal
German Aerospace Center
Oldenburg, Germany

ISSN 0302-9743   ISSN 1611-3349 (electronic)
Lecture Notes in Computer Science
ISBN 978-3-031-97536-3   ISBN 978-3-031-97537-0 (eBook)
https://doi.org/10.1007/978-3-031-97537-0

This Springer imprint is published by the registered company Springer Nature Switzerland AG
The registered company address is: Gewerbestrasse 11, 6330 Cham, Switzerland

If disposing of this product, please recycle the paper.

Preface

We are happy to present this Festschrift volume, edited for the occasion of the 71st birthday of our dear colleague Werner Damm. It contains 18 contributions written by 55 authors, all of them colleagues with whom Werner has cooperated intensely in the vast variety of contexts his academic career has covered.

Werner is known to all of us as a passionate scientist, an influential research manager, an entrepreneur, and an academic teacher and mentor. In different forms, this Festschrift strives to touch on all these perspectives, trying to match Werner's exceptionally broad interests: it provides industrial reports as well as original research covering a variety of topics in computer science and cyber-physical systems, and it gives the word to several of Werner's academic descendants. The contributions exemplify the breadth of Werner's academic interests and show the appreciation he receives in the scientific community. As part of our work on the volume, we organised a review process that involved a program committee consisting of all contributing authors.

Werner had and has a mission, namely to render digitally controlled systems safe and societally acceptable and beneficial. For this, he has diligently worked on developing formal methods of computer science in general, and automatic verification, analysis, and testing of embedded and cyber-physical systems in particular, from an academic curiosity into an industrial reality. However, he has always embedded these technical contributions into a much broader context, acknowledging that technical reliability, while being a cornerstone of trustworthy digitalization, addresses the issues of societal acceptability and responsibility only fragmentarily, and he has consequently actively promoted bridging computer science with the humanities.

We are glad to know that Werner's scientific activities do not end yet. Werner still is very active in various contexts ranging from industry cooperations and joint research initiatives with industry over political and academic consultancy to basic and applied research — and he is always open to fresh research challenges.

We thank Werner for inspiring all of us and congratulate him on the occasion of his 71st birthday!

November 2023

Martin Fränzle
Jürgen Niehaus
Bernd Westphal

Organization

Program Committee

Martin Fränzle	Carl von Ossietzky Universität Oldenburg
Jürgen Niehaus	SafeTRANS
Bernd Westphal	DLR Systems Engineering for Future Mobility Institute

Additional Reviewers

Rasmus Adler	Hardi Hungar
Albert Benveniste	Inigo Incer
Tom Bienmüller	Franz Korf
Jens Braband	Sebastian Lehnhoff
Manfred Broy	Dejan Nickovic
Felix Brüning	Jens Oehlerking
Bernd Finkbeiner	Ernst-Ruediger Olderog
Radu Grosu	Jan Peleska
Willem Hagemann	Alexander Pretschner
Axel Hahn	Astrid Rakow
Paul Hannibal	Jan Reich
Thomas Henzinger	Daniel Schneider
Holger Hermanns	Viorica Sofronie-Stokkermans
Wen-Ling Huang	Ingo Stierand

Contents

BTC Embedded Systems – Bringing Formal Methods from Norddeutschland to the World

Tom Bienmüller[(✉)], Matthias Büker, Udo Brockmeyer, Jürgen Bohn, Hans J. Holberg, Tobe Toben, and Hartmut Wittke

BTC Embedded Systems AG, Oldenburg, Germany
{tom.bienmuller,matthias.buker,udo.brockmeyer,jurgen.bohn,
hans.holberg,tobe.toben,hartmut.wittke}@btc-embedded.com
http://www.btc-embedded.com

Abstract. In 1998 a visionary dreamt a dream of turning academic formal methods into production-ready solutions and off-the-shelf tools applicable for industry. Equipped with this mission, together with some early birds of his university department, he approached this business as part of OSC GmbH, which was founded in 1999. This paper tells the history and the evolution of this quest, and describes the fields in which its products and solutions based on formal methods are integrated in today's industrial applications.

1 Introduction

OSC GmbH[1] was founded in 1999 in Oldenburg, Germany, as a collection of three spin-offs of the OFFIS[2] research institute (namely "Embedded Systems", "Information Management" and "Low Power Design"). Based on academic work previously done at OFFIS, OSC GmbH – for its formal methods part "Embedded Systems" – approached the gap between academic research and industrial use by focusing on abstract system-level designs specified using STATEMATE statecharts and their formal verification including automatic test vector generation (ATG). Additionally, for the UML-tool Rhapsody, test-case specification (and later also ATG) was promoted. With initially mixed success, in particular with respect to the STATEMATE add-ons – not only because of lacking experience in building off-the-shelf products but also due to the decreasing relevance of STATEMATE with a continuously reducing market size – this set the foundation of the first own product "EmbeddedValidator". This tool was integrated with the code generator TargetLink from dSPACE and was embedded into the MATLAB Simulink/Stateflow world[3]. Timely correlated to this episode, OSC Embedded

---

[1] the acronym "OSC" stands for "OFFIS Systems and Consulting".

[2] www.offis.de.

[3] for all tools and add-ons developed for Rhapsody-UML, any marketing and sales activities have been done by the owner of the Rhapsody tools (from I-Logix, to Telelogic, to IBM).

M. Fränzle et al. (Eds.): Werner Damm Festschrift, LNCS 15471, pp. 1–8, 2026.
https://doi.org/10.1007/978-3-031-97537-0_1

Systems (OSC-ES) was extracted as an independent company out of the OSC GmbH federation.

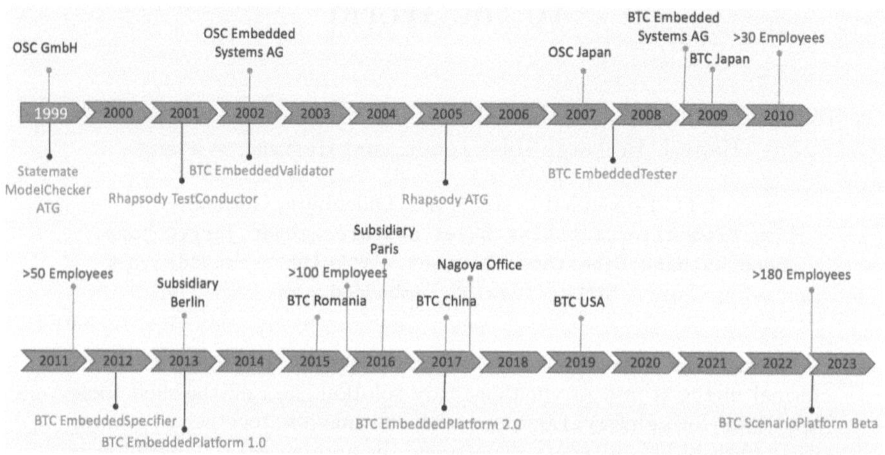

Fig. 1. Historical timeline of BTC Embedded Systems AG from 1999 to 2023

Raise and Fall and Raise

Gaining experience with EmbeddedValidator and getting some reputation in the field of verification of automatically generated code for Simulink/Stateflow led to a joined project together with MAN Germany, and the Japanese companies NISSAN and HITACHI. This project aimed at automating the review process for proving correct translation of a Simulink/Stateflow design to C-code by applying back-to-back testing together with structural automatic test case generation. On the technical side, this project led to the product "EmbeddedTester", which gained very good market acceptance and was essential for setting up a healthy company in the years thereafter. Anyway, before that healthy state, the economics of OSC-ES were not that well, leading to close insolvency of OSC-ES in 2007, and, subsequently, to the integration of OSC-ES into the BTC Group[4], headed by BTC AG. As a consequence, OSC Embedded Systems was renamed to BTC Embedded Systems (BTC-ES), coming along with a 30% reduction of the team size to "18+1", where "+1" stands for one member of the Japanese branch. Even though dramatic, this event led to, first, a recovery which lasts until today – between 2008 and 2023 BTC-ES is a healthy company with very positive contributions to the overall BTC Group results. Second, the company experienced the power of focus as motivated in [5], which, retrospectively, was also key to success.

Closely aligned with the advent of the automotive functional safety standard ISO 26262 [2] new tools for (semi-)formal specification were released by

---

[4] the acronym "BTC" stands for "Business Technology Consulting".

BTC Embedded Systems in 2012, followed by the introduction of today's major product "EmbeddedPlatform" in 2013, which combines different use-cases of formal methods into one single tool. Today, this tool has over 2.000 active licenses world-wide and is used by over 200 customers.

Since 2017, BTC-ES performs a project together with and for MAN Truck & Bus with the goal of introducing and applying (virtual) scenario-based testing as essential ingredient to homologate an SAE-L4 autonomous truck for german highways. Over 70 people were working in 2022 in this project to achieve the goals, half of them employed at BTC-ES, the other half working subcontracted from BTC-ES at partners from the field of artificial intelligence, statistics and law. The project is expected to reach final results in the second half of the 2020 s. From this project, similar to the evolution for the off-the-shelf product EmbeddedTester, another product "ScenarioPlatform" for scenario-based testing highly automated vehicles reached beta-state beginning of 2023.

From Norddeutschland to the World

Obviously, over 25 million Euros revenues in 2022 were not possible without different branches and subsidiaries of BTC-ES. Today, BTC-ES has sales and support branches and companies in Japan, China, France and the US. Product development and support is assisted by several teams employed at BTC Embedded Systems Romania, another daughter company of BTC Embedded Systems. Currently, BTC-ES employs over 180 people in total who are working either in product, customer-specific solution or innovation development, BTC-ES-tool application for customers, marketing, sales or product support. Anyway, from the beginning to today, the Oldenburg headquarter is in the center of new innovations and evolutions for the whole company, and the majority of the team members reside here. With increasing size, new headquarter locations were needed over the years (see Fig. 2).

(a) Industriestraße 1999-2008

(b) Buschstraße 2009-2011

(c) An der Schmiede 2011-2013

(d) Gerhard-Stalling Straße 2014-today

Fig. 2. BTC-ES headquarter locations in Oldenburg

The following sections are taken from publication [10] which was originally released at the Formal Methods Symposium in 2021. It describes the application areas of formal methods used for verification and testing as implemented in the aforementioned tools of BTC Embedded Systems for the past two decades. Some enhancements to [10] have been added where appropriate, in particular, on the topic of virtual validation of autonomous vehicles.

In general, the major challenge for the company was to demystify complex and very mathematical techniques and make them available to non-experts in the field of formal methods [1]. Over the years we learned, on the one hand, that such "magic rocket-science technology" with all its countless applications actually attracts large interest in industry. On the other hand, a loose coupling of rocket-science academic tools for expert users was not the key to our success. Moreover, to convince customers to use formal methods in their projects, it is equally important to design an easy-to-use tool for non-expert users that focus on solving the problem of the customer in a "push-button" manner. Thus, one of our top priorities is to make formal methods available in the most user-friendly way.

To keep pace with the recent academic trends, we maintain a strong academic network, e.g. with the universities of Oldenburg, Freiburg, and Oxford, and participate in large international research projects[5]. In doing so, the focus of our industrial research is always driven by finding solutions for the problems of our customers: for us, research is not an end in itself but must always be a means to an end. With this mission statement in mind, we were able to contribute to the success of our customers during the last two decades.[6] The maturity and effectiveness of BTC-ES verification solutions compared to academic tools are also recognized by the academic world [12]. In [9], we give some more insights into our customer-centric development process for the application of formal methods.

2 Formal Methods in BTC-ES Products

In this section, we present the application of formal methods in our test and verification tools: the specification languages, the backend solving tools, and, finally, the use cases based upon formal methods – with ⓘ referring to Fig. 3.

Specification Languages. By our long-standing cooperation with our customers, we designed an intuitive ① graphical specification language for the formalization of natural-language requirements based on a temporal trigger-action relationship called *universal pattern* (UP) [11], which is easy to use for non-expert users. Scenario-based virtual testing will play a key role in the validation of autonomous driving systems, for which we developed an intuitive ② *graphical traffic scenario language* [6]. Figure 4 shows examples for both.

---

[5] More details can be found at https://www.btc-embedded.com/research/.

[6] Some selected success stories of our customers can be found at https://www.btc-embedded.com/references/.

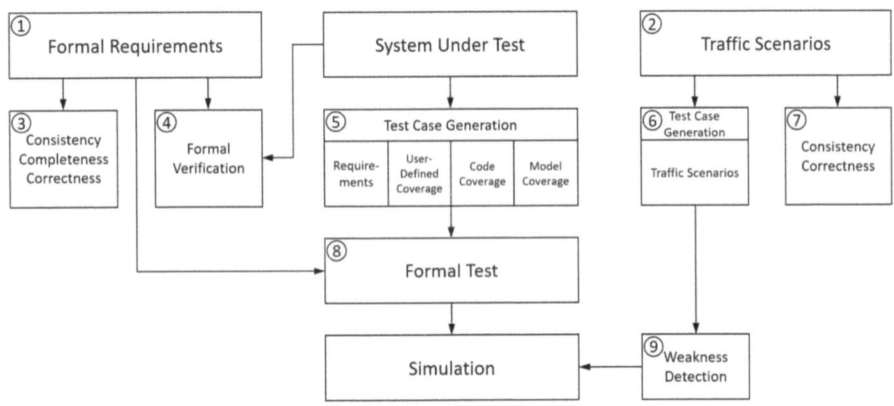

Fig. 3. Fields of application of formal methods within BTC-ES products

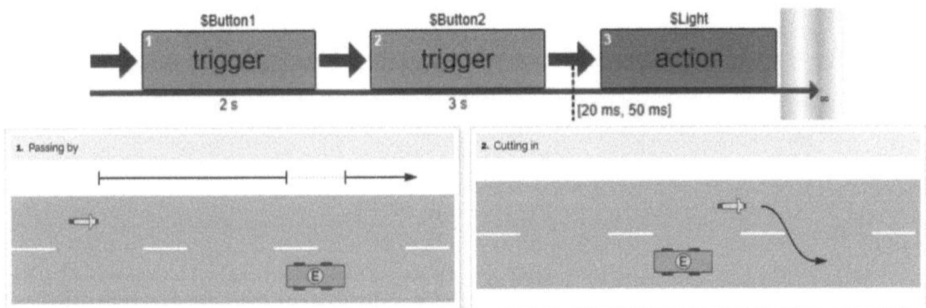

Fig. 4. A UP describing that if button 1 is pressed for 2 s and thereafter button 2 is pressed for 3 s then the light is swichted on within 20 ms to 50 ms (top), and a traffic scenario specifying a passing-by maneuver followed by a cut-in (bottom).

Backend Solving Tools. Most of the analysis tasks shown in Fig. 3 can be reduced to the *state reachability problem.* In our products, we employ a diverse set of state-of-the-art *model checking* tools like iSAT3 [7], CBMC [8], and some other BDD and SAT-based solvers – in particular, to deal with *floating-point* code efficiently. The progress of the SMT community is also of special interest for us, in particular whether promising SMT solvers can be integrated into our products in the future [4]. We further employ *non-exhaustive analysis methods* based on random sampling and mathematical optimization like genetic algorithms or swarm intelligence. Moreover, for some special adaptive control problem, as explained later on, we use techniques from control theory like *model predictive control.*

Fields of Applications. The most prominent use case of our customers is ⑤ *automatic test case generation* for the system under test (SUT), in particular to achieve a high level of code and model coverage. But also test cases derived from the requirements or from user-defined coverage goals are of increasing interest. For that use case, random sampling and model checking are currently the most successful techniques, though recent experiments suggest that also optimization algorithms could be a very promising complement.

The classical ④ *formal verification* problem, i.e. whether the model or code satisfies the formal specification, is handled by the aforementioned model checkers. Due to the well-known state-space explosion problem of model checking that might occur at customer site, our tool suite also provides a more scalable though incomplete simulation-based testing approach, called ⑧ *formal test*: executable monitors are automatically derived from the formal requirements that judge the SUT when simulated using manual or automatically derived test cases.

In a very early design stage, namely before system development starts, it is of utmost importance to ensure a high quality of the system requirements. Our products support that effort by automatic ③ *consistency*, i.e. whether no contradiction exists within the set of requirements, *completeness*, i.e. whether no requirement was missed, and *correctness* checks, i.e. whether the requirements actually express the intended behavior [11]. The algorithmic core of this requirement analysis use case is based on model checking technology.

For the use case of scenario-based testing, we follow a similar approach. Specified traffic scenarios can be checked for ⑦ *consistency*, i.e. whether the abstract traffic scenario is concretizable, and *correctness*, i.e. whether the scenario specification expresses the intended traffic situation. For that purpose, we implemented a special scenario-concretization solver derived from iSAT3 [6].

To actually test the autonomous system within the simulation using, e.g., formal test, ⑥ *test case generation* from traffic scenarios calls for a special treatment: the autonomous and complex behavior of the SUT can hardly be predicted a priori since any howsoever small detail in the concrete virtual simulation environment like slope of the road or conditions of illumination potentially influences internal decision making of the SUT. Therefore, the actual behavior of the surrounding traffic needs to be adapted reactively during simulation according to the behavior of the SUT, in order to successfully execute the scenario specification. For solving that complex control task, we combined techniques from control theory like model predictive control and from model checking.

One of the most challenging problems in scenario-based testing is to find critical edge cases of an abstract traffic scenario, i.e. preferably the most critical concrete scenario execution matching it. Given some criticality metrics like time-to-collision, we conduct ⑨ *weakness detection*, realized by optimization algorithms, to find the most critical concrete scenario instance within the space of variation parameters (like velocities or distances) of the abstract traffic scenario.

All these techniques are embedded into a scenario-based testing framework for the virtual homologation of highly automated driving functions that is compliant to the new ISO standard SOTIF (Safety of the Intended Functionality) [3]. This framework enables to automatically derive test scenarios that cover the Operational Design Domain (ODD) and to automatically judge whether the SUT behaves reasonably safe, conform to its specification (intended behavior) and to regulatory rules and requirements. Reasonably safe in this context means that the safety requirements and the acceptance criteria defined during the SOTIF development process are satisfied. An example for an acceptance criterion is to show that the SUT performs better (to a certain amount) than a skilled and attentive human driver in the same scenario. Finally, all relevant aspects are combined into a verdict function to provide an automatic overall assessment of the validation results.

3 Conclusion

Formal methods play an essential role in what we do: from specification languages with clear formal semantics that allow e.g. the precise translation into observer code, via model checkers and solvers that allow finding even the most obscure corner cases that would be overlooked by humans and that are unlikely to be detected by mere random guessing, to mixtures with approximative methods like our formal test that combines the scalability of a simulation-based method with the precision of a formal requirements specification. The key to success from our experience is that these methods must be embedded into and surrounded by intuitive tools and processes that are easy to use and understand. This was and is vital for BTC Embedded Systems for being successful in the market. Formalisms must be precise without being unnecessarily complex, tools must be powerful without being overly hard to use, and results must be translated back to the problem space of the user such that they can be understood and assessed correctly depending on their context. No easy task, but certainly one remaining interesting and challenging for decades to come.

Acknowledgments.. We would like to thank all our colleagues for their various contributions to the tools which we briefly discussed in this paper. Additionally, we thank the authors of [10] for their great input. Last not least we thank Prof. Dr. Werner Damm for his deep expertise in formal methods, being our supervisor during studies and PhD times, and, in particular, for his entrepreneurship regarding BTC Embedded Systems, of which he is still a supervisory board member. Without him this paper and also the success of BTC Embedded Systems would not have been possible.

References

1. Bienmüller, T., Damm, W., Wittke, H.: The STATEMATE verification environment - making it real. In: 12th International Conference on Computer Aided Verification LNCS, vol. 1855, pp. 561–567. Springer (2000)
2. ISO: road vehicles – functional safety (2018)
3. ISO: road vehicles – safety of the intended functionality (2022)
4. Mentel, L., Scheibler, K., Winterer, F., Becker, B., Teige, T.: Benchmarking SMT solvers on automotive code. In: MBMV (2021)
5. Moore, G.: Crossing the chasm: marketing and selling disruptive products to mainstream customers. Collins Business Essentials, HarperCollins (2002). https:// books.google.de/books?id=yJXHUDSaJgsC

6. Scheibler, K., Eggers, A., Teige, T., Walz, M., Bienmüller, T., Brockmeyer, U.: Solving constraint systems from traffic scenarios for the validation of autonomous driving. In: SC^2 (2019)
7. Scheibler, K., et al.: Accurate ICP-based floating-point reasoning. In: Formal Methods in Computer-Aided Design (FMCAD) (2016)
8. Schrammel, P., Kroening, D., Brain, M., Martins, R., Teige, T., Bienmüller, T.: Incremental bounded model checking for embedded software. Formal Aspects Comput. **29**(5), 911–931 (2017). https://doi.org/10.1007/s00165-017-0419-1
9. Teige, T.: The power of focus: how to optimize a model checker for embedded software (2020). https://www.btc-es.de/en/blog/the-power-of-focus-how-to-optimize-a-model-checker-for-embedded-software.html
10. Teige, T., et al.: Two decades of formal methods in industrial products at btc embedded systems. In: Huisman, M., Păsăreanu, C., Zhan, N. (eds.) FM 2021. LNCS, vol. 13047, pp. 725–729. Springer, Cham (2021). https://doi.org/10.1007/978-3-030-90870-6_40
11. Teige, T., Meincke, W., Brockmeyer, U.: Applying automated formal CCC checks for complex systems development. In: Embedded World Conference (2021)
12. Westhofen, L., Berger, P., Katoen, J.-P.: Benchmarking Software Model Checkers on Automotive Code. In: Lee, R., Jha, S., Mavridou, A., Giannakopoulou, D. (eds.) NFM 2020. LNCS, vol. 12229, pp. 133–150. Springer, Cham (2020). https://doi.org/10.1007/978-3-030-55754-6_8

Some Algebraic Aspects of Assume-Guarantee Reasoning

Inigo Incer[1]([✉]), Albert Benveniste[2], and Alberto Sangiovanni-Vincentelli[3]

[1] University of Michigan, Ann Arbor, MI, USA
iir@umich.edu
[2] INRIA/IRISA, Rennes, France
[3] University of California, Berkeley, Berkeley, CA, USA

Abstract. We present the algebra of assume-guarantee (AG) contracts. We define contracts, provide new as well as known operations, and show how these operations are related. Contracts are functorial: any Boolean algebra has an associated contract algebra. We study monoid and semiring structures in contract algebra—and the mappings between such structures. We discuss the actions of a Boolean algebra on its contract algebra.

1 Introduction

The design of complex cyber-physical systems (CPS) involves fundamental challenges in modeling, specification, and integration. Among the modeling challenges, the need to model the interconnection of components modeled using discrete transitions with those using differential equations is fundamental. It is the opposition between the discrete and the continuous, which René Thom calls "the fundamental aporia of mathematics." Specification is well known to be hard in system engineering: practitioners consistently rank the generation of specifications for a project among the top challenges in system design [23,27]. Finally, integration has seen steep increases in magnitude in the 20th century, bringing technical challenges to engineering (how do we design, build, and maintain such systems?)[1] and organizational challenges to the business landscape (how does an organization change to support the development, construction, and mainte-

---

[1] We read, for example: "We are increasingly experiencing a new type of accident that arises in the interactions among components (electromechanical, digital, and human) rather than in the failure of individual components" [19]. "Almost all SI [system integration] failures occur at interfaces primarily due to incomplete, inconsistent, or misunderstood specifications" [20].

This paper is based on Chapter 6 of [14].

© The Author(s), under exclusive license to Springer Nature Switzerland AG 2026
M. Fränzle et al. (Eds.): Werner Damm Festschrift, LNCS 15471, pp. 9–35, 2026.
https://doi.org/10.1007/978-3-031-97537-0_2

nance of such systems?)[2]. The organizational challenges of system engineering are so salient that some authors categorize the field as a branch of management [7,17].

Werner Damm has championed two key concepts to address system design challenges: rich components and contracts. In [5], he observes that "the design of components in complex systems inherently involves multi-site, multi-domain and cross-organizational design teams" and that "in spite of well-defined and enforced process models, such as the V-model, a multitude of 'disturbances' may lead to undesirable design iterations. 'Disturbances' in this context take a variety of forms, such as late or incomplete sets of requirements, late requirement changes, unspecified assumptions, unexpected disruption in a supply chain sub-process, the failure to meet non-functional constraints such as communication latencies, implicit interdependencies, etc." One way to deal with these disturbances is by characterizing the domains of validity of our models. In [5], a rich component is proposed as a classical component upgraded by

> (1) extending component specifications to cover all viewpoints necessary for electronic system design; (2) explicating the dependency of such specifications on assumptions on the context of a component; (3) providing classifiers to such assumptions, relating both the positioning in a layered design space (horizontal, up, down), as well as to their confidence level.

The idea of representing components using assume-guarantee (AG) contracts launched a major research effort in system engineering [3,4,12,25,28]. We wanted to streamline two key aspects of complex system design: relationships between a system integrator and its suppliers (cross-organizational design) and the interactions between multiple engineering organizations within the same company (cross-domain design). In contract-based design, to each component in our system we assign an assume-guarantee specification—or contract. A rich contract algebra [4,14] allows us to relate global system properties to the local properties of the components it comprises. This algebra provides mathematical support for key aspects of the system design process. In this paper, we present this algebra and discuss its role in system design.

Contracts. AG contracts can be understood as formal specifications split in two parts: (i) assumptions made on the environment, and (ii) responsibilities assigned to the object satisfying the specification when it is instantiated in an environment which meets the assumptions of the contract. Contracts were introduced to streamline the integration of complex systems and to support concurrent design.

---

[2] Hobday et al. [11] argue that "systems integration has evolved beyond its original technical and operational tasks to encompass a strategic business dimension becoming, therefore, a core capability of many high-technology corporations...The more complex, high-technology, and high cost the product, the more significant systems integration becomes to the productive activity of the firm...Systems integration capabilities are inextricably linked to decisions on whether to make in-house, outsource, or collaborate in production and competition." Davies et al. [6] claim that " the traditional advantages of the vertically-integrated systems seller offering single-vendor designed systems is no longer a major source of competitive advantage in many industries."

System integration pertains to the composition of multiple design objects into a coherent whole. For example, suppose a company wishes to implement a system with a specification C; designers may realize that there are two sub-specifications C_1 and C_2 such that the composition of their implementations always yields an implementation for the top level specification. In the language of AG contracts, we would say that the composition of C_1 and C_2, written $C_1 \parallel C_2$, *refines* C. This company may now develop an implementation M_1 for C_1, and assign C_2 to a third-party OEM to deliver an implementation M_2. If M_2 is an implementation for C_2, the original company knows that M_1 and M_2 can be composed and that this composition meets the top-level specification C. In this setting, frictions in the supply chain are alleviated as companies exchange formal specifications expressed as contracts.

An additional use of contracts is as follows. Suppose our company wants to implement a system with a specification C using a component M_1 with specification C_1 that is not sufficient to implement C. Contracts provide an operation called *quotient* which yields the specification whose implementation is *exactly* the component M' such that M_1 composed with M' meets the specification C. The operation of quotient has uses in synthesis (when we have made incremental progress towards meeting a goal) and in every situation where we need to find *missing components*.

To say that contracts support concurrent design refers to another aspect of the design process. The design of some components involves multiple engineers working on different aspects of the same object. For example, a team may work on the functionality aspects of an integrated circuit, while another works on its timing characterization. If the functionality team generates a specification C_f, and the timing team generates a specification C_t, the two teams can combine their specs into a single contract object C through an operation called *merging*. In contract theory, these various aspects of a component are called *viewpoints*.

The work on the assume-guarantee reasoning of Floyd-Hoare logic [8,10], on assume-guarantee specifications of Lamport and Abadi [1,18], on design by contract by Meyer [22], on interface automata by de Alfaro and Henzinger [2], and applications of formal specifications to cyber-physical systems by Damm [5], yielded that formal assume-guarantee specifications could be used to design and analyze any cyber-physical system, including their discrete and continuous aspects. Assume-guarantee contracts were thus introduced for this purpose by Benveniste et al. in [3]. Cyber-physical-system design methodologies using contracts were described in [25,29].

Two questions of practical relevance (the related discussions are written in blue) inform our discussion below. *Question 1: how can multiple specifications be combined to generate a system specification?* This leads us to consider binary operations on assume-guarantee contracts and their uses in the system design process. The question has a corollary: *when a system is decomposed across multiple suppliers, as well as across multiple viewpoints, is there a right order for applying the contract operations? Does the result depend on this order?* This question will lead us to consider how the various operations interact among themselves.

The second question has to do with computational complexity. We know that abstracting specifications tends to yield computationally-friendlier semidecision procedures. *Question 2: how can we compute contract abstractions?* Part of our answer to this question will make use of semiring actions.

The structure of the paper is as follows. Section 2 begins to address Question 1 by covering the standard definitions of contracts [4] and all known contract operations. The content of this section is a review, except for the operations of *implication* and *coimplication,* which, to the best of our knowledge, have not been published before. Section 3 treats contracts as an algebra associated with any Boolean algebra and presents a study of monoid and semiring structures within a contract algebra. In other words, this section deals with how the various contract operations interact with each other, yielding insight into compositional design methodologies using contracts. We use these results to define contract actions, which play a role in answering Question 2. Section 4 covers two ways in which a Boolean algebra can act on its contract algebra, and Sect. 5 discusses contract abstractions. Sections 3, 4 and 5 summarize the contributions of this paper.

2   Assume-Guarantee Contracts

This section defines assume guarantee contracts and addresses Question 1: how can multiple specifications be combined to generate a system specification?

We discuss four binary operations that allow us to combine contracts. We also explore adjoint operations that allow us to carry out contract decompositions optimally. The content in this section borrows from [4,24], and the references in the text. Its contributions are the closed-form expressions for implication and coimplication and the use of duality to unify the presentation of the contract operations.

Let $\mathcal{B}$ be a set called the *universe of behaviors.* Its elements are called *behaviors.* The universe of behaviors fixes the modeling formalism we have chosen in our application. A property is defined as a subset of $\mathcal{B}$. A component is also a subset of $\mathcal{B}$. The difference is semantics: we think of components as the set of behaviors they can display. Properties contain behaviors meeting a certain criterion. We say that a component M satisfies a property P, written $M \models P$ if $M \subseteq P$. Component composition is given by set intersection, i.e., given components M and M', their composition is $M \parallel M' = M \cap M'$. Assume-guarantee contracts are pairs of properties.

Definition 1. *A contract $\mathcal{C}$ is a pair of properties $\mathcal{C} = (A, G)$. We call A* assumptions, *and G* guarantees.

Components can have two types of relationships with respect to a contract.

Definition 2. *Let $\mathcal{C} = (A, G)$ be a contract. We say that a component E is an* environment *for $\mathcal{C}$, written $E \models^E \mathcal{C}$, if $E \models A$.*

Environments are those components which meet the assumptions of a contract. Implementations are those which meet the guarantees of the contract when operating in an environment accepted by the contract.

Definition 3. *Let $C = (A, G)$ be a contract. We say that a component M is an implementation for C, written $M \models^M C$, if $M \parallel E \models G$ for every environment E of C.*

Now that we have definitions for environments and implementations, we define a relation on contracts that declares two contracts equivalent when they have the same environments and the same implementations:

Definition 4. *Let C and C' be two contracts. We say they are equivalent when they have the same environments and the same implementations.*

This means that for $C = (A, G)$ and $C' = (A', G')$ to be equivalent, we must have $A = A'$ and $G \cap A = G' \cap A' = G' \cap A$ (because $A' = A$). The largest G' meeting this condition is $G' = G \cup \neg A$ (the complement is taken with respect to $\mathcal{B}$). Enforcing this constraint for a contract allows us to have a unique mathematical object for each set of environments and implementations. We thus define an AG contract in canonical form as follows:

Definition 5. *A contract in canonical form is a contract $C = (A, G)$ satisfying $A \cup G = \mathcal{B}$.*

From now on, we assume all contracts are in canonical form.

2.1 Duality

There is a unary operation which is helpful in revealing structure for AG contracts.

Definition 6. *Let $C = (A, G)$ be a contract. We define a unary operation called reciprocal as follows: $C^{-1} = (G, A)$.*

This operation flips environments and implementations, i.e., it gives us the "environment view" of the specification C. Note that the reciprocal respects canonicity.

Definition 7. *Let $\circ$ and $\star$ be two binary operations on AG contracts, we say that the operations are dual when $(C_a \circ C_b)^{-1} = C_a^{-1} \star C_b^{-1}$.*

2.2 Order

Definition 8. *Suppose C and C' are two contracts. We say that C is a refinement of C', written $C \leq C'$, when all implementations of C are implementations of C' and all environments of C' are environments of C.*

The association we make of a specification being a refinement is that it is harder to meet than another. This is why we say that a specification accepting more environments is a refinement of one accepting less. We can express this order relation using assumptions and guarantees.

Proposition 1 (Theorem 5.2 of [4]). *Let $C = (A, G)$ and $C' = (A', G')$ be two contracts. Then $C \leq C'$ when $G \subseteq G'$ and $A' \subseteq A$.*

2.3   Conjunction and Disjunction

The notion of order provides a lattice structure to AG contracts in canonical form. Given contracts $\mathcal{C} = (A, G)$ and $\mathcal{C}' = (A', G')$, their meet (GLB) and join (LUB) are given by

$$\mathcal{C} \wedge \mathcal{C}' = (A \cup A', G \cap G') \quad \text{and} \quad \mathcal{C} \vee \mathcal{C}' = (A \cap A', G \cup G').$$

We leave it to the reader to verify that conjunction and disjunction are monotonic with respect to the refinement order. Also, conjunction and disjunction furnish our first example of dual operations: $\mathcal{C} \wedge \mathcal{C}' = (G \cap G', A \cup A')^{-1} = ((G, A) \vee (G', A'))^{-1} = (\mathcal{C}^{-1} \vee \mathcal{C}'^{-1})^{-1}$.

If we interpret a contract as the entailment $A \Rightarrow G$, then contract conjunction is the conjunction of such entailments. Conjunction can be used to combine viewpoints. Disjunction has an application in product lines.

2.4   Composition

The notion of composition of AG contracts yields the specification of systems obtained from composing implementations of each of the contracts being composed. Contract composition formalizes how contracts with suppliers result in a system level contract. This operation is defined by axiom as follows:

Suppose $\mathcal{C}_1$ and $\mathcal{C}_2$ are two specifications to be composed. Call $\mathcal{C}$ the composite specification. Let M_1 and M_2 be arbitrary implementations of $\mathcal{C}_1$ and $\mathcal{C}_2$, respectively, and let E be any environment of $\mathcal{C}$. We define $\mathcal{C}$ to be the smallest contract satisfying the following constraints: the composite $M_1 \parallel M_2$ is an implementation of $\mathcal{C}$; the composite $M_1 \parallel E$ is an environment of $\mathcal{C}_2$; and the composite $M_2 \parallel E$ is an environment of $\mathcal{C}_1$.

The first requirement states that composing implementations of the specs being composed yields an implementation of the composite specification. The second requirement states that instantiating an implementation of $\mathcal{C}_1$ in an environment of the composite specification yields an environment for $\mathcal{C}_2$. And the last requirement is the analogous statement for $\mathcal{C}_1$. This principle, which states how to compose specifications split between environment and implementation requirements, was stated for the first time by M. Abadi and L. Lamport [1]. We can obtain a closed-form expression of this principle for AG contracts:

Proposition 2 (Theorem 5.2 of [4]). *Let* $\mathcal{C}_1 = (A_1, G_1)$ *and* $\mathcal{C}_2 = (A_2, G_2)$ *be two AG contracts. Their composition, denoted* $\mathcal{C}_1 \parallel \mathcal{C}_2$, *is given by* $\mathcal{C}_1 \parallel \mathcal{C}_2 = (A_1 \cap A_2 \cup \neg(G_1 \cap G_2), G_1 \cap G_2)$.

We state without proof an important property of composition:

Proposition 3. *Composition of AG contracts is monotonic with respect to the refinement order.*

2.5   Strong Merging (or Merging)

We said that AG contracts are used to handle the specifications of the various viewpoints of the same design element. Suppose $\mathcal{C}_1$ and $\mathcal{C}_2$ are specifications corresponding to different aspects to the same design object, e.g., functionality and power. We define their merger, denoted $C_1 \bullet C_2$, to be the contract which guarantees the guarantees of both specifications when the assumptions of both specifications are respected: $C_1 \bullet C_2 = (A_1 \cap A_2, G_1 \cap G_2 \cup \neg(A_1 \cap A_2))$. This contract is equivalent to contract $(A_1 \cap A_2, G_1 \cap G_2)$, which is exactly what we defined merging to be. Merging can be used to combine viewpoints. Merging and composition are duals, as pointed out in [24].

2.6   Adjoints

We have introduced four operations on AG contracts: two were obtained from the partial order, and two by axiom. Now we obtain the adjoints of these operations. Adjoints are used to compute optimal decompositions of contracts.

Quotient (or Residual). The adjoint of composition is called *quotient*. Let C and C' be two AG contracts. The quotient (also called residual in the literature), denoted C/C', is defined as the largest AG contract C'' satisfying $C' \parallel C'' \leq C$.

Due to the fact that the quotient is the largest contract with this property, Proposition 3 tells us that any of its refinements has this property.

If we interpret C as a top-level specification that our system has to meet (e.g., the specification of a vehicle), and C' as the specification of a subset of the design for which we already have an implementation (e.g., a powertrain), then the quotient is the specification whose implementations are exactly those components that, if added to our partial design, would yield a system meeting the top-level specification. The following proposition gives us a closed-form expression for the quotient of AG contracts:

Proposition 4 (Theorem 3.5 of [13]). *Let $C = (A, G)$ and $C' = (A', G')$ be two AG contracts. The quotient, denoted C/C', is given by*

$$C/C' = (A \cap G', G \cap A' \cup \neg(A \cap G')).$$

For an in-depth study of the notion of a quotient across several compositional theories, see [16]. We can readily show that

$$C/C' = C \bullet (C')^{-1}. \tag{1}$$

Separation. Just like composition has an adjoint operation (the quotient), merging has an adjoint. For contracts C and C', we define the operation of *separation*, denoted $C \div C'$, as the smallest contract C'' satisfying $C \leq C' \bullet C''$. This operation has a closed-form solution:

Proposition 5 (Theorem 3.12 of [26]) *Let $C = (A, G)$ and $C' = (A', G')$ be two AG contracts. Then $C \div C' = (A \cap G' \cup \neg(G \cap A'), G \cap A')$.*

Separation obeys $C \div C' = C \parallel (C')^{-1}$. From this identity and (1), it follows that quotient and separation are duals. For examples of merging and separation, see [26].

Implication and Coimplication. Given contracts $\mathcal{C}$ and $\mathcal{C}'$, the definition of implication, denoted $\mathcal{C}' \to \mathcal{C}$, in a lattice is $\forall \mathcal{C}''.\ \mathcal{C}'' \wedge \mathcal{C}' \leq \mathcal{C} \Leftrightarrow \mathcal{C}'' \leq (\mathcal{C}' \to \mathcal{C})$. In other words, $\mathcal{C}' \to \mathcal{C}$ is the largest contract $\mathcal{C}''$ satisfying $\mathcal{C}'' \wedge \mathcal{C}' \leq \mathcal{C}$. The following proposition tells us how to compute this object:

Proposition 6. *Let $\mathcal{C} = (A, G)$ and $\mathcal{C}' = (A', G')$ be two contracts. Implication has the closed form expression $\mathcal{C}' \to \mathcal{C} = ((A \cap \neg A') \cup (G' \cap \neg G), G \cup \neg G')$.*

Dually, we can ask what is the smallest contract $\mathcal{C}''$ satisfying $\mathcal{C}'' \vee \mathcal{C}' \geq \mathcal{C}$. We will denote this object $\mathcal{C}' \nrightarrow \mathcal{C}$. A similar proof yields the following proposition.

Proposition 7. *Let $\mathcal{C} = (A, G)$ and $\mathcal{C}' = (A', G')$ be two contracts. The smallest contract $\mathcal{C}''$ satisfying $\mathcal{C}'' \vee \mathcal{C}' \geq \mathcal{C}$ has the closed form expression $\mathcal{C}' \nrightarrow \mathcal{C} = (A \cup \neg A', (G \cap \neg G') \cup (A' \cap \neg A))$.*

We observe that

$$\mathcal{C}' \to \mathcal{C} = (G \cup \neg G', (A \cap \neg A') \cup (G' \cap \neg G))^{-1} = ((\mathcal{C}')^{-1} \nrightarrow \mathcal{C}^{-1})^{-1},$$

which shows that implication and coimplication are duals.

2.7   Summary of Binary Operations

The following diagram shows how all AG contract operations are related.

$$\text{Order} \begin{array}{c} \nearrow \\ \Rightarrow \end{array} \begin{array}{c} \text{Conjunction} \xrightarrow{\text{Right adjoint}} \text{Implication} \\ \text{Dual} \updownarrow \qquad\qquad \updownarrow \text{Dual} \\ \text{Disjunction} \xrightarrow{\text{Left adjoint}} \text{Coimplication} \end{array} \qquad \text{Axiom} \begin{array}{c} \nearrow \\ \Rightarrow \end{array} \begin{array}{c} \text{Composition} \xrightarrow{\text{Right adjoint}} \text{Quotient} \\ \text{Dual} \updownarrow \qquad\qquad \updownarrow \text{Dual} \\ \text{Merging} \xrightarrow{\text{Left adjoint}} \text{Separation} \end{array}$$

Two operations—conjunction and disjunction—come from the definition of order, and two—composition and merging—are defined by axiom. The rest of the operations are adjoints of these four.

3   Algebraic Structures Within Contracts

In this section, we investigate the corollary of Question 1: Since both viewpoints and subsystems specifications need to be combined, does the order among these operations influence the result? If yes, is there a best order?

Inspection of the various binary formulas for AG contracts suggests that contracts can be defined over any Boolean algebra, not just that corresponding to properties over a set of behaviors. From now on, we deal with contracts defined over an arbitrary Boolean algebra and embark on a study of the relations between the various contract operations.

Let B be a Boolean algebra with bottom and top elements 0_B and 1_B, respectively. We form the contract algebra $\mathbf{C}(B)$ associated with B. The elements of $\mathbf{C}(B)$ are all pairs $(a, b) \in B^2$ such that $a \vee b = 1_B$. The notions of order and

Table 1. Closed-form expressions of operations for contracts over a Boolean algebra

Conjunction	Disjunction
$\mathcal{C} \wedge \mathcal{C}' = (a \vee a', g \wedge g')$	$\mathcal{C} \vee \mathcal{C}' = (a \wedge a', g \vee g')$
Composition	Merging
$\mathcal{C}_1 \parallel \mathcal{C}_2 = (a_1 \wedge a_2 \vee \neg(g_1 \wedge g_2), g_1 \wedge g_2)$	$\mathcal{C}_1 \bullet \mathcal{C}_2 = (a_1 \wedge a_2, g_1 \wedge g_2 \vee \neg(a_1 \wedge a_2))$
Quotient	Separation
$\mathcal{C}/\mathcal{C}' = (a \wedge g', g \wedge a' \vee \neg(a \wedge g'))$	$\mathcal{C} \div \mathcal{C}' = (a \wedge g' \vee \neg(g \wedge a'), g \wedge a')$
Implication	Coimplication
$\mathcal{C}' \rightarrow \mathcal{C} = ((a \wedge \neg a') \vee (g' \wedge \neg g), g \vee \neg g')$	$\mathcal{C}' \nrightarrow \mathcal{C} = (a \vee \neg a', (g \wedge \neg g') \vee (a' \wedge \neg a))$

the binary operations work exactly the same as for AG contracts over sets of behaviors. Table 1 summarizes these operations.

The contract $1 = (0_B, 1_B)$ is larger than any contract. $0 = (1_B, 0_B)$ is smaller than any contract. The contract $e = (1_B, 1_B)$ is an identity for composition and merging. 1 is an identity for conjunction, and 0 for disjunction. Table 2 shows how various operations behave with respect to the distinguished elements.

Table 2. Contract operations and the distinguished elements

0	1	e
$\mathcal{C} \wedge 0 = 0$	$\mathcal{C} \wedge 1 = \mathcal{C}$	$(a, g) \wedge e = (1_B, g)$
$\mathcal{C} \vee 0 = \mathcal{C}$	$\mathcal{C} \vee 1 = 1$	$(a, g) \vee e = (a, 1_B)$
$\mathcal{C} \parallel 0 = 0$	$(a, g) \parallel 1 = (\neg g, g)$	$\mathcal{C} \parallel e = \mathcal{C}$
$(a, g) \bullet 0 = (a, \neg a)$	$\mathcal{C} \bullet 1 = 1$	$\mathcal{C} \bullet e = \mathcal{C}$
$\mathcal{C}/0 = 1$	$(a, g)/1 = (a, \neg a)$	$\mathcal{C}/e = \mathcal{C}$
$0/(a, g) = (g, \neg g)$	$1/\mathcal{C} = 1$	$e/\mathcal{C} = \mathcal{C}^{-1}$
$(a, g) \div 0 = (\neg g, g)$	$\mathcal{C} \div 1 = 0$	$\mathcal{C} \div e = \mathcal{C}$
$0 \div \mathcal{C} = 0$	$1 \div (a, g) = (\neg a, a)$	$e \div \mathcal{C} = \mathcal{C}^{-1}$
$(a, g) \rightarrow 0 = (g, \neg g)$	$\mathcal{C} \rightarrow 1 = 1$	$(a, g) \rightarrow e = (\neg a, 1_B)$
$0 \rightarrow \mathcal{C} = 1$	$1 \rightarrow \mathcal{C} = \mathcal{C}$	$e \rightarrow (a, g) = (\neg g, g)$
$\mathcal{C} \nrightarrow 0 = 0$	$(a, g) \nrightarrow 1 = (\neg a, a)$	$(a, g) \nrightarrow e = (1_B, \neg g)$
$0 \nrightarrow \mathcal{C} = \mathcal{C}$	$1 \nrightarrow \mathcal{C} = 0$	$e \nrightarrow (a, g) = (a, \neg a)$

3.1   Monoids

We begin to study the interactions among the various operations. We recall that a monoid is a semigroup with identity, i.e., a set together with an associative binary operation and an identity element for that operation. A contract algebra contains several monoids:

Proposition 8. $\mathbf{C}_\wedge^M(B) = (\mathbf{C}(B), \wedge, 1_B)$, $\mathbf{C}_\vee^M(B) = (\mathbf{C}(B), \vee, 0_B)$, $\mathbf{C}_\|^M(B) = (\mathbf{C}(B), \|, e)$, and $\mathbf{C}_\bullet^M(B) = (\mathbf{C}(B), \bullet, e)$ are idempotent, commutative monoids.

It turns out these monoids are isomorphic:

Proposition 9. The monoids $\mathbf{C}_\wedge^M(B)$, $\mathbf{C}_\vee^M(B)$, $\mathbf{C}_\|^M(B)$, and $\mathbf{C}_\bullet^M(B)$ are isomorphic. Moreover, the following diagram commutes, where $\theta_g(a, g) = (\neg(a \wedge g), g)$ and $\theta_a(a, g) = (a, \neg(a \wedge g))$:

$$
\begin{array}{ccc}
\mathbf{C}_\wedge^M(B) & \xleftarrow{\;(\cdot)^{-1}\;}_{\simeq} & \mathbf{C}_\vee^M(B) \\[4pt]
\theta_g \big\uparrow\wr & & \wr\big\uparrow \theta_a \\[4pt]
\mathbf{C}_\|^M(B) & \xrightarrow[\;(\cdot)^{-1}\;]{\simeq} & \mathbf{C}_\bullet^M(B)
\end{array}
\tag{2}
$$

3.2   Maps Between Monoids

The previous result showed how to express contract operations in terms of others. Now we explore how to map contract monoids across their underlying Boolean algebras.

Suppose we have two Boolean algebras, B and B'. Due to Proposition 9, it is sufficient to study the structure of the maps between the contract monoids $\mathbf{C}_\|^M(B)$ and $\mathbf{C}_\|^M(B')$ in order to understand the structure of the maps between all contract monoids associated with each Boolean algebra. First we study maps that allow us to construct and split contracts. Then we consider the general maps.

Let $\mathbf{M}_\wedge(B)$ and $\mathbf{M}_\vee(B)$ be the monoids $\mathbf{M}_\wedge(B) = (B, \wedge, 1_B)$ and $\mathbf{M}_\vee(B) = (B, \vee, 0_B)$. We define the two monoid maps

$$
\iota_a \colon \mathbf{M}_\wedge(B) \to \mathbf{C}_\|^M(B) \quad a \mapsto (a, 1_B)
$$
$$
\iota_g \colon \mathbf{M}_\wedge(B) \to \mathbf{C}_\|^M(B) \quad g \mapsto (1_B, g).
$$

These maps generate an epic monoid map $\pi \colon \mathbf{M}_\wedge(B) \times \mathbf{M}_\wedge(B) \to \mathbf{C}_\|^M(B)$ defined as

$$
(a, g) \mapsto \iota_a(a) \,\|\, \iota_g(g) = (g \to a, g).
$$

Similarly, we have monoid maps that allow us to split a contract:

$$
\pi_g \colon \mathbf{C}_\|^M(B) \to \mathbf{M}_\wedge(B) \quad \pi_g(a, g) = g
$$
$$
\pi_a \colon \mathbf{C}_\wedge^M(B) \to \mathbf{M}_\vee(B) \quad \pi_a(a, g) = a.
$$

We use the monoid isomorphisms (2) to obtain a map $\mathbf{C}_\wedge^M(B) \to \mathbf{M}_\wedge(B)$ from the last morphism:

$$
\neg \circ \pi_a \circ \theta_g \colon \mathbf{C}_\|^M(B) \to \mathbf{M}_\wedge(B)
$$
$$
(a, g) \mapsto a \wedge g.
$$

The two maps $\mathbf{C}_{\|}^{M}(B) \to \mathbf{M}_{\wedge}(B)$ yield the monic monoid map

$$\iota \colon \mathbf{C}_{\|}^{M}(B) \to \mathbf{M}_{\wedge}(B) \times \mathbf{M}_{\wedge}(B)$$

$$(a, g) \mapsto (a \wedge g, g).$$

This map is left-invertible:

$$\pi \circ \iota = e.$$

The elementary maps just described enable us to find the general structure between the monoid maps between the parallel monoids corresponding to two Boolean algebras.

Theorem 1. *Let* $f \colon \mathbf{C}_{\|}^{M}(B) \to \mathbf{C}_{\|}^{M}(B')$. *Then we can write* f *as*

$$f = \pi \circ (l_a(ag)l_g(g)r_a(ag)r_g(g), r_a(ag)r_g(g)) \circ \iota,$$

where $l_a, l_g, r_a, r_g \colon \mathbf{M}_{\wedge}(B) \to \mathbf{M}_{\wedge}(B')$ *are monoid morphisms.*

3.3   Semirings

Now that we have four isomorphic monoids, we look for additional algebraic structure within the contract algebra, namely, the existence of semirings. This will capture the interactions between algebraic operations. First we study the distributivity of the binary operations. Distributivity and semi-distributivity answer the corollary of Question 1: Does the order between subspecification composition and viewpoint combination matter?

Table 3. Distributivity of contract operations

	Conjunction	Disjunction	Composition	Merging
Conjunction	$\mathcal{C} \wedge (\mathcal{C}' \wedge \mathcal{C}'') =$ $(\mathcal{C} \wedge \mathcal{C}') \wedge (\mathcal{C} \wedge \mathcal{C}'')$	$\mathcal{C} \wedge (\mathcal{C}' \vee \mathcal{C}'') =$ $(\mathcal{C} \wedge \mathcal{C}') \vee (\mathcal{C} \wedge \mathcal{C}'')$	$\mathcal{C} \wedge (\mathcal{C}' \| \mathcal{C}'') =$ $(\mathcal{C} \wedge \mathcal{C}') \| (\mathcal{C} \wedge \mathcal{C}'')$	$e \wedge (1 \bullet 0) \neq$ $(e \wedge 1) \bullet (e \wedge 0)$
Disjunction	$\mathcal{C} \vee (\mathcal{C}' \wedge \mathcal{C}'') =$ $(\mathcal{C} \vee \mathcal{C}') \wedge (\mathcal{C} \vee \mathcal{C}'')$	$\mathcal{C} \vee (\mathcal{C}' \vee \mathcal{C}'') =$ $(\mathcal{C} \vee \mathcal{C}') \vee (\mathcal{C} \vee \mathcal{C}'')$	$e \vee (1 \| 0) \neq$ $(e \vee 1) \| (e \vee 0)$	$\mathcal{C} \vee (\mathcal{C}' \bullet \mathcal{C}'') =$ $(\mathcal{C} \vee \mathcal{C}') \bullet (\mathcal{C} \vee \mathcal{C}'')$
Composition	$\mathcal{C} \| (\mathcal{C}' \wedge \mathcal{C}'') =$ $(\mathcal{C} \| \mathcal{C}') \wedge (\mathcal{C} \| \mathcal{C}'')$	$\mathcal{C} \| (\mathcal{C}' \vee \mathcal{C}'') =$ $(\mathcal{C} \| \mathcal{C}') \vee (\mathcal{C} \| \mathcal{C}'')$	$\mathcal{C} \| (\mathcal{C}' \| \mathcal{C}'') =$ $(\mathcal{C} \| \mathcal{C}') \| (\mathcal{C} \| \mathcal{C}'')$	$1 \| (0 \bullet e) \neq$ $(1 \| 0) \bullet (1 \| e)$
Merging	$\mathcal{C} \bullet (\mathcal{C}' \wedge \mathcal{C}'') =$ $(\mathcal{C} \bullet \mathcal{C}') \wedge (\mathcal{C} \bullet \mathcal{C}'')$	$\mathcal{C} \bullet (\mathcal{C}' \vee \mathcal{C}'') =$ $(\mathcal{C} \bullet \mathcal{C}') \vee (\mathcal{C} \bullet \mathcal{C}'')$	$0 \bullet (1 \| e) \neq$ $(0 \bullet 1) \| (0 \bullet e)$	$\mathcal{C} \bullet (\mathcal{C}' \bullet \mathcal{C}'') =$ $(\mathcal{C} \bullet \mathcal{C}') \bullet (\mathcal{C} \bullet \mathcal{C}'')$

Proposition 10. *Table 3 shows whether the binary operations displayed in the rows distribute over the binary operations in the columns.*

If we use conjunction to combine viewpoints, then the distributivity of composition over conjunction says that combining viewpoints and composing subspecifications can be performed at will, in any order. This is wrong if we use merging to combine viewpoints. Nevertheless, the following proposition expresses that, if we use merging to combine viewpoints, it is preferred to combine viewpoints first, and then compose subspecifications, than the converse.

Proposition 11. *Let C_i be a contract for $1 \leq i \leq 4$. We have the following semi-distributions:*

$$(C_1 \wedge C_2) \star (C_3 \wedge C_4) \leq (C_1 \star C_3) \wedge (C_2 \star C_4) \quad and$$
$$(C_1 \vee C_2) \star (C_3 \vee C_4) \geq (C_1 \star C_3) \vee (C_2 \star C_4),$$

where $\star$ is any of the operations conjunction, disjunction, composition, or merging.

The sub-distributivity of parallel composition over conjunction is proved in Chap. 4 of [4]. We will now use the distributivity results to look for semiring structure within the algebra of contracts. We recall the definition of a semiring (see, e.g., [9]):

Definition 9. *A semiring $(R, \cdot, +, 1_R, 0_R)$ is a nonempty set R where (a) $(R, +, 0_R)$ is a commutative monoid; (b) $(R, \cdot, 1_R)$ is a monoid; (c) $r(s + t) = rs + rt$ and $(s + t)r = sr + tr$ for all $r, s, t \in R$; (d) $r \cdot 0_R = 0_R \cdot r = 0_R$ for all $r \in R$; and (e) $0_R \neq 1_R$.*
A map of semirings $f : (R, \cdot, +, 1_R, 0_R) \rightarrow (R', \cdot, +, 1_{R'}, 0_{R'})$ satisfies (a) $f(0_R) = f(0_{R'})$, (b) $f(1_R) = f(1_{R'})$, (c) $f(r+s) = f(r)+f(s)$, and (d) $f(r \cdot s) = f(r) \cdot f(s)$.

The following result provides the semiring structures available in the contract algebra.

Proposition 12. *Using the operations of conjunction, disjunction, composition, and merging, we have exactly four semirings: (a) the conjunction semiring $\mathbf{C}_\wedge^S(B) = (\mathbf{C}(B), \wedge, \vee, 1, 0)$, (b) the disjunction semiring $\mathbf{C}_\vee^S(B) = (\mathbf{C}(B), \vee, \wedge, 0, 1)$, (c) the composition semiring $\mathbf{C}_\parallel^S(B) = (\mathbf{C}(B), \parallel, \vee, e, 0)$, and (d) the merging semiring $\mathbf{C}_\bullet^S(B) = (\mathbf{C}(B), \bullet, \wedge, e, 1)$.*

These four semirings have two isomorphisms.

Proposition 13. *We have the isomorphisms $\mathbf{C}_\wedge^S(B) \cong \mathbf{C}_\vee^S(B)$ and $\mathbf{C}_\parallel^S(B) \cong \mathbf{C}_\bullet^S(B)$. There are no isomorphisms between these two pairs.*

3.4   Some Semiring Maps

Let $\mathbf{S}_\wedge(B)$ and $\mathbf{S}_\vee(B)$ be, respectively, the semirings $(B, \wedge, \vee, 1, 0)$ and $(B, \vee, \wedge, 0, 1)$. We first observe that complementation is a semiring isomorphism for $\mathbf{S}_\wedge(B)$ and $\mathbf{S}_\vee(B)$. We define maps $\Delta_g : \mathbf{S}_\wedge(B) \rightarrow \mathbf{C}_\wedge^S(B)$ and $\iota_g : \mathbf{S}_\wedge(B) \rightarrow \mathbf{C}_\parallel^S(B)$ as follows: $\Delta_g(b) = (\neg b, b)$ and $\iota_g(b) = (1_B, \, b)$.

Proposition 14. *Δ_g and ι_g are semiring homomorphisms.*

Observe that Δ_g can be used to obtain a semiring map from $\mathbf{S}_\vee(B)$ to $\mathbf{C}_\vee^S(B)$ using the semiring isomorphisms $\neg : \mathbf{S}_\wedge(B) \xrightarrow{\sim} \mathbf{S}_\vee(B)$ and $(\cdot)^{-1} : \mathbf{C}_\wedge^S(B) \xrightarrow{\sim} \mathbf{C}_\vee^S(B)$ as follows:

$$\mathbf{S}_\vee(B) \xrightarrow{\ \neg\ } \mathbf{S}_\wedge(B) \xrightarrow{\ \Delta_g\ } \mathbf{C}^S_\wedge(B) \xrightarrow{\ (\cdot)^{-1}\ } \mathbf{C}^S_\vee(B)$$

$$b \longmapsto \neg b \longmapsto (b, \neg b) \longmapsto (\neg b, b)$$

This means that Δ_g is also a semiring homomorphism $\mathbf{S}_\vee(B) \xrightarrow{\ \Delta_g\ } \mathbf{C}^S_\vee(B)$. The following diagram commutes in the category of semirings:

$$\begin{array}{ccc}
\mathbf{S}_\wedge(B) & \overset{\neg}{\underset{\simeq}{\longleftrightarrow}} & \mathbf{S}_\vee(B) \\
\Delta_g \downarrow & \searrow \Delta_a \nearrow & \downarrow \Delta_g \\
\mathbf{C}^S_\wedge(B) & \underset{(\cdot)^{-1}}{\overset{\simeq}{\longleftrightarrow}} & \mathbf{C}^S_\vee(B)
\end{array}$$

The commutativity of the diagram gives rise to the diagonal arrow $\Delta_a = (\cdot)^{-1} \circ \Delta_g = \Delta_g \circ \neg$. This map is a semiring homomorphism from $\mathbf{S}_\wedge(B)$ to $\mathbf{C}^S_\vee(B)$ and from $\mathbf{S}_\vee(B)$ to $\mathbf{C}^S_\wedge(B)$. Explicitly, this map is

$$\Delta_a b = (\Delta_g(b))^{-1} = (\neg b, b)^{-1} = (b, \neg b) \quad (b \in B).$$

If we use the map ι_g, we can obtain $\iota'_a : \mathbf{S}_\vee(B) \to \mathbf{C}^S_\bullet(B)$ as follows:

$$\mathbf{S}_\vee(B) \xrightarrow{\ \neg\ } \mathbf{S}_\wedge(B) \xrightarrow{\ \iota_g\ } \mathbf{C}^S_\|(B) \xrightarrow{\ (\cdot)^{-1}\ } \mathbf{C}^S_\vee(B)$$

$$b \longmapsto \neg b \longmapsto (1_B, \neg b) \longmapsto (\neg b, 1_B)$$

We obtain the diagram below.

$$\begin{array}{ccc}
\mathbf{S}_\wedge(B) & \overset{\neg}{\underset{\simeq}{\longleftrightarrow}} & \mathbf{S}_\vee(B) \\
\iota_g \downarrow \quad \iota_a & \times & \iota'_g \quad \downarrow \iota'_a \\
\mathbf{C}^S_\|(B) & \underset{(\cdot)^{-1}}{\overset{\simeq}{\longleftrightarrow}} & \mathbf{C}^S_\bullet(B)
\end{array}$$

The commutativity of the diagram provides the semiring maps $\mathbf{S}_\wedge(B) \xrightarrow{\iota_a} \mathbf{C}^S_\bullet(B)$ and $\mathbf{S}_\vee(B) \xrightarrow{\iota'_g} \mathbf{C}^S_\|(B)$ given by $\iota_a = (\cdot)^{-1} \circ \iota_g$ and $\iota'_g = \iota_g \circ \neg$.

Now we consider maps to $\mathbf{S}_\wedge(B)$. The map $\pi_g : (a, g) \mapsto g$ is a semiring homomorphism from $\mathbf{C}^S_\wedge(B)$ to $\mathbf{S}_\wedge(B)$ and from $\mathbf{C}^S_\|(B)$ to $\mathbf{S}_\wedge(B)$. Similarly, the map $\pi'_a : (a, g) \mapsto \neg a$ is a semiring homomorphism from $\mathbf{C}^S_\wedge(B)$ to $\mathbf{S}_\wedge(B)$. The following diagrams commute and define the maps not specified before.

$$\begin{array}{ccc}
\mathbf{S}_\wedge(B) \xrightleftharpoons[\simeq]{\neg} \mathbf{S}_\vee(B) &
\mathbf{S}_\wedge(B) \xrightleftharpoons[\simeq]{\neg} \mathbf{S}_\vee(B) &
\mathbf{S}_\wedge(B) \xrightleftharpoons[\simeq]{\neg} \mathbf{S}_\vee(B) \\
\mathbf{C}_\|^S(B) \xrightleftharpoons[(\cdot)^{-1}]{\simeq} \mathbf{C}_\bullet^S(B) &
\mathbf{C}_\wedge^S(B) \xrightleftharpoons[(\cdot)^{-1}]{\simeq} \mathbf{C}_\vee^S(B) &
\mathbf{C}_\wedge^S(B) \xrightleftharpoons[(\cdot)^{-1}]{\simeq} \mathbf{C}_\vee^S(B)
\end{array}$$

(First diagram: vertical maps π_g, π_g', π_a, π_a'. Second diagram: π_g, π_g', π_a, π_a'. Third diagram: π_a', π_a, π_g', π_g.)

4 Actions

One of the questions we sought to answer was: how can we compute contract abstractions? Abstractions are useful to carry out refinement verification in less costly computational environments. We will define abstractions through contract actions.

The semiring maps just described can be used to generate actions of the semirings $\mathbf{S}_\wedge(B)$ and $\mathbf{S}_\vee(B)$ over the contract semirings. Consider, for example the map $\Delta_g : \mathbf{S}_\wedge(B) \to \mathbf{C}_\wedge^S(B)$. For a contract $\mathcal{C} = (a, g)$, we have $\Delta_g(b) \wedge \mathcal{C} = (\neg b, b) \wedge (a, g) = (b \to a, b \wedge g)$.

Now consider the map $\iota_g : \mathbf{S}_\wedge(B) \to \mathbf{C}_\|^S(B)$: $\iota_g(b) \parallel \mathcal{C} = (1_B, b) \parallel (a, g) = ((b \wedge g) \to a, b \wedge g) = (b \to a, b \wedge g)$. We observe that $\mathbf{S}_\wedge(B)$ acts in the same way on the semirings $\mathbf{C}_\wedge^S(B)$ and $\mathbf{C}_\|^S(B)$. This leads us to

Definition 10. *The right action of B on $\mathbf{C}(B)$ is $(a, g) \cdot b = (b \to a, b \wedge g)$.*

Similarly, the semiring map $\Delta_a : \mathbf{S}_\wedge(B) \to \mathbf{C}_\vee^S(B)$ yields $\Delta_a(b) \vee \mathcal{C} = (b, \neg b) \vee (a, g) = (b \wedge a, b \to g)$, and the semiring map $\iota_a : \mathbf{S}_\wedge(B) \to \mathbf{C}_\bullet^S(B)$ results in $\iota_a(b) \bullet \mathcal{C} = (b, 1_B) \bullet (a, g) = (b \wedge a, b \to g)$. $\mathbf{S}_\wedge(B)$ acts in the same way on the semirings $\mathbf{C}_\vee^S(B)$ and $\mathbf{C}_\bullet^S(B)$. We thus obtain

Definition 11. *The left action of B on $\mathbf{C}(B)$ is given by $b \cdot (a, g) = (b \wedge a, b \to g)$.*

The left and right actions have the practical meaning of adding assumptions or guarantees, respectively, to a contract. The following proposition shows several properties of these actions.

Proposition 15. *The identities for the left and right actions shown in Table 4 hold.*

The left action on contracts can be used to generate an abstraction for the guarantees of a contract. Given a contract $\mathcal{C}$ and $b \in B$, we know from Table 4 that $\mathcal{C} \leq b \cdot \mathcal{C}$. This tells us that we can obtain a more relaxed contract by adding assumptions. Subsequently, we may make use of this additional assumption to coarsen the guarantees of the contract. This operation is rich in algebraic properties, as shown in Table 4. For algorithmic manipulations of contracts, see Chapter 7 of [14].

Table 4. Identities for the left and right actions of a Boolean algebra B over its contract algebra ($b, b' \in B$ and $\mathcal{C}, \mathcal{C}' \in \mathbf{C}(B)$)

Order	
$b \cdot \mathcal{C} \geq \mathcal{C}$	$\mathcal{C} \cdot b \leq \mathcal{C}$
$\mathcal{C} \leq \mathcal{C}' \Rightarrow b \cdot \mathcal{C} \leq b \cdot \mathcal{C}'$	$\mathcal{C} \leq \mathcal{C}' \Rightarrow \mathcal{C} \cdot b \leq \mathcal{C}' \cdot b$
Reciprocal	
$(b \cdot \mathcal{C})^{-1} = \mathcal{C}^{-1} \cdot b$	
Associativity	
$(b \wedge b') \cdot \mathcal{C} = b \cdot (b' \cdot \mathcal{C})$	$\mathcal{C} \cdot (b \wedge b') = (\mathcal{C} \cdot b) \cdot b'$
Distributivity over the Boolean algebra	
$(b \vee b') \cdot \mathcal{C} = (b \cdot \mathcal{C}) \wedge (b' \cdot \mathcal{C})$	$\mathcal{C} \cdot (b \vee b') = (\mathcal{C} \cdot b) \vee (\mathcal{C} \cdot b')$
Actions and the contract operations	
$b \cdot (\mathcal{C} \wedge \mathcal{C}') = b \cdot \mathcal{C} \wedge b \cdot \mathcal{C}'$	$(\mathcal{C} \wedge \mathcal{C}') \cdot b = \mathcal{C} \cdot b \wedge \mathcal{C}'$
$b \cdot (\mathcal{C} \vee \mathcal{C}') = b \cdot \mathcal{C} \vee \mathcal{C}'$	$(\mathcal{C} \vee \mathcal{C}') \cdot b = \mathcal{C} \cdot b \vee \mathcal{C}' \cdot b$
$b \cdot (\mathcal{C} \parallel \mathcal{C}') = b \cdot \mathcal{C} \parallel b \cdot \mathcal{C}'$	$(\mathcal{C} \parallel \mathcal{C}') \cdot b = \mathcal{C} \cdot b \parallel \mathcal{C}'$
$b \cdot (\mathcal{C} \bullet \mathcal{C}') = b \cdot \mathcal{C} \bullet \mathcal{C}'$	$(\mathcal{C} \bullet \mathcal{C}') \cdot b = (\mathcal{C} \cdot b) \bullet (\mathcal{C}' \cdot b)$
Actions and the adjoint operations	
$b \cdot (\mathcal{C}/\mathcal{C}') = \mathcal{C}/(\mathcal{C}' \cdot b) = (b \cdot \mathcal{C})/\mathcal{C}'$	$(\mathcal{C}/\mathcal{C}') \cdot b = (\mathcal{C} \cdot b)/(b \cdot \mathcal{C}')$
$b \cdot (\mathcal{C} \div \mathcal{C}') = (b \cdot \mathcal{C}) \div (\mathcal{C}' \cdot b)$	$(\mathcal{C} \div \mathcal{C}') \cdot b = (\mathcal{C} \cdot b) \div \mathcal{C}' = \mathcal{C} \div (b \cdot \mathcal{C}')$
$b \cdot (\mathcal{C}' \rightarrow \mathcal{C}) = \mathcal{C}' \rightarrow b \cdot \mathcal{C} = \mathcal{C}' \cdot b \rightarrow \mathcal{C}$	$(\mathcal{C}' \rightarrow \mathcal{C}) \cdot b = b \cdot \mathcal{C}' \rightarrow \mathcal{C} \cdot b$
$b \cdot (\mathcal{C}' \nrightarrow \mathcal{C}) = \mathcal{C}' \cdot b \nrightarrow b \cdot \mathcal{C}$	$(\mathcal{C}' \nrightarrow \mathcal{C}) \cdot b = \mathcal{C}' \nrightarrow \mathcal{C} \cdot b = b \cdot \mathcal{C}' \nrightarrow \mathcal{C}$

5   Contract Abstractions

It is often useful to vary the level of detail used to represent objects. More detail may be needed to carry out some analysis tasks, but too much detail may hinder computation. We will explore Question 2: how can we compute contract abstractions?

5.1   The Galois-Connection Abstraction

Suppose we have a monotone operator $\alpha \colon B_c \rightarrow B_a$, where B_a and B_c are Boolean algebras called abstract and concrete domains, respectively. Suppose that α is also a monoid map $\alpha \colon \mathbf{M}_\vee(B_c) \rightarrow \mathbf{M}_\vee(B_a)$, i.e., it commutes with disjunction and maps 1_{B_c} to 1_{B_a}. We can define a map $\bar{\alpha} \colon \mathbf{C}(B_c) \rightarrow \mathbf{C}(B_a)$ as

$$\bar{\alpha} \colon (a, g) \mapsto (\alpha(a), \alpha(g)). \tag{3}$$

The map is well-defined since $\alpha(a) \vee \alpha(g) = \alpha(a \vee g) = \alpha(1_{B_c}) = 1_{B_a}$. Moreover, the monotonicity of $\bar{\alpha}$ follows from the monotonicity of α. This abstraction is a slight generalization of that proposed in Chapter 5 of [4].

How can such a map α be obtained? Suppose that α is a monotone map that maps 1_{B_c} to 1_{B_a}. If α is a left element of a Galois connection pair, then it commutes with disjunction (because left adjoints commute with colimits). Thus, such an α generates a valid contract abstraction (3).

5.2 Contract Galois Connections from Boolean Algebra Maps

Let $f : B \to B'$ be a Boolean algebra map. f induces a map $f^* : \mathbf{C}(B) \to \mathbf{C}(B')$ between their contract algebras given by $f^*(a, g) = (f(a),\ \mathrm{f(g)})$. Observe that $f(a) \vee f(g) = f(a \vee g) = f(1_B) = 1_{B'}$. As f commutes with the Boolean algebra operations, f^* commutes with the contract operations. Thus, f^* is well-defined.

Let $\gamma : B_a \to B_c$ and $\alpha : B_c \to B_a$ be Boolean algebra maps. These maps generate contract maps $\mathbf{C}(B_a) \xrightarrow{\ \gamma^*\ } \mathbf{C}(B_c)$ and $\mathbf{C}(B_c) \xrightarrow{\ \alpha^*\ } \mathbf{C}(B_a)$,   as described before.

We are interested in exploring the conditions needed for these maps to form a Galois connection. Specifically, for $\mathcal{C}_a = (a_a, g_a) \in \mathbf{C}(B_a)$ and $\mathcal{C}_c = (a_c, g_c) \in \mathbf{C}(B_c)$, we want $\alpha^*(\mathcal{C}_c) \leq \mathcal{C}_a$ if and only if $\mathcal{C}_c \leq \gamma^*(\mathcal{C}_a)$.

This means that $(\alpha a_c, \alpha g_c) \leq (a_a, g_a)$ if and only if $(a_c, g_c) \leq (\gamma a_a, \gamma g_a)$. If we set $a_c = 1_{B_c}$ and $a_a = 1_{B_a}$, we get $\alpha g_c \leq g_a$ if and only if $g_c \leq \gamma g_a$, and if we set $g_c = 1_{B_c}$ and $g_a = 1_{B_a}$, we obtain $a_a \leq \alpha a_c$ if and only if $\gamma a_a \leq a_c$.

This means that α and γ must be simultaneously the left and right adjoints of each other. By setting $g_c = \gamma g_a$ and $a_c = \gamma a_a$ in the equations above, we obtain that $\alpha \circ \gamma$ is the identity map. Similarly, by setting $g_a = \alpha g_c$ and $a_a = \alpha a_c$, we get that $\gamma \circ \alpha$ is the identity map. This means that B_a and B_c are isomorphic. We conclude that contract Galois connections generated from Boolean algebra maps impose very rigid constraints on the Boolean algebras over which the contracts are defined.

6 Conclusions

Assume-guarantee contracts provide effective tools in system engineering design. In this paper, we described the algebra of contracts that can provide a framework for manipulating contracts in a structured way.

We explored in-depth this algebraic structure *per se*, while establishing a mathematically sound methodology for contract-based system design. The rich algebraic structure of assume-guarantee contracts provides sound support for a number of operations in system design: combining viewpoints, composing sub-specifications, and patching a design. At the same time, it raises a number of open methodological issues: does a designer have freedom in ordering the different design steps? Alternatively, does the theory impose or recommend an ordering? Abstractions are a well established way to provide semi-decision procedures at a lower computational cost. In this paper, we answered the following question: is there a systematic way to lift abstractions of properties to abstractions of contracts?

While writing assume-guarantee contracts is intuitive in applications, the computation of the algebraic operations is often not. Pacti[3] was recently introduced [15] to support the design tasks that are backed by the algebra of contracts. Pacti is able to automatically compute several of the operations we discussed.

An important question remains open: is there a sound way to lift testing from properties—where this is well-developed—to assume-guarantee contracts? Our algebra of assume-guarantee contracts does not provide answers to this question as yet.

A   Proofs

Proof of Proposition 6. Let $\mathcal{C}_i$ be the contract stated in the proposition. Observe that $\mathcal{C}_i \wedge \mathcal{C}' = \mathcal{C} \wedge \mathcal{C}'$. Thus, by the monotonicity of conjunction, $\mathcal{C}'' \leq \mathcal{C}_i$ implies that $\mathcal{C}'' \wedge \mathcal{C}' \leq \mathcal{C}$.

Now write $\mathcal{C}''$ as $\mathcal{C}'' = (A'', G'')$ and assume that $\mathcal{C}'' \wedge \mathcal{C}' \leq \mathcal{C}$. Then

$$G'' \cap G' \leq G \text{ and}$$
$$A'' \cup A' \geq A.$$

From this we conclude that

$$G'' \leq G \cup \neg G' \text{ and} \tag{4}$$
$$A'' \geq A \cap \neg A'. \tag{5}$$

From (4) and the fact that $A'' \geq \neg G''$ (which follows from the definition of AG contracts), we obtain $A'' \geq G' \cap \neg G$. This result and (5) yield

$$A'' \geq (A \cap \neg A') \cup (G' \cap \neg G).$$

This expression and (4) mean that $\mathcal{C}'' \leq \mathcal{C}_i$, completing the proof. $\square$

Proof of Proposition 8. We already know that 1 and 0 are, respectively, the identity elements of conjunction and disjunction. e is the identity for composition and merging. The idempotence of these operations follows immediately from their definitions. It remains to show that these operations are associative.

Let $\mathcal{C} = (a, g)$, $\mathcal{C}' = (a', g')$, and $\mathcal{C}'' = (a'', g'')$ be contracts.

– Conjunction.

$$\begin{aligned}
\mathcal{C} \wedge (\mathcal{C}' \wedge \mathcal{C}'') &= (a, g) \wedge (a' \vee a'', g' \wedge g'') \\
&= (a \vee (a' \vee a''), g \wedge (g' \wedge g'')) \\
&= ((a \vee a') \vee a'', (g \wedge g') \wedge g'') \\
&= (a \vee a', g \wedge g') \wedge \mathcal{C}'' = (\mathcal{C} \wedge \mathcal{C}') \wedge \mathcal{C}''
\end{aligned}$$

---

[3] https://www.pacti.org.

– Disjunction.

$$\mathcal{C} \vee (\mathcal{C}' \vee \mathcal{C}'') = \left(\mathcal{C}^{-1} \wedge \left((\mathcal{C}')^{-1} \wedge (\mathcal{C}'')^{-1}\right)\right)^{-1}$$
$$= \left((\mathcal{C}^{-1} \wedge (\mathcal{C}')^{-1}) \wedge (\mathcal{C}'')^{-1}\right)^{-1} = (\mathcal{C} \vee \mathcal{C}') \vee \mathcal{C}''$$

– Composition.

$$\mathcal{C} \parallel (\mathcal{C}' \parallel \mathcal{C}'')$$
$$= (a, g) \parallel (\neg g' \vee \neg g'' \vee (a' \wedge a''), g' \wedge g'')$$
$$= (\neg g \vee \neg g' \vee \neg g'' \vee (a \wedge a' \wedge a''), g \wedge (g' \wedge g''))$$
$$= (\neg g \vee \neg g' \vee \neg g'' \vee ((a \wedge a') \wedge a''), (g \wedge g') \wedge g'')$$
$$= (\neg g \vee \neg g' \vee (a \wedge a'), g \wedge g') \parallel \mathcal{C}'' = (\mathcal{C} \parallel \mathcal{C}') \parallel \mathcal{C}''$$

– Merging.

$$\mathcal{C} \bullet (\mathcal{C}' \bullet \mathcal{C}'') = \left(\mathcal{C}^{-1} \parallel \left((\mathcal{C}')^{-1} \parallel (\mathcal{C}'')^{-1}\right)\right)^{-1}$$
$$= \left((\mathcal{C}^{-1} \parallel (\mathcal{C}')^{-1}) \parallel (\mathcal{C}'')^{-1}\right)^{-1} = (\mathcal{C} \bullet \mathcal{C}') \bullet \mathcal{C}'' \qquad \square$$

Proof of Proposition 9. Due to the duality relations between conjunction and disjunction and between composition and merging, the reciprocal map provides monoid isomorphisms between $(\mathbf{C}(B), \wedge, 1)$ and $(\mathbf{C}(B), \vee, 0)$ and between $(\mathbf{C}(B), \parallel, e)$ and $(\mathbf{C}(B), \bullet, e)$.

We now show that the map $\theta_g : \mathbf{C}_{\parallel}^M(B) \to \mathbf{C}_{\wedge}^M(B)$ defined as

$$\theta_g : (a, g) \mapsto (\neg(a \wedge g), g)$$

is a monoid isomorphism.

Observe that $\theta_g^2(a, g) = \theta_g(\neg(a \wedge g), g) = (\neg(\neg(a \wedge g) \wedge g), g) = (a, g)$, so θ_g is an involution. We proceed to show it is a monoid homomorphism. Let $\mathcal{C} = (a, g)$ and $\mathcal{C}' = (a', g')$.

– $\theta_g(e) = \theta_g(1, 1) = (0, 1) = 1$
– We verify whether θ_g commutes with the multiplications:

$$\theta_g(\mathcal{C} \parallel \mathcal{C}') = \theta_g(\neg(g \wedge g') \vee (a \wedge a'), g \wedge g')$$
$$= (\neg(a \wedge g \wedge a' \wedge g'), g \wedge g')$$
$$= (\neg(a \wedge g) \vee \neg(a' \wedge g'), g \wedge g')$$
$$= (\neg(a \wedge g), g) \wedge (\neg(a' \wedge g'), g') = \theta_g(\mathcal{C}) \wedge \theta_g(\mathcal{C}').$$

As θ_g is an involution, we have to check that it is a monoid map from $\mathbf{C}_{\wedge}^M(B)$ to $\mathbf{C}_{\parallel}^M(B)$.

$$\theta_g(\mathcal{C} \wedge \mathcal{C}')$$
$$= \theta_g(a \vee a', g \wedge g') = (\neg(g \wedge g' \wedge (a \vee a')), g \wedge g')$$
$$= (\neg(g \wedge g') \vee (\neg a \wedge \neg a'), g \wedge g')$$
$$= (\neg(g \wedge g') \vee (\neg(a \wedge g) \wedge \neg(a' \wedge g')), g \wedge g')$$
$$= (\neg(a \wedge g), g) \parallel (\neg(a' \wedge g'), g') = \theta_g(\mathcal{C}) \parallel \theta_g(\mathcal{C}')$$

The isomorphism θ_a is defined using the diagram (2), i.e.,

$$\theta_a(a, g) = \left(\theta_g(a, g)^{-1}\right)^{-1} = (\theta_g(g, a))^{-1} = (\neg(a \wedge g), a)^{-1} = (a, \neg(a \wedge g)). \quad \square$$

Proof of Theorem 1. Because ι is monic, f generates a unique monoid map $f^{\#}$:

$$
\begin{array}{ccc}
\mathbf{M}_{\wedge}(B) \times \mathbf{M}_{\wedge}(B) & & \mathbf{M}_{\wedge}(B') \times \mathbf{M}_{\wedge}(B') \\
\downarrow{\scriptstyle \pi} & {\scriptstyle f^{\#}} & \uparrow{\scriptstyle \iota} \\
\mathbf{C}_{\|}^{M}(B) & \xrightarrow{\ \ f\ \ } & \mathbf{C}_{\|}^{M}(B')
\end{array}
$$

Because π is epic, $f^{\#}$ generates a unique monoid map $f^{\flat}$

$$
\begin{array}{ccc}
\mathbf{M}_{\wedge}(B) \times \mathbf{M}_{\wedge}(B) & \xrightarrow{\ f^{\flat}\ } & \mathbf{M}_{\wedge}(B') \times \mathbf{M}_{\wedge}(B') \\
\downarrow{\scriptstyle \pi} & {\scriptstyle f^{\#}} & \uparrow{\scriptstyle \iota} \\
\mathbf{C}_{\|}^{M}(B) & \xrightarrow{\ \ f\ \ } & \mathbf{C}_{\|}^{M}(B')
\end{array}
$$

Thus, we have the diagram

$$
\begin{array}{ccc}
\mathbf{M}_{\wedge}(B) \times \mathbf{M}_{\wedge}(B) & \xrightarrow{\ f^{\flat}\ } & \mathbf{M}_{\wedge}(B') \times \mathbf{M}_{\wedge}(B') \\
\iota\uparrow\downarrow{\scriptstyle \pi} & {\scriptstyle f^{\#}} & \pi\downarrow\uparrow{\scriptstyle \iota} \\
\mathbf{C}_{\|}^{M}(B) & \xrightarrow{\ \ f\ \ } & \mathbf{C}_{\|}^{M}(B')
\end{array}
\qquad (6)
$$

$f^{\flat}$ can be factored as the product of two maps

$$\mathbf{M}_{\wedge}(B) \times \mathbf{M}_{\wedge}(B) \to \mathbf{M}_{\wedge}(B').$$

We also observe that $f^{\flat}(a, g) = f^{\flat}((a, 1) \wedge (1, g)) = f^{\flat}(a, 1) \wedge f^{\flat}(1, g)$. This means there are monoid maps $l_a, l_g, r_a, r_g \colon \mathbf{M}_{\wedge}(B) \to \mathbf{M}_{\wedge}(B')$ such that

$$f^{\flat}(a, g) = (l_a(a)l_g(g), r_a(a)r_g(g)).$$

To obtain further restrictions on these maps, we use (6): $f^{\flat}(a, g) = \iota \circ f \circ \pi(a, g) = \iota \circ \pi \circ f^{\flat} \circ \iota \circ \pi(a, g) = (l_a(ag)l_g(g)r_a(ag)r_g(g), r_a(ag)r_g(g)).$ $\quad \square$

Proof of Proposition 10. Let $\mathcal{C} = (a, g)$, $\mathcal{C}' = (a', g')$, and $\mathcal{C}'' = (a'', g'')$ be contracts.

- Conjunction.

$$\mathcal{C} \wedge (\mathcal{C}' \vee \mathcal{C}'') = (a, g) \wedge (a' \wedge a'', g' \vee g'')$$
$$= ((a \vee a') \wedge (a \vee a''), (g \wedge g') \vee (g \wedge g''))$$
$$= (\mathcal{C} \wedge \mathcal{C}') \vee (\mathcal{C} \wedge \mathcal{C}'')$$
$$\mathcal{C} \wedge (\mathcal{C}' \parallel \mathcal{C}'')$$
$$= (a, g) \wedge ((g' \wedge g'') \rightarrow (a' \wedge a''), g' \wedge g'')$$
$$= (a \vee (a' \wedge a'') \vee \neg g' \vee \neg g'', g \wedge g' \wedge g'')$$
$$= (a \vee \neg g \vee (a' \wedge a'') \vee \neg g' \vee \neg g'', g \wedge g' \wedge g'')$$
$$= ((g \wedge g' \wedge g'') \rightarrow ((a \vee a') \wedge (a \vee a'')), g \wedge g' \wedge g'')$$
$$= (a \vee a', g \wedge g') \parallel (a \vee a'', g \wedge g'')$$
$$= (\mathcal{C} \wedge \mathcal{C}') \parallel (\mathcal{C} \wedge \mathcal{C}'')$$
$$e \wedge (1 \bullet 0) = e \wedge 1 = e \neq 0$$
$$= e \bullet 0 = (e \wedge 1) \bullet (e \wedge 0)$$

- Disjunction.

$$\mathcal{C} \vee (\mathcal{C}' \wedge \mathcal{C}'') = \left(\mathcal{C}^{-1} \wedge ((\mathcal{C}')^{-1} \vee (\mathcal{C}'')^{-1})\right)^{-1}$$
$$= \left((\mathcal{C}^{-1} \wedge (\mathcal{C}')^{-1}) \vee (\mathcal{C}^{-1} \wedge (\mathcal{C}'')^{-1})\right)^{-1}$$
$$= (\mathcal{C} \vee \mathcal{C}') \wedge (\mathcal{C} \vee \mathcal{C}'')$$
$$\mathcal{C} \vee (\mathcal{C}' \bullet \mathcal{C}'') = \left(\mathcal{C}^{-1} \wedge ((\mathcal{C}')^{-1} \parallel (\mathcal{C}'')^{-1})\right)^{-1}$$
$$= \left((\mathcal{C}^{-1} \wedge (\mathcal{C}')^{-1}) \parallel (\mathcal{C}^{-1} \wedge (\mathcal{C}'')^{-1})\right)^{-1}$$
$$= (\mathcal{C} \vee \mathcal{C}') \bullet (\mathcal{C} \vee \mathcal{C}'')$$
$$e \vee (1 \parallel 0) = e \vee 0 = e \neq 1$$
$$= 1 \parallel e = (e \vee 1) \parallel (e \vee 0)$$

- Composition.

$$\mathcal{C} \parallel (\mathcal{C}' \wedge \mathcal{C}'') = (a, g) \parallel (a' \vee a'', g' \wedge g'')$$
$$= \begin{pmatrix} \neg(g \wedge g') \vee \neg(g \wedge g'') \vee (a \wedge a') \vee (a \wedge a''), \\ (g \wedge g') \wedge (g \wedge g'') \end{pmatrix}$$
$$= ((g \wedge g') \rightarrow (a \wedge a'), (g \wedge g')) \wedge$$
$$((g \wedge g'') \rightarrow (a \wedge a''), (g \wedge g''))$$
$$= (\mathcal{C} \parallel \mathcal{C}') \wedge (\mathcal{C} \parallel \mathcal{C}'')$$
$$\mathcal{C} \parallel (\mathcal{C}' \vee \mathcal{C}'') = (a, g) \parallel (a' \wedge a'', g' \vee g'')$$
$$= \begin{pmatrix} \neg(g \wedge g') \wedge \neg(g \wedge g'') \vee ((a \wedge a') \wedge (a \wedge a'')), \\ (g \wedge g') \vee (g \wedge g'') \end{pmatrix}$$
$$= \begin{pmatrix} (\neg(g \wedge g') \vee (a \wedge a')) \wedge (\neg(g \wedge g'') \vee (a \wedge a'')), \\ (g \wedge g') \vee (g \wedge g'') \end{pmatrix}$$

$$= ((g \wedge g') \to (a \wedge a'), (g \wedge g'))$$
$$\vee ((g \wedge g'') \to (a \wedge a''), (g \wedge g''))$$
$$= (\mathcal{C} \parallel \mathcal{C}') \vee (\mathcal{C} \parallel \mathcal{C}'')$$
$$1 \parallel (0 \bullet e) = 1 \parallel 0 = 0 \neq 1 = 0 \bullet 1$$
$$= (1 \parallel 0) \bullet (1 \parallel e)$$

The distributivity of composition over conjunction was shown in [21].

– Merging.

$$\mathcal{C} \bullet (\mathcal{C}' \wedge \mathcal{C}'') = \left(\mathcal{C}^{-1} \parallel ((\mathcal{C}')^{-1} \vee (\mathcal{C}'')^{-1}) \right)^{-1}$$
$$= \left((\mathcal{C}^{-1} \parallel (\mathcal{C}')^{-1}) \vee (\mathcal{C}^{-1} \parallel (\mathcal{C}'')^{-1}) \right)^{-1}$$
$$= (\mathcal{C} \bullet \mathcal{C}') \wedge (\mathcal{C} \bullet \mathcal{C}'')$$
$$\mathcal{C} \bullet (\mathcal{C}' \vee \mathcal{C}'') = \left(\mathcal{C}^{-1} \parallel ((\mathcal{C}')^{-1} \wedge (\mathcal{C}'')^{-1}) \right)^{-1}$$
$$= \left((\mathcal{C}^{-1} \parallel (\mathcal{C}')^{-1}) \wedge (\mathcal{C}^{-1} \parallel (\mathcal{C}'')^{-1}) \right)^{-1}$$
$$= (\mathcal{C} \bullet \mathcal{C}') \vee (\mathcal{C} \bullet \mathcal{C}'')$$
$$0 \bullet (1 \parallel e) = 0 \bullet 1 = 1 \neq 0 = 1 \parallel 0$$
$$= (0 \bullet 1) \parallel (0 \bullet e)$$

$\square$

Proof of Proposition 11. Due to the distributivity of $\star$ over conjunction and disjunction (Proposition 10), we have $(C_1 \wedge C_2) \star (C_3 \wedge C_4) = (C_1 \star C_3) \wedge (C_1 \star C_4) \wedge (C_2 \star C_3) \wedge (C_2 \star C_4) \leq (C_1 \star C_3) \wedge (C_2 \star C_4)$ and $(C_1 \vee C_2) \star (C_3 \vee C_4) = (C_1 \star C_3) \vee (C_1 \star C_4) \vee (C_2 \star C_3) \vee (C_2 \star C_4) \geq (C_1 \star C_3) \vee (C_2 \star C_4)$. $\square$

Proof of Proposition 12. Tables 2 and 3 tell, respectively, how operations behave with respect to the distinguished elements and how operations distribute.

Suppose conjunction is the multiplication operation. Since $\mathcal{C} \wedge e \neq e$, neither merging nor composition can be the addition operations. On the other hand, $\mathcal{C} \wedge 0 = 0$, and conjunction distributes over disjunction. Thus, $(\mathbf{C}(B), \wedge, \vee, 1, 0)$ is a semiring.

Now we assume disjunction is the multiplication operation. Since $\mathcal{C} \vee e \neq e$, neither merging nor composition can be the addition operations. However, $\mathcal{C} \vee 1 = 1$, and disjunction distributes over conjunction. Thus, $(\mathbf{C}(B), \vee, \wedge, 0, 1)$ is a semiring.

Suppose composition is the multiplication operation. Since composition does not distribute over merging, merging cannot be addition. Since $\mathcal{C} \parallel 1 \neq 1$, conjunction cannot be addition. However, $\mathcal{C} \parallel 0 = 0$ and composition distributes over disjunction. Thus, $(\mathbf{C}(B), \parallel, \vee, e, 0)$ is a semiring.

Now suppose that merging is the multiplication. Since merging does not distribute over composition, composition cannot be addition. Also, since $\mathcal{C} \bullet 0 \neq 0$, conjunction cannot be addition. However, $\mathcal{C} \bullet 1 = 1$ and merging distributes over conjunction. Thus, $(\mathbf{C}(B), \bullet, \wedge, e, 1)$ is a semiring. $\square$

Proof of Proposition 13. The two semiring isomorphisms are given by the reciprocal map. Suppose there is a semiring map $\beta \colon \mathbf{C}_{\parallel}^{S}(B) \to \mathbf{C}_{\wedge}^{S}(B)$. Then

$$\beta(a, 1_B) = \beta((a, 1_B) \vee e) = \beta(a, g) \vee 1 = 1,$$

which means that β is not invertible. $\qquad\square$

Proof of Proposition 14. Let $b, b' \in B$.

- $\Delta_g(0_B) = (1_B, 0_B) = 0$ and $\Delta_g(1_B) = (0_B, 1_B) = 1$
- $\Delta_g(b \wedge b') = (\neg(b \wedge b'), b \wedge b') = (\neg b \vee \neg b', b \wedge b') = \Delta_g(b) \wedge \Delta_g(b')$
- $\Delta_g(b \vee b') = (\neg(b \vee b'), b \vee b') = (\neg b \wedge \neg b', b \vee b') = \Delta_g(b) \vee \Delta_g(b')$

This shows that Δ_g is a semiring homomorphism. Now we study ι_g:

- $\iota_g(0_B) = (1_B, 0_B) = 0$ and $\iota_g(1_B) = (1_B, 1_B) = e$
- $\iota_g(b \wedge b') = (1_B, b \wedge b') = (1_B, b) \parallel (1_B, b') = \iota_g(b) \parallel \iota_g(b')$
- $\iota_g(b \vee b') = (1, b \vee b') = (1_B, b) \vee (1_B, b') = \iota_g(b) \vee \iota_g(b')$

We conclude that ι_g is a semiring homomorphism as well. $\qquad\square$

Proof of Proposition 15. Let $b, b' \in B$, $\mathcal{C} = (a, g)$, and $\mathcal{C}' = (a', g')$. We have the following properties:

- Order.

$$b \cdot \mathcal{C} = b \cdot (a, g) = (b \wedge a, b \to g) \geq (a, g) = \mathcal{C}$$
$$\mathcal{C} \cdot b = (a, g) \cdot b = (b \to a, b \wedge g) \leq (a, g) = \mathcal{C}$$

  Now suppose $\mathcal{C} = (a, g) \leq \mathcal{C}' = (a', g')$. We have $a' \leq a$ and $g \leq g'$. Since the operations $b \wedge (\cdot)$ and $b \to (\cdot)$ are monotonic, we have $b \cdot \mathcal{C} \leq b \cdot \mathcal{C}'$ and $\mathcal{C} \cdot b \leq \mathcal{C}' \cdot b$.
- Reciprocal.

$$(b \cdot \mathcal{C})^{-1} = (b \wedge a, b \to g)^{-1} = (b \to g, b \wedge a)$$
$$= (g, a) \cdot b = \mathcal{C}^{-1} \cdot b$$

- Associativity.

$$(b \wedge b') \cdot \mathcal{C} = ((b \wedge b') \wedge a, (b \wedge b') \to g)$$
$$= (b \wedge (b' \wedge a), b \to (b' \to g))$$
$$= b \cdot (b' \cdot (a, g)) = b \cdot (b' \cdot \mathcal{C})$$
$$\mathcal{C} \cdot (b \wedge b') = ((b \wedge b') \cdot \mathcal{C}^{-1})^{-1} = ((b' \wedge b) \cdot \mathcal{C}^{-1})^{-1}$$
$$= (b' \cdot (b \cdot \mathcal{C}^{-1}))^{-1} = (b \cdot \mathcal{C}^{-1})^{-1} \cdot b' = (\mathcal{C} \cdot b) \cdot b'$$

– Distributivity over the Boolean algebra B.

$$(b \vee b') \cdot \mathcal{C} = ((b \vee b') \wedge a, (b \vee b') \to g)$$
$$= ((b \wedge a) \vee (b' \wedge a), (b \to g) \wedge (b' \to g))$$
$$= (b \cdot \mathcal{C}) \wedge (b' \cdot \mathcal{C})$$
$$\mathcal{C} \cdot (b \vee b') = \left((b \vee b') \cdot \mathcal{C}^{-1}\right)^{-1}$$
$$= \left(b \cdot \mathcal{C}^{-1} \wedge b' \cdot \mathcal{C}^{-1}\right)^{-1}$$
$$= \mathcal{C} \cdot b \vee \mathcal{C} \cdot b'$$

– Distributivity over the contract operations.
  • Conjunction.

$$b \cdot (\mathcal{C} \wedge \mathcal{C}') = (b \wedge (a \vee a'), b \to (g \wedge g'))$$
$$= ((b \wedge a) \vee (b \wedge a'), (b \to g) \wedge (b \to g'))$$
$$= b \cdot \mathcal{C} \wedge b \cdot \mathcal{C}'$$
$$(\mathcal{C} \wedge \mathcal{C}') \cdot b = (b \to (a \vee a'), b \wedge (g \wedge g'))$$
$$= ((b \to a) \vee a', (b \wedge g) \wedge g')$$
$$= (\mathcal{C} \cdot b) \wedge \mathcal{C}'$$

  • Disjunction.

$$b \cdot (\mathcal{C} \vee \mathcal{C}') = \left((\mathcal{C}^{-1} \wedge (\mathcal{C}')^{-1}) \cdot b\right)^{-1}$$
$$= \left(\mathcal{C}^{-1} \cdot b \wedge (\mathcal{C}')^{-1}\right)^{-1}$$
$$= b \cdot \mathcal{C} \vee \mathcal{C}'$$
$$(\mathcal{C} \vee \mathcal{C}') \cdot b = \left(b \cdot (\mathcal{C}^{-1} \wedge (\mathcal{C}')^{-1})\right)^{-1}$$
$$= \left(b \cdot \mathcal{C}^{-1} \wedge b \cdot (\mathcal{C}')^{-1}\right)^{-1}$$
$$= \mathcal{C} \cdot b \vee \mathcal{C}' \cdot b$$

  • Composition.

$$b \cdot (\mathcal{C} \parallel \mathcal{C}')$$
$$= (b \wedge ((g \wedge g') \to (a \wedge a')), b \to (g \wedge g'))$$
$$= \begin{pmatrix} (b \to (g \wedge g')) \to (b \wedge a \wedge a'), \\ b \to (g \wedge g') \end{pmatrix}$$
$$= \begin{pmatrix} ((b \to g) \wedge (b \to g')) \to ((b \wedge a) \wedge (b \wedge a')), \\ (b \to g) \wedge (b \to g') \end{pmatrix}$$
$$= b \cdot \mathcal{C} \parallel b \cdot \mathcal{C}'$$

$$(\mathcal{C} \parallel \mathcal{C}') \cdot b$$
$$= (b \rightarrow ((g \wedge g') \rightarrow (a \wedge a')), b \wedge g \wedge g')$$
$$= ((b \wedge g \wedge g') \rightarrow (a \wedge a'), b \wedge g \wedge g')$$
$$= ((b \wedge g \wedge g') \rightarrow ((b \rightarrow a) \wedge a'), b \wedge g \wedge g')$$
$$= (\mathcal{C} \cdot b) \parallel \mathcal{C}'$$

- Merging.

$$b \cdot (\mathcal{C} \bullet \mathcal{C}') = ((\mathcal{C}^{-1} \parallel (\mathcal{C}')^{-1}) \cdot b)^{-1}$$
$$= (\mathcal{C}^{-1} \cdot b \parallel (\mathcal{C}')^{-1})^{-1} = b \cdot \mathcal{C} \bullet \mathcal{C}'$$
$$(\mathcal{C} \bullet \mathcal{C}') \cdot b = (b \cdot (\mathcal{C}^{-1} \parallel (\mathcal{C}')^{-1}))^{-1}$$
$$= (b \cdot \mathcal{C}^{-1} \parallel b \cdot (\mathcal{C}')^{-1})^{-1} = \mathcal{C} \cdot b \bullet \mathcal{C}' \cdot b$$

– Distributivity over the adjoint operations.
  - Quotient.

$$b \cdot (\mathcal{C}/\mathcal{C}') = b \cdot (\mathcal{C} \bullet (\mathcal{C}')^{-1}) = b \cdot \mathcal{C} \bullet (\mathcal{C}')^{-1}$$
$$= (b \cdot \mathcal{C})/(\mathcal{C}')$$
$$b \cdot (\mathcal{C}/\mathcal{C}') = b \cdot (\mathcal{C} \bullet (\mathcal{C}')^{-1}) = \mathcal{C} \bullet b \cdot (\mathcal{C}')^{-1}$$
$$= \mathcal{C} \bullet (\mathcal{C}' \cdot b)^{-1} = \mathcal{C}/(\mathcal{C}' \cdot b)$$
$$(\mathcal{C}/\mathcal{C}') \cdot b = (\mathcal{C} \bullet (\mathcal{C}')^{-1}) \cdot b = \mathcal{C} \cdot b \bullet (\mathcal{C}')^{-1} \cdot b$$
$$= \mathcal{C} \cdot b \bullet (b \cdot \mathcal{C}')^{-1} = (\mathcal{C} \cdot b)/(b \cdot \mathcal{C}')$$

  - Separation.

$$b \cdot (\mathcal{C} \div \mathcal{C}') = b \cdot (\mathcal{C} \parallel (\mathcal{C}')^{-1}) = b \cdot \mathcal{C} \parallel b \cdot (\mathcal{C}')^{-1}$$
$$= b \cdot \mathcal{C} \parallel (\mathcal{C}' \cdot b)^{-1} = (b \cdot \mathcal{C}) \div (\mathcal{C}' \cdot b)$$
$$(\mathcal{C} \div \mathcal{C}') \cdot b = (\mathcal{C} \parallel (\mathcal{C}')^{-1}) \cdot b = \mathcal{C} \cdot b \parallel (\mathcal{C}')^{-1}$$
$$= (\mathcal{C} \cdot b) \div \mathcal{C}'$$
$$(\mathcal{C} \div \mathcal{C}') \cdot b = (\mathcal{C} \parallel (\mathcal{C}')^{-1}) \cdot b$$
$$= \mathcal{C} \parallel (\mathcal{C}')^{-1} \cdot b = \mathcal{C} \div (b \cdot \mathcal{C}')$$

  - Implication.

$$b \cdot (\mathcal{C}' \rightarrow \mathcal{C}) = b \cdot ((a \wedge \neg a') \vee (g' \wedge \neg g), g \vee \neg g')$$
$$= (b \wedge (a \wedge \neg a') \vee (b \wedge g' \wedge \neg g), (b \rightarrow g) \vee \neg g')$$
$$= \left(\begin{array}{l} ((b \wedge a) \wedge \neg a') \vee (g' \wedge \neg (b \rightarrow g)), \\ (b \rightarrow g) \vee \neg g' \end{array} \right)$$
$$= \mathcal{C}' \rightarrow b \cdot \mathcal{C}$$

$$b \cdot (\mathcal{C}' \to \mathcal{C}) = b \cdot ((a \wedge \neg a') \vee (g' \wedge \neg g), g \vee \neg g')$$

$$= \begin{pmatrix} (a \wedge \neg(b \to a')) \vee (b \wedge g' \wedge \neg g), \\ (b \wedge g') \to g \end{pmatrix}$$

$$= \mathcal{C}' \cdot b \to \mathcal{C}$$

$$(\mathcal{C}' \to \mathcal{C}) \cdot b = ((a \wedge \neg a') \vee (g' \wedge \neg g), g \vee \neg g') \cdot b$$

$$= \begin{pmatrix} (b \to (a \wedge \neg a')) \vee (b \to (g' \wedge \neg g)), \\ (b \wedge g) \vee \neg(b \to g') \end{pmatrix}$$

$$= \begin{pmatrix} (b \to a) \wedge \neg(b \wedge a') \vee (b \to g') \wedge \neg(b \wedge g), \\ (b \wedge g) \vee \neg(b \to g') \end{pmatrix}$$

$$= b \cdot \mathcal{C}' \to \mathcal{C} \cdot b$$

- Coimplication.

$$b \cdot (\mathcal{C}' \nrightarrow \mathcal{C}) = \left(((\mathcal{C}')^{-1} \to \mathcal{C}^{-1}) \cdot b \right)^{-1}$$

$$= \left(b \cdot (\mathcal{C}')^{-1} \to \mathcal{C}^{-1} \cdot b \right)^{-1} = \mathcal{C}' \cdot b \nrightarrow b \cdot \mathcal{C}$$

$$(\mathcal{C}' \nrightarrow \mathcal{C}) \cdot b = \left(b \cdot ((\mathcal{C}')^{-1} \to \mathcal{C}^{-1}) \right)^{-1}$$

$$= \left((\mathcal{C}')^{-1} \to b \cdot \mathcal{C}^{-1} \right)^{-1} = \mathcal{C}' \nrightarrow \mathcal{C} \cdot b$$

$$= \left((\mathcal{C}')^{-1} \cdot b \to \mathcal{C}^{-1} \right)^{-1} = b \cdot \mathcal{C}' \nrightarrow \mathcal{C}$$

$$\square$$

References

1. Abadi, M., Lamport, L.: Composing specifications. ACM Trans. Program. Lang. Syst. **15**(1), 73–132 (1993). https://doi.org/10.1145/151646.151649
2. de Alfaro, L., Henzinger, T.A.: Interface automata. In: Proceedings of the 8th European Software Engineering Conference Held Jointly with 9th ACM SIGSOFT International Symposium on Foundations of Software Engineering. ESEC/FSE-9, pp. 109–120. Association for Computing Machinery, New York (2001). https://doi.org/10.1145/503209.503226
3. Benveniste, A., Caillaud, B., Ferrari, A., Mangeruca, L., Passerone, R., Sofronis, C.: Multiple viewpoint contract-based specification and design. In: de Boer, F.S., Bonsangue, M.M., Graf, S., Willem-Paul de Roever (eds.) Formal Methods for Components and Objects, 6th International Symposium (FMCO 2007), Amsterdam, The Netherlands, October 24–26, 2007, Revised Papers, Lecture Notes in Computer Science, vol. 5382, pp. 200–225. Springer Verlag, Berlin Heidelberg (2008). https://doi.org/10.1007/978-3-540-92188-2
4. Benveniste, A., et al.: Contracts for system design. Found. Trends® Electr. Des. Autom. **12**(2-3), 124–400 (2018)
5. Damm, W.: Controlling speculative design processes using rich component models. In: Fifth International Conference on Application of Concurrency to System Design (ACSD'05), pp. 118–119 (2005). https://doi.org/10.1109/ACSD.2005.35

6. Davies, A., Brady, T., Hobday, M.: Organizing for solutions: systems seller vs. systems integrator. Ind. Mark. Manage. **36**(2), 183–193 (2007)
7. Emes, M., Smith, A., Cowper, D.: Confronting an identity crisis-how to "brand" systems engineering. Syst. Eng. **8**(2), 164–186 (2005). https://doi.org/10.1002/sys.20028
8. Floyd, R.W.: Assigning meanings to programs. Mathematical Aspects of Computer Science, ed. Schwartz, JT, Amer. Math. Soc (1967)
9. Golan, J.S.: Semirings and their Applications, 1st edn. Springer, Dordrecht (1999). https://doi.org/10.1007/978-94-015-9333-5
10. Hoare, C.A.R.: An axiomatic basis for computer programming. Commun. ACM **12**(10), 576–580 (1969). https://doi.org/10.1145/363235.363259
11. Hobday, M., Davies, A., Prencipe, A.: Systems integration: a core capability of the modern corporation. Ind. Corp. Chang. **14**(6), 1109–1143 (2005)
12. Iannopollo, A., Nuzzo, P., Tripakis, S., Sangiovanni-Vincentelli, A.: Library-based scalable refinement checking for contract-based design. In: 2014 Design, Automation & Test in Europe Conference & Exhibition (DATE), pp. 1–6. IEEE (2014)
13. Incer, I., Sangiovanni-Vincentelli, A.L., Lin, C.W., Kang, E.: Quotient for assume-guarantee contracts. In: 16th ACM-IEEE International Conference on Formal Methods and Models for System Design, MEMOCODE'18, pp. 67–77 (2018). https://doi.org/10.1109/MEMCOD.2018.8556872
14. Incer, I.: The Algebra of Contracts. Ph.D. thesis, EECS Department, University of California, Berkeley (2022)
15. Incer, I., et al.: Pacti: scaling assume-guarantee reasoning for system analysis and design. arXiv preprint: arXiv:2303.17751 (2023)
16. Incer, I., Mangeruca, L., Villa, T., Sangiovanni-Vincentelli, A.L.: The quotient in preorder theories. In: Raskin, J.F., Bresolin, D. (eds.) Proceedings 11th International Symposium on Games, Automata, Logics, and Formal Verification, Brussels, Belgium, September 21-22, 2020. Electronic Proceedings in Theoretical Computer Science, vol. 326, pp. 216–233. Open Publishing Association, Brussels, Belgium (2020). https://doi.org/10.4204/EPTCS.326.14
17. Jackson, S.: Memo to industry: the crisis in systems engineering. Insight **13**(1), 43–43 (2010). https://doi.org/10.1002/inst.201013143
18. Lamport, L.: A simple approach to specifying concurrent systems. Commun. ACM **32**(1), 32–45 (1989). https://doi.org/10.1145/63238.63240
19. Leveson, N.: A new accident model for engineering safer systems. Saf. Sci. **42**(4), 237–270 (2004). https://doi.org/10.1016/S0925-7535(03)00047-X
20. Madni, A.M., Sievers, M.: Systems integration: key perspectives, experiences, and challenges. Syst. Eng. **17**(1), 37–51 (2014). https://doi.org/10.1002/sys.21249
21. Mallozzi, P., Nuzzo, P., Pelliccione, P.: Incremental refinement of goal models with contracts. In: Hojjat, H., Massink, M. (eds.) Fundamentals of Software Engineering, pp. 35–50. Springer International Publishing, Cham (2021)
22. Meyer, B.: Applying "design by contract". IEEE Comput. **25**(10), 40–51 (1992). https://doi.org/10.1109/2.161279
23. National Defense Industrial Association and others: Top systems engineering issues in US defense industry. Systems Engineering Division Task Group Report (2016). https://www.ndia.org/-/media/sites/ndia/divisions/systems-engineering/studies-and-reports/ndia-top-se-issues-2016-report-v7c.ashx
24. Negulescu, R.: Process spacess. Tech. Rep. CS-95-48, University of Waterloo (1995)
25. Nuzzo, P., Sangiovanni-Vincentelli, A.L., Bresolin, D., Geretti, L., Villa, T.: A platform-based design methodology with contracts and related tools for the design

of cyber-physical systems. Proc. IEEE **103**(11), 2104–2132 (2015). https://doi.org/10.1109/JPROC.2015.2453253

26. Passerone, R., Incer, I., Sangiovanni-Vincentelli, A.L.: Coherent extension, composition, and merging operators in contract models for system design. ACM Trans. Embed. Comput. Syst. **18**(5s) (2019). https://doi.org/10.1145/3358216

27. Sage, A.P., Lynch, C.L.: Systems integration and architecting: an overview of principles, practices, and perspectives. Syst. Eng.: J. Int. Counc. Syst. Eng. **1**(3), 176–227 (1998)

28. Sangiovanni-Vincentelli, A., Damm, W., Passerone, R.: Taming Dr. Frankenstein: contract-based design for cyber-physical systems. Eur. J. Control **18**(3), 217–238 (2012)

29. Sangiovanni-Vincentelli, A.L., Damm, W., Passerone, R.: Taming Dr. Frankenstein: contract-based design for cyber-physical systems. Eur. J. Control **18**(3), 217 – 238 (2012). https://doi.org/10.3166/ejc.18.217-238, http://www.sciencedirect.com/science/article/pii/S0947358012709433

Understanding Integration Testing

Alexander Pretschner and Lena Gregor$^{(\boxtimes)}$

Chair of Software and Systems Engineering, TUM School of Computation,
Information and Technology, Technical University of Munich, Munich, Germany
{alexander.pretschner,lena.gregor}@tum.de

Abstract. What is integration testing? There are multiple definitions in the literature and relevant standards. The notion of integration testing also is understood and implemented differently across and even within organizations. We define integration testing intensionally, delineate it from unit, subsystem and system testing, and summarize analytical and constructive approaches at integration testing. We argue that *partial negative specifications* which reflect recurring integration faults can help detect and even avoid integration problems.

1 Introduction

Software is the most powerful driver of innovation today. Powered by communication over the internet, software enables the integration of things, data, concepts, and systems. Because of the enormous complexity of the integrated systems and data we build, integration is becoming ever more relevant and difficult.

In addition to conceptual integrity, the integration process requires to match hardware and software interfaces. In practice, this turns out to be a very difficult task: What does it mean for interfaces to match? How can we make sure they match? How do we check if they match? In this essay, we focus on matching software interfaces and consider hardware interfaces only in passing. While many ideas we discuss can and probably should also be used constructively, we will specifically focus on the activity of integration testing.

The literature roughly suggests two ways to understand integration tests. One understanding is that integration tests aim at testing the *combined functionality* of some subset of components, or units. The other understanding is that integration tests aim at evaluating the *interaction* in-between components, most often without saying what exactly this means or entails. In fact, one standard [1] uses both definitions within the same standard, where "integration tests" are the former and "integration testing" is the latter.

Academic distinctions may not be so relevant to practitioners. However, our own anecdotal experience is that different companies and even different teams within one company have different understandings of integration tests, and do integration tests to vastly varying extents. Some observe in hindsight that unit and integration tests seemingly tested for the same "thing;" some observe that problems found by integration tests could have been found using unit tests; and some realize that unit tests fundamentally fail to identify relevant problems.

M. Fränzle et al. (Eds.): Werner Damm Festschrift, LNCS 15471, pp. 36–53, 2026.
https://doi.org/10.1007/978-3-031-97537-0_3

Test pyramid. This is in line with differing perspectives on the test pyramid. The idea is to have many automated unit tests; fewer (possibly automated) integration tests; and as few end-to-end tests as possible. At the same time, different organizations today observe that the structure of their tests rather resembles ice cones, honeycombs, sand glasses, or any other shape.

The Influence of Architecture. The architecture of the system matters, of course. If we think about a system written in Java, we may have at least a clear understanding of what a unit test is. If we then integrate units and test these, this intuitively becomes an integration test. But if we test the composition of two units, what are the consequences for the *intention* of tests? If we re-use the unit tests and simply replace drivers or stubs (mocks) with the actual components, does this turn unit tests into integration tests? If we consider a microservice architecture, what is the difference between a unit and an integration test?

Granularity. Integration testing happens at several levels [2]. "Units" to be integrated include classes, components, things, (micro) services, full-fledged enterprise systems as well as electronic control units (ECUs) in cars. We will use examples from these domains throughout our essay, but our treatise aims at an understanding that is independent of the specific notion of unit. We do, therefore, not think that the conceptual ambiguities about integration tests are a result of different notions of "unit."

Development Process. In addition to the architecture and granularity, the development process clearly has an impact on the understanding and necessity of integration tests. If software is developed by distributed teams, possibly under different governance, the integration problem seems different from one single agile team that develops a specific piece of software: Intuitively, it matters if the testing process is intertwined with or even precedes development, or if it is performed after development, possibly by a different team.

Why is it difficult? While there may be no consensus on what exactly integration testing is, integration *problems* are widely acknowledged. Why do we have these problems? When system components are developed, there is both explicit knowledge and implicit assumptions about the technical context of the system: Among others, APIs and database schemas constitute explicit knowledge while the interpretation of numbers or the communication format, among many others listed in Sect. 4.1, often remain implicit. If two component implementations then make conflicting assumptions, integration problems are unavoidable.

Overview. The remainder of this essay is organized as follows. Section 2 introduces our formalism, the way it can be used to model development processes, and recaps the fundamentals of testing. Section 3 introduces different test levels. We argue that there are two kinds of integration faults that relate to (1) inadequate design decisions and (2) to inadequate connections at the implementation level. Section 4 argues that integration tests should address integration faults, defines the latter, and provides examples. A key insight will be that we cannot expect specifications for integration testing, and that lists of integration faults can be used as negative specifications for integration testing. Section 5 uses these insights pragmatically, and Sect. 6 concludes.

2 Preliminaries

2.1 Behavior Descriptions

Throughout this paper, we gently make use of formalism where we believe it can help clarify concepts or lead to more succinct descriptions. We see the development process as a transformation of *behavior descriptions* [3] in the spirit of Broy's work on using streams for formally founded model-based development [4]. In the simplest form, behavior descriptions relate input to output. We extend this notion to nondeterministic as well as underspecified behavior descriptions that relate input to sets of output. Where appropriate, we may assume that the formalism includes notions of logical or real time, in order to capture real-time behavior or the performance of a system. We may also assume that the formalism captures special security properties such as confidentiality expressed as hyper properties. Because the precise nature of the formalism is immaterial to our argumentation, we refrain from incorporating the latter properties and stick to the intuition that behavior descriptions are predicates whose semantics are a mapping from streams of input $\mathbb{I}$ to streams of sets of output $\mathbb{O}$,

$$(\mathbb{N} \to \mathbb{I}) \to (\mathbb{N} \to 2^{\mathbb{O}}).$$

Software and systems development is based on user requirements $\mathbb{U}$ and then possibly systems requirements $\mathbb{R}$. At least in principle, requirements are turned into *system specifications*, which are turned into *subsystem* specifications, turned into *unit specifications* and ultimately *implemented* by code. All these artifacts can be interpreted as behavior descriptions, but they usually do not exist as formal documents. The notion of subsystem is recursive where units are implemented but otherwise not further refined. Subsystems hence include units; we also use the word *component* as a synonym. We generalize each of these development steps into a transformation $\rightsquigarrow$ between behavior descriptions, where the left behavior description is turned into the right behavior description.

Nondeterminism is one way to express underspecification: if for a given input, any output is possible, this amounts to leaving open or *not* specifying the reaction to that input. The behavior description of a specification is meant to reflect the requirements—but it may take some additional decisions that were left open by the requirements. Similarly, the specification of subsystems should reflect the system behavior, but may add additional behavior. Finally, implementations of units are meant to reflect the specifications of the latter but themselves are allowed to take further decisions. If this is the case, our development steps are stepwise refinements in a formal sense. Refinement hence amounts to reducing non-determinism, commonly expressed as logical implication between behavior descriptions, or as trace inclusion at the semantic level. In order not to clutter formalism, we will not distinguish between syntax and semantics: For requirements, system specifications, subsystem specifications and implementations P, Q, $P \rightsquigarrow Q$ is a refinement step iff $Q \Rightarrow P$ or, semantically, iff

$$\forall i \in \mathbb{N} \to \mathbb{I}, t \in \mathbb{N} : (Q(i))(t) \subseteq (P(i))(t).$$

We do not think that stepwise refinement is or should be a realistic development methodology. We also do not advocate that formalized behavior descriptions should drive a development process. We are perfectly fine if behavior descriptions only exist virtually. In our context, which is independent of a specific development methodology, we use formalism solely to characterize the testing activities, not the development activities.

2.2   Development

Many design steps are functional decompositions. Because once again an exact formal definition of the composition is immaterial to our paper, we assume some composition operator $\oplus$ with a well-defined semantics that type-correctly connects inputs and outputs of the subsystems. One example for such an operator is provided by the Focus formalism [4], where the semantics simply is the logical conjunction $\wedge$. We will write $S \rightsquigarrow S_1 \oplus \ldots \oplus S_n$ to denote the decomposition of S into n subsystems and their respective connections. We will use the same symbol $S \rightsquigarrow I$ to denote an implementation step.

An idealized development process starts with a set of user requirements $\mathbb{U}$, refined into system requirements $\mathbb{R}$, in turn refined into an overall system specification S with $\mathbb{U} \rightsquigarrow \mathbb{R} \rightsquigarrow S$. The system S is functionally decomposed via $S \rightsquigarrow S_1 \oplus \ldots \oplus S_n$. Each subsystem S_i can recursively be decomposed into smaller subsystems. When the design process comes to an end, subsystems S_i are realized by implementations I_i in implementation steps $S_i \rightsquigarrow I_i$. If a subsystem S was decomposed by $S_1 \oplus \ldots \oplus S_n$ and realized by implementations $I_1, \ldots, I_n$, then $I_1 \oplus \ldots \oplus I_n \rightsquigarrow I$ is the integration step that leads to the implementation I of S. A subsystem's implementation I may be used in further integration steps to form higher-level subsystems.

While we focus on software systems in this essay, our conceptualization also works for cyber-physical systems (CPS). The behavior of CPS usually cannot be reduced to I/O relations only as it includes attributes such as weight, energy consumption, temperature, etc. Clearly lacking elegance, we can model these as part of the system output. At the level of specifications, we then need to include a subset relationship between ranges of values in order to cater to refinements. At the level of implementations, we do not consider refinements between implementations, but only between implementations and possibly hypothetical specifications.

Moreover, CPS development includes a further software-hardware integration step, concerned with the mapping of software components (tasks, communication) to hardware components (processors, memory, busses). We acknowledge the potential for confusion because of the terminology and consider this step to be a mix of *design* and *implementation* steps which include the partitioning of tasks, the choice of schedules, the exploration of and decisions about hardware design trade-offs related to cost, reliability, availability, weight, energy consumption, etc., and the deployment of software components to hardware resources.

There is a long tradition of mathematically modeling development steps as logical implications or, semantically, as trace inclusions, and thus understanding them as refinement steps in the above sense (the only exception is the integration step, which is rather the opposite: $I \rightsquigarrow I_1 \oplus \ldots \oplus I_n$ is or should be a refinement, not $I_1 \oplus \ldots \oplus I_n \rightsquigarrow I$—but we need to implement before we can integrate, and in agile processes, we continuously integrate).

From a technical perspective, there is no need for testing in an ideal world. If during design it can be shown that $\mathbb{R} \Rightarrow \mathbb{U}$, $S \Rightarrow \mathbb{R}$, recursively that $S_1 \wedge \ldots \wedge S_n \Rightarrow S$ for all subsystems and that $I_i \Rightarrow S_i$, then the implementation I of the system satisfies $I \Rightarrow \mathbb{U}$ by construction, under the assumption that the composition of implementations reflects the semantics of the dual decomposition steps and also includes specifications of all the infrastructure components, including libraries, middleware, virtual machines, etc. Note the similarity with the communication and computation infrastructure for CPS mentioned above.

2.3   Testing

In reality, of course, things are more complex. Requirements change; specifications are inexistent, incomplete, wrong, or bad; design steps cannot be shown to be refinements; and implementations cannot be proved to be refinements either. This motivates the need for incomplete verification steps: checking the correctness of implementations $I_i \Rightarrow S_i$ is approximated using unit testing; checking the correctness $I_{j1} \wedge \ldots \wedge I_{jn_j} \Rightarrow S_j$ of subsystem S_j implemented by $I_{j1}, \ldots, I_{jn_j}$ is approximated using subsystem testing; checking if the top level implementation I implements the requirements $\mathbb{R}$ is approximated with system testing; and $I \Rightarrow \mathbb{U}$ is verified during acceptance testing.

Let us assume a (sub)system S has been decomposed into $S_1 \oplus \ldots \oplus S_n$ and the S_i are implemented by I_i such that unit testing suggests $I_i \Rightarrow S_i$. Then the correctness of at least two composed unit implementations I_i with $i \in \mathbb{J}, \mathbb{J} \subseteq \{1, \ldots, n\}, |\mathbb{J}| \geq 2$ against a *hypothetical* specification $H_{\mathbb{J}}$ is checked using *integration testing* by approximating (i.e., testing) $\bigwedge_{i \in \mathbb{J}} I_i \Rightarrow H_{\mathbb{J}}$ such that $H_{\mathbb{J}} \oplus X = S$ for some equally unknown X. In case $|\mathbb{J}| = n$, we have $H_{\mathbb{J}} = S$; this is the last level of integration testing, different from system testing because the latter checks $\bigwedge I_i \Rightarrow \mathbb{R}$ rather than $\bigwedge I_i \Rightarrow S$ (cf. §§3.2 and 3.3). Jorgenson finds this distinction academic because in reality, the distinction between a system specification and a system requirements specification is not always clear-cut [5]. In practice, indeed, testing the last integration stage and system testing cannot always be clearly separated. In any case, we argue that the hypothetical specification $H_{\mathbb{J}}$ is the key to understanding what integration testing really is and should be, an insight missing in prior work [6].

The above formalization of design and testing steps suggests a development process that reflects the V model. Yet, our perspective is agnostic to the concrete development process. The activities are the same in other development processes, e.g., agile processes. The difference is the temporal ordering of the steps (when is integration taking place, when are requirements refined) and the nature of the specifications: In an agile world, specifications come as user stories, test cases, or acceptance tests. These are user *requirements* specifications rather than *(sub)system* specifications in a traditional sense; and yet they all give rise to behavior descriptions. Hypothetical specifications $H_{\mathbb{J}}$ for integration testing are often captured by the previously implemented user stories. They unlikely exist as *explicit (sub)system specifications* in either development process, possibly with the exception of the last stage where $H_{\mathbb{J}} = S$.

Test cases, or tests for short, consist of input streams and expected output streams, and of environment conditions. Testing is the process of applying input to a system and comparing the system's actual output to its expected output. Test cases reflect specifications and are themselves partial behavior descriptions. Sometimes, test cases *are* the the relevant specifications, which is typical for test-first approaches such as test-driven development.

Testing in most cases cannot prove if one behavior description refines another but usually provides partial evidence or falsifies a refinement relationship. That is, testing in practice usually cannot show $\bigwedge I_i \Rightarrow H_{\mathbb{J}}$ but at most $\neg(\bigwedge I_i \wedge H_{\mathbb{J}})$. With the exception of code and tests, we use behavior descriptions only for conceptualizing our perspective. They are unlikely to be amenable to formal treatment in any realistic development process anyway, so this is not a huge practical concern—and the approaches in §5 do not require formalized specifications. Yet, for gaining confidence that the implementation of a system satisfies its specification and/or addresses the users' needs, the choice of relevant, or good, test cases is crucial.

Testing aims both at gaining confidence if requirements are implemented correctly and at revealing faults. In terms of the latter, the idea that good test cases are usually defect-based is spelt out in earlier work [3], getting back to early considerations by Weyuker and Jeng [7]. The starting point of our understanding is the text book definition of test cases being good when they reveal faults. As intuitive as this definition may be from a management perspective, it immediately leads to the odd conclusion that there cannot be good test cases for a hypothetical perfect system. It is then a short step to extend the definition of good tests to reveal potential faults, and marry that to the economical or risk-based intuition that not all faults are equally problematic, tests easy to debug, or cheap to run. In this vein, we consider tests to be good if they reveal potential faults with good cost effectiveness and will, in this paper, concentrate on faults being potential.

Requirements-based tests usually do not target specific faults beyond "requirement incorrectly implemented." If requirements are prioritized w.r.t. importance, respective tests can still be "good" as they address specified (and therefore probably relevant) use cases, which makes their incorrect implementation naturally risk-based and hence helps design tests that are cost-effective.

3   Test Levels

3.1   Unit Tests: $I_i \Rightarrow S_i$?

Unit tests are meant to check if the implementation I_i of a unit specification S_i is correct, that is, if $I_i \Rightarrow S_i$ (once again note that in most cases tests do not provide proof, but we may content ourselves with falsification). In order to test such a unit independently, in practice we need to *mock*, or *stub*, those units that are used by I_i, be that by calls to other functions, messages to other services, or signals sent to other ECUs in a car. Depending on the development context, there is a chance that there is no explicit specification S_i for I_i, or that the unit tests themselves constitute S_i. In practice, even without specifications, unit tests are written by using knowledge about requirements and context.

Unit tests reveal *unit faults* that are usually rooted and fixed in the unit itself. In the V model, unit tests immediately follow the implementation step. In agile contexts, unit tests are written before or during development in a sprint, and are executed whenever code is checked in to the continuous integration environment. Within each sprint, a subsystem consisting of several features is built.

3.2   Integration Tests: $\bigwedge I_i \Rightarrow H_{\mathbb{J}}$?

Integration takes place whenever components (that may be units) are connected to each other. "Connecting" here means that previously mocked functions are now replaced by the respective real implementations. The way functions from other units are called depends on the system: In a simple system, these may be synchronous function calls. In more complex systems, this may be asynchronous inter-process communication; the sending and receiving of messages within some middleware; or the exchange of messages via the Internet in REST-based (micro)service architectures.

Assume we integrate a subset of at least two implementations I_j of components S_j with $j \in \mathbb{J}$ and $\mathbb{J} \subseteq \{1, \ldots, n\} \wedge |\mathbb{J}| \geq 2$ for a (sub)system S with $S \rightsquigarrow S_1 \oplus \ldots \oplus S_n$. We further assume that unit tests suggest that the I_j are correct w.r.t. S_j. Testing includes the comparison with expected results, which is why we explained in §2.3 that we need to assume a hypothetical specification $H_{\mathbb{J}}$ for this subset of components to be given—which, because in most cases $|\mathbb{J}| < n$ and only in one case $|\mathbb{J}| = n$, almost certainly does not explicitly exist. However, in agile processes, $H_{\mathbb{J}}$ can be seen to capture all user stories that were built in earlier sprints, and regardless of the development process, *it comes as the integration test suite and hence reflects the testers' intentions*.

Remember that we have defined $\wedge$ as the semantics of the composition operator $\oplus$, without making a distinction between $\wedge_\alpha$ at the level of specifications and $\wedge_\gamma$ at the level of implementations. Now assume that integration tests reveal $\bigwedge_{j \in \mathbb{J}} I_j \not\Rightarrow H_{\mathbb{J}}$. The respective design, implementation, and integration steps preceding this integration test are $S \rightsquigarrow S_1 \oplus \ldots \oplus S_n$; $S_j \rightsquigarrow I_j$; and $I_1 \oplus \ldots \oplus I_\ell \rightsquigarrow I$ for some $\ell \leq n$. Because the I_j are assumed to be correct by virtue of unit testing, and because we assume the semantics $\wedge_\alpha$ of $\oplus$ to be adequate by design of the formalism, there are two possible causes for the integration test to fail: the implementation of $\wedge_\gamma$ at the implementation level or the decomposition at the specification level.

Interaction Faults. The implementation of the "connection" (abstractly, $\oplus$ with semantics $\wedge_\gamma$) of at least two subsystems I_{j1}, I_{j2} does not work as expected by the design: The connection at the implementation level does not have the same properties as the connection at the design level. At the level of *abstract* *behavior descriptions*, including both specifications and implementations, this mismatch cannot materialize: a logical $\wedge$ is an $\wedge$! However, this connection is implemented by some technical mechanism that transfers control or data. At the level of decomposing systems into subsystem specifications, we may just assume that the connection $\wedge_\alpha$ "works." The concrete implementation-level connection mechanism $\wedge_\gamma$, in contrast, is implemented by a simple synchronous method call or a remote procedure call in a full-fledged middleware (in which case we could also see the system as composed of another component, the middleware itself). It includes characteristics of the communication that are outside the control of the interacting partners, including jitter, latency, message loss, etc., and if these are instrumental for system functionality, they are a cause for integration faults. $\wedge_\gamma$ also captures the full complexities of hardware-software integration as introduced in §1.

We have defined tests to include environment conditions in §2.3. The common understanding is the configuration of the hardware and software stacks. However, one may also see characteristics of the transfer of control and data to be such environment conditions. This becomes apparent when considering electromagnetic interferences or cosmic rays as a source for a system's malfunctioning: These evidently are environment conditions.

The connection mechanism does *not* include the content, encoding, format of the messages or function calls or the way the recipient is addressed. These are the responsibility of the components that interact. At the level of the implementation, $\wedge_\gamma$ may be defective or not share all properties with its abstract counterpart $\wedge_\alpha$: It may be that $S_{j1} \wedge_\alpha S_{j2} \Rightarrow S_j$, $I_{j1} \Rightarrow S_{j1}$ and $I_{j2} \Rightarrow S_{j2}$, but $I_{j1} \wedge_\gamma I_{j2} \not\Rightarrow S_j$. This can happen, for instance, if messages are simply not forwarded by the communication infrastructure. We call these integration faults *interaction faults*. They are typically meant when integration testing is said to target the interaction between subsystems, cf. §1.

For pure software systems, the majority of integration problems does not seem to be due to the interaction itself. In contrast, interaction problems are more relevant, if not prevalent, when considering hardware components.

Design Faults. The decomposition $S \rightsquigarrow S_1 \oplus \ldots \oplus S_n$ was incorrect. That is to say, $S_1 \wedge \ldots \wedge S_n \not\Rightarrow S$ already at the specification level, but this *design fault* was not or could not be detected at the level of specifications, for instance, simply because of a lack of respective specification documents. At this level, we usually assume that the connection $\wedge_\alpha$ between the subsystems simply works correctly. However, the (possibly implicit) component specifications may have left open the communication format, message frequency, encoding, interpretation of numbers, and so on. Yet, these are the responsibility of individual components, not of the connection $\wedge_\alpha$. Implementations of the subsystems each took their own decisions. These were possibly correct w.r.t. their own possibly implicit specifications, but later turn out to be in conflict with each other. For instance, one subsystem implementation I_1 took the decision to use XML, in line with specification S_1, and another subsystem implementation I_2 to use JSON, in line with specification S_2. These integration faults could, *in principle*, have been avoided and detected at the level of specifications during design step $S \rightsquigarrow S_1 \oplus \ldots \oplus S_n$, but have not; and therefore constitute *design faults*.

Without considering non-functional properties, Reiter calls these faults *architecture faults* [8]. We prefer the term design fault because not all design faults are architecture faults, see §4.3. Performance and security problems, for which integration testing probably is the wrong technique, are discussed in §4.5.

In practice, tests may also incorrectly suggest that the hypothetical specification H_J is satisfied, because they are incomplete. Yet, *at a later integration stage*, the system may not work as expected. This may also be because of the possibility of a fourth cause, incompleteness (or incorrectness) of the hypothetical specification H_J used during integration testing.

If H_J does not contain or contains inconsistent information about the transfer of control or data between two components, including format, timing, units, data types and data model, as spelt out in §4.1, incorrect implementation choices may be impossible to detect at the level of unit testing. In practice, the same holds for subsystem specifications S_i that are used for subsystem testing: If these do not contain information about timing, for instance, then two implementations are free to pick their own timing, and this choice may be in conflict with other choices, a recurring problem in the context of flaky tests, cf. §4.1. As mentioned in the introduction, integration faults are based on conflicting assumptions.

Depending on the development context and the need for standard conformance, there is a chance that there are specifications for full subsystems. In contrast, when integration testing is performed, the only explicit representation of H_J usually is the test cases themselves, containing input and expected output, written to reflect the testers' intentions. In practice, at this test level, engineers perform rather explorative testing to see if the interfaces of the subsystems "match." We will therefore suggest in §4.7 to approximate H_J by formulating such a specification negatively, based on commonly recurring integration faults.

3.3   System, Subsystem, and Acceptance Tests: $I \Rightarrow \mathbb{R}$? $I \Rightarrow \mathbb{U}$?

In a V modellish process, system tests are performed after system integration has taken place. Text books sometimes additionally require that the system's infrastructure be identical to the prospective deployment infrastructure. The goal of these tests is to show that a system's implementation I implements the (stipulated) requirements $\mathbb{R}$, $I \Rightarrow \mathbb{R}$. System tests are sometimes called end-to-end tests because they allow for user interactions that mimic those in the real deployment context. Good system test cases address important use cases and risky potential defects. In agile development, system tests are run continuously, where the integrated system consists of all those components that have hitherto been developed and those that need to be mocked. As there often are no explicit system requirements, checking is instead done directly against a subset of the user requirements that often come as user stories.

In addition to the perspective that system testing aims at verifying if $I \Rightarrow \mathbb{R}$, it is not uncommon to see system testing as the last stage of integration testing, or at least not to insist on a clear-cut distinction [5]. As explained above, in this case the hypothetical specification $H_\mathbb{J}$ is not hypothetical anymore but instead is given by S, i.e., $H_\mathbb{J} = S$. We then test if the development $S \rightsquigarrow I_1 \oplus \ldots \oplus I_n$ actually was a refinement.

The idea of system testing recursively applies to *subsystems* S_i of a system S with $S_i \rightsquigarrow S_{i1} \oplus \ldots \oplus S_{in}$ and an implementation $I_i = I_{i1} \oplus \ldots \oplus I_{in}$. Subsystem testing checks if $I_i \Rightarrow \mathbb{R}_i$ holds for subsystem requirements $\mathbb{R}_i$, as long as these are available. We may also take again the perspective that a subsystem test for S_i actually is the last stage of the integration tests for the implementation I_i of S_i. In this case, the specification of S_i is actual rather than hypothetical.

Subsystem and system testing are done w.r.t. requirements. Integration testing is done w.r.t. possibly hypothetical system specifications.

Finally, acceptance tests verify if a system implementation I satisfies the user requirements, $I \Rightarrow \mathbb{U}$. Requirements specifications are likely incomplete, and respective faults may incorrectly be considered to be integration faults. We will consider three examples in §4.4.

4   Integration Faults

If good test cases target potential faults, cf. §2.3, then it is natural to require integration tests to target specific integration faults. As it turns out, there is a huge body of literature and practice on recurring faults in different system types. Interestingly, very few specifically target integration faults.

4.1   Examples

There is a wide range of known integration faults, including different expectations regarding units, encoding, data models, formats, data types, value ranges, frequency, quantization, message synchronicity, message or call ordering, schedule, information freshness and timeouts. Popular examples include the Mars climate orbiter that crashed due to unit inconsistencies where one component expected data in metric units of Newton-seconds and another component sent data in Imperial units of pound-seconds. Ariane-V exploded because two connected components operated with different representations, and hence value ranges, of integers. Buffer overflows can be seen as a further example of integration problems related to value ranges: input strings in one component are longer than expected by the memory allocated to buffer variables in other components. In microservice architectures, mismatches between communication formats JSON and XML are frequently observed. Moreover, as microservices are meant to maintain their own data and thus data model, changes to the model in one service are not always consistently implemented in other services. Respective inconsistencies are not uncommon in contexts where the data model frequently changes, as seen in the insurance sector with frequently changing regulations. Finally, flaky tests can be understood as integration problems. One source of test flakiness are timeouts, where testers specify timeouts that have not been specified at all during design and implementation of the system to be tested.

We do not claim this list to be complete. Indeed, there is a huge body of literature regarding recurring faults for different system types. Some of these works include integration-relevant faults considering various aspects of integration, among them integration faults in monolithic (e.g. [9]), object-oriented (e.g. [2]), or cyber-physical systems (e.g. [10,11]), interface faults (e.g. [12–14]), and faults in the composition of services (e.g. [14,15]).

4.2 Integration Faults as Design Faults

In this essay, the lack of space makes us focus on design rather than interaction faults. We are fully aware of the importance of interaction faults specifically in CPS. That said, let us turn our attention to why *design faults* occur, and why they recur. We have already observed that in practice, the boundaries between the last stage of integration testing and system testing are blurred. Textbooks indicate that system testing is requirements-based and tests if $I_S \Rightarrow \mathbb{R}_S$. We have seen that system testing sometimes also is understood as the last step of integration testing where it is checked if $I_S \Rightarrow S$. Integration testing takes some subset of components and checks their composition against a hypothetical specification. In agile settings, that specification corresponds to previously implemented user requirements. We may speculate that in agile settings, integration testing corresponds to the flavor of integration testing that targets correct functionality rather than correct interactions between components.

We have argued that specifically software integration faults often are a special form of design faults. For simplicity's sake, let us consider the interaction of only two correct implementations I_1 and I_2 with specifications S_1, S_2 in $S \rightsquigarrow S_1 \oplus \ldots \oplus S_n$. Integration tests are executed against a hypothetical specification $H_{\{1,2\}}$ and fail: $I_1 \wedge_\gamma I_2 \not\Rightarrow H_{\{1,2\}}$. If the implementation of the connection $\wedge_\gamma$ between components is correct, the problem must be the *design* with $S_1 \wedge_\alpha \ldots \wedge_\alpha S_n \not\Rightarrow S$.

That the problem is a specification problem does not contradict the intuition that it can be *repaired* at the implementation level: It will often be possible to modify implementations I_i such that $\bigwedge I_i \Rightarrow H_{\mathbb{J}}$, independent of the specifications, and as long as $H_{\mathbb{J}}$ is consistent. Note that we use the term modify rather than refine here: implementations usually are fully refined. After this fix, it may be that $I_i \not\Rightarrow S_i$ which means that we would need to update the specification.

It is important to remember that $S_1 \wedge_\alpha S_2 \not\Rightarrow S$ can be due to the following situation: Both S_1 and S_2 leave open the same implementation choice, e.g., the communication format JSON or XML. This means that implementations I_1, I_2 that refine these specifications can decide differently, thus introducing an integration fault. And yet, this is a *design fault*: When designing $S \rightsquigarrow S_1 \oplus \ldots \oplus S_n$, the communication format should have been chosen consistently! However, as we lack ways to understand $S_1 \wedge S_2 \not\Rightarrow S$ at *design time*, we are likely to detect this problem only *after implementation.*

In sum, the problem can often be fixed in the implementations, but the root cause is the design. In order to understand integration faults, we therefore study them in terms of what can go wrong at the design level.

- If in order to repair the design fault, both S_1 and S_2 *must* and can consistently be refined into S_1' and S_2' such that $S_1' \wedge S_2' \wedge S_3 \wedge \ldots \wedge S_n \Rightarrow S$, then both S_1 and S_2 were underspecified. The implementations took conflicting decisions, probably inadvertently. This is the case that we studied above: *both* specifications leave open the communication format. Note that there is no unique component to blame for this integration fault.
- If it is sufficient to refine *only one* component specification S_1 into S_1' (or symmetrically, S_2 into S_2') and thus ensure that $I_1' \oplus I_2 \oplus \ldots \oplus I_n \Rightarrow S$, then S_1 initially offered implementation choices that *could* conflict with the design of the rest of the system, and I_1 happened to precisely take a decision that did conflict. This happens, for instance, if an implementation I_2 of S_2 is re-used (and thus equated with S_2) and performs computations with 32-bit numbers while S_1 *didn't specify* if numbers were 32 or 64 bits long, but some implementation I_1 decided to do 64 bits. It also happens when the integration of components leads to a buffer overflow.

  In this case, there is a conflict between two implementations that could have been avoided by fixing one component specification and making sure that it is correctly implemented.
- It may also be *impossible* to simultaneously refine both S_1 and S_2 into a consistent subsystem specification, i.e. for all refinements $S_i' \Rightarrow S_i$ and correct implementations $I_i \Rightarrow S_i$, we have $S_1' \oplus \ldots \oplus S_n' \not\Rightarrow S$. This is the case if component specifications took conflicting design decisions, for instance, if S_1 requires its implementations to communicate with I_2 at $100\,\text{Hz}$ and S_2 requires its implementations to communicate with I_1 at $33\,\text{Hz}$. In this case, either one or both specifications (and their respective implementations) need to be modified, *not refined,* in order to conform with the behavior of the respective other component.
- Finally, one may be tempted to think that in addition to the above possibilities, it is possible that *any* of the two specifications be refined in order to fix the integration problem, but not necessarily both: In this case, $S_1 \rightsquigarrow S_1'$ with $S_1' \Rightarrow S_1 \wedge S_2' = S_2$ *or* $S_2 \rightsquigarrow S_2'$ with $S_1' = S_1 \wedge S_2' \Rightarrow S_2$ fix one specific observed integration problem.

  That is, both $S_1 \oplus S_2 \rightsquigarrow S_1' \oplus S_2$ and $S_1 \oplus S_2 \rightsquigarrow S_1 \oplus S_2'$ are possible refinements that fix the integration problem. However, this means that there is underspecification in both specifications for some inputs. If we remove this underspecification in one component, then implementation choices in the other component may, once again, lead to the same integration problem. In practice, this is unlikely to happen because only one implementation is actually changed. We have seen before that it is possible to resolve the integration problem at the level of either *implementation.* However, if the unchanged implementation is re-used with its unchanged specification in a different context, then this may lead to future integration problems.

For simplicity's sake, we have argued with two rather than more components. It is of course possible that an integration fails when three but not two components are integrated. Then, for instance, it may be possible to refine S_1 and S_2

as well as S_2 and S_3, but there still is an inconsistency in $S_1' \oplus S_2' \oplus S_3'$. The characterization of different integration faults, however, remains the same.

4.3 Other Design Faults

Are there design faults that are *not* integration faults? Intuitively, this is the case, for instance, if the intended functionality of one single component S_i happens to be incorrectly specified. This, however, begs the question of the reference w.r.t. which S_i is incorrect. There seems to be one special case where we do not need this external reference: a component S_1 is obviously problematic regardless of the rest of the system if it is *impossible* to refine (or implement) it in a way such that for *any* refinement of the other units $S_2' \oplus \ldots \oplus S_n'$, we have $S_1 \wedge S_2' \ldots S_n' \not\Rightarrow S$. In this case, the design of S_1 is "broken," must be modified, and cannot be repaired by the implementation of the remaining system.

Unfortunately, this case is of rather theoretical interest, as in practice the misbehavior of one unit often can be compensated by other units (exception handling, error correction, redundancy). However, we may accept the practical intuition that there is a design fault in the definition of S_1 if "the other units *could but should not* fix the problem in S_1" and refrain from attempting to formalize this idea.

Design faults which are not integration faults can also lead to the violation of non-functional requirements, including performance and security, cf. §4.3.

4.4 Requirements Rather Than Integration Faults

Some faults may seem like integration faults but really are due to incorrect or incomplete requirements for a system at a higher level.

To start with, consider a car2infrastructure component that is meant to help optimize traffic. In order to do so, it provides information to vehicles about the situation behind a turn at a crossing: pedestrians, children, other cars, buses, etc. The input to such a component are signals derived from camera or infrared or other sensors; the output is a representation of the real world behind the turn. Acceptance testing such a system may then lead to challenging situations for the recognition capabilities of the controller—but what really matters is the safety of traffic, the throughput of the overall system, and so on. In other words, the relevant properties are system-level properties that transcend the properties of the controller subsystem. The controller is a subsystem, not the top-level system.

Higher-level properties often do not make sense at a lower level. This is the case for drone swarms, for instance, where one part of the system's desired behavior is to maintain a minimum distance to other drones. This is *controlled* by each individual drone but can only be observed if there is at least a second drone, i.e., after the top-level system "drone swarm" has been put in place.

Now consider an automated cruise control component, or ACC. Phantom jams occur when changes in the (usually decelerating) movement of one car are reflected and amplified by the following car, leading to a further amplification of the movement by the next following car, and so on. This phenomenon, called string instability, is provoked by many ACCs on the market [16]. Possibly overlooked by the designers of the respective control algorithms, it is however relatively easy to fix in the control algorithms themselves [17]. The point here is that the phenomenon cannot be observed by (acceptance) testing one individual ACC alone: String stability is a property of the higher-level system within which the ACC is embedded: traffic management. If string stability is ignored during the design of the ACC, it is unlikely that the ACC induces string-stable behavior. However, upon integrating it into cars and then into traffic, the property becomes relevant. From the perspective of the overall traffic, we may—incorrectly—see a lack of string stability as an interaction problem.

Finally, let us consider the problem of feature interactions [18], well studied in telecommunication systems, and also common in other distributed systems such as cars. Feature interactions occur when the combination of one or more features lead to undesired or at least unspecified system behavior. For instance, if n_1 enables call forwarding to n_2, which in turn enables call forwarding to n_3, should calls to n_1 be redirected to n_3, even though n_1 didn't explicitly say so? Similarly, if n_1 has blocked outgoing calls to n_2, but n_3 has enabled call forwarding to n_2, should a call from n_1 to n_3 be redirected to n_2?

These situations share the commonality that systems are embedded within other systems, and acceptance tests at a lower component level—the traffic controller, drone swarms, the ACC, the individual feature—naturally cannot reveal the problems at a higher level. Seeing them as *interaction problems* from the higher level is incorrect: They actually are *design* problems in the higher-level systems. We are aware that avoiding this kind of situations without knowing the potential problems upfront is a daunting task.

4.5 Non-Functional Properties

Performance, a system-level property, is influenced by architecture, be that the result of too many indirections in a multi-tiered architecture, be that the result of an inadequate network topology. Performance also is influenced by unit-level choices of adequate data structures—in which case bad performance could be traced back to unit faults at the level of design or implementation.

Similarly, security is a system-level property. Its violation can result from both inadequate implementations and inadequate architectures. We have discussed buffer overflows as integration problems above; we would argue along similar lines for most injection attacks. We also know that some definitions of confidentiality such as non-interference are non-compositional: individual systems may be secure, but their composition is not. Moreover, system-level security of course is influenced by the mechanisms that transfer control, possibly attacked by system call interposition or in-memory process patching, or data: communication channels compromised by a variety of man-in-the-middle attacks.

Non-functional properties like security and performance cross-cut all possible executions of a system and hence naturally matter when (sub)systems are integrated. Security and performance problems are rooted in protocol design, architecture, interaction, and implementation and can hence be seen as all three design, interaction, and implementation faults. Yet, because they manifest at the level of the overall (sub)system, integration testing is not the right level nor the right technique for checking these properties. Specifically, reviews or static analysis tools may be more adequate for the identification of security problems.

At this stage, it is of course natural to consider the notion of *emergent properties*, a topic, however, that we see outside the scope of this paper.

4.6   Locating Integration Faults

The above integration problems between implementations result from inconsistent design choices made during development of the integrated components. The reason for the possibility of inconsistency often is underspecification. Later design choices that, taken together, led to inconsistent behavior, have been taken inadvertently or because incorrect assumptions on the implementation of the respective other components were made. We have already seen that the integration problem can be repaired in the *implementation of any component*: Both modifications $I_1 \oplus I_2 \rightsquigarrow I_1' \oplus I_2$ and $I_1 \oplus I_2 \rightsquigarrow I_1 \oplus I_2'$ solve the problem, which can cater to both design and interaction faults.

This raises the question of where to locate integration faults. At the level of the *implementations*, there is no unique component to blame. As an example, consider two connected components where one receives an input that the other component processes and then fails with a buffer overflow. Truncation of the string could be done in either unit; and usually performance considerations lead to one choice or the other. It is a typical problem that the second unit later is re-used but assumes that the string has been truncated to the length of the buffer. The underlying problem is that the specifications of the two units did not agree on a maximum length for the input.

Even if we can fix an integration problem in implementations, our considerations show that the root cause often is the *design*, not the implementation. As we have seen, integration faults at the design level can be repaired by refining only one or by refining or modifying both specifications. If the implementation is not modified accordingly, this process may entail that one of the implementations is inconsistent with its specification after this refinement or modification.

Unit tests, in contrast, immediately indicate the location of the fault. And of course, if a specification had been refined in order to avoid later integration problems, this refined specification could have been used for unit testing, and unit tests would possibly have revealed an implementation fault. This explains the common intuition in practice that some integration problems could and should have been detected with unit tests, as introduced in §1.

It is tempting to include the notion of "fault for which there is no canonical location to fix it" into a definition of integration faults. Indeed, we believe that this is a useful perspective in terms of understanding integration testing. In general, unfortunately, we cannot see, *beforehand*, if a test targets a fault that can be fixed by touching several alternative components. This seems, however, possible if we restrict ourselves to the above class of specific integration faults.

4.7   Integration Faults for Negative Specifications

The above argumentation essentially entails the following definition of integration testing: the verification if a composition of components—that often do not form a "full" subsystem which may come with an explicit specification—matches the implicit specification of this composition. We agreed that these specifications usually do not actually exist but are hypothetical instead.

In practice, testers write integration tests that themselves constitute this very partial specification. We now suggest to simply see the above list of integration faults as negative specifications, of course tailored to a specific context. $H_\mathbb{J}$ then may come as "absence of unit mismatch" or "absence of format mismatch," and the tester needs to find a way to instantiate these statements in a given context. This leads to a mostly methodological approach at integration testing: checklists of both design and interaction faults, some of which have been published in the literature [2,9,12–15], thus yielding an extensional but not an intensional definition of integration tests, which is the point of this essay. Obviously, it is even preferable to use these checklists during design and implementation of the system, in order to avoid rather than to identify them at a later stage.

5   Constructive and Analytical Approaches

Checklists. The use of checklists with known integration faults, tailored to a specific context, has been discussed in §4.7. While this of course cannot guarantee the absence of integration faults, awareness of respective problems is a powerful first step, specifically when systems are developed in distributed teams.

Contract-Driven Design. One recurring attempt at embedding precise partial specifications into the design process is the use of contracts [4,19,20]. We would argue that while the idea has been around for sixty years now, there are few places where it is implemented in practice. Let us venture to speculate that the reason is of a methodological rather than a technical nature: engineers need guidance w.r.t. what constitutes an adequate level of abstraction for these contracts. It seems like the "absence of recurring faults," adequately contextualized and formalized, can be a realistic level of abstraction for writing contracts.

Contract-Based Testing. One way to target rather simple integration faults such as data type, format, or range of values mismatch, is contract-based testing for service-oriented testing, which is more "lightweight" than other integration tests because it allows to test the interaction of two services in isolation: For consumer-driven contract testing, the developers of the consumer manually write tests to specify simple conditions to the interface of their provider. The provider team can then use these tests to verify if their service actually fulfills these conditions. Provider-driven contract testing works in the opposite order.

Fault Injection. Approaches for test adequacy assessment through integration fault injection into software systems or their interfaces are rather sparse, e.g. interface mutation for programs in C [21–23]. The above taxonomies and catalogues of integration faults can be used to inject realistic integration faults into component interfaces, which is the subject of our own current work.

6   Conclusions

Integration tests target integration faults. Integration testing is the process of verifying if some composition of components satisfies its specification. In contrast to subsystem tests, it is almost certain that such specifications do not exist. Instead, they come as the very integration tests that testers write. In addition to explorative testing if component interfaces match at this level, we have suggested to use the absence of integration faults as negative specifications and use this information to design integration tests.

We have argued that integration faults usually are *design faults*, at least for pure software systems without complex transfers of control or data. If the characteristics of the communication between two components, including latency, packet loss, etc., can impact the correct functioning of the system, integration testing also needs to target *interaction faults*. This is the case for hardware components where the nature of the communication channels very much matters and often constitutes the source of integration problems, e.g., network, wiring, voltage, or electromagnetic interference problems.

The methodological challenge behind integration faults is that during development, design decisions are taken that are based on assumptions on the technical context of a set of components. Integration faults materialize during implementation when these assumptions turn out to be incorrect.

Our conceptualization intensionally defines integration tests, is agnostic to the development process, and may help explain some of the confusion about integration testing: When practitioners feel in hindsight that unit tests could have detected integration problems, this is likely because shared assumptions were not understood, refined, and made explicit. When integration tests seemingly yield the same results as unit tests, then there likely is awareness of shared assumptions, but these are not explicit part of specifications and thus unit tests. If integration tests fail to reveal integration faults, they were probably not designed with potential integration faults in mind.

Acknowledgments. Tiziano Munaro provided valuable insights into the relationship between integration tests and software/hardware integration. David Marson observed the relationship between interaction faults and environment conditions in the definition of tests. Lars Alvincz, Daniel Elsner, Joachim Fröhlich, Anja Hentschel, Silke Reimer, Matthias Saft, and Horst Sauer helped us understand why there are so different perceptions of integration testing. Manfred Broy was instrumental in relating the abstract notion of refinement to practical development processes and in his skepticism about emergent properties. The anonymous reviewers provided invaluable feedback.

References

1. "ISO/IEC/IEEE International Standard - Systems and software engineering–Vocabulary," ISO/IEC/IEEE 24765:2017(E), pp. 1–541 (2017)
2. Winter, M., Ekssir-Monfared, M., Sneed, H.M., Seidl, R., Borner, L.: Der Integrationstest: Von Entwurf und Architektur zur Komponenten-und Systemint (2012)

3. Pretschner, A.: Defect-based testing. In: Dependable Software Systems Engineering, vol. 50 of NATO Science for Peace and Security, pp. 141–163. IOS (2017)
4. Broy, M., Stølen, K.: Specification and Development of Interactive Systems - Focus on Streams, Interfaces, and Refinement. Springer (2001)
5. Jorgenson, P.: Software Testing—A Craftman's Approach. CRC Press (2002)
6. Broy, M., Pretschner, A.: A model-based view onto testing. In: Model-Based Testing for Embedded Systems, CRC Press (2011)
7. Weyuker, E.J., Jeng, B.: Analyzing partition testing strategies. IEEE Trans. Software Eng. **17**(7), 703–711 (1991)
8. Reiter, H.: Reduktion von Integrationsproblemen für SW im Auto durch frühzeitige Erkennung und Vermeidung von Architekturfehlern. PhD thesis, TUM (2010)
9. Leung, H., White, L.: A study of integration testing and software regression at the integration level. In: Proc. Conf. on Software Maintenance, pp. 290–301 (1990)
10. Madni, A.M., Sievers, M.: Systems integration: key perspectives, experiences, and challenges. Syst. Eng. **17**(1), 37–51 (2014)
11. Leveson, N.G.: Role of software in spacecraft accidents. J. Spacecr. Rocket. **41**(4), 564–575 (2004)
12. Duraes, J.A., Madeira, H.S.: Emulation of SW faults: a field data study and a practical approach. IEEE Trans. SW Eng. **32**(11), 849–867 (2006)
13. Nakajo, T., Kume, H.: A case history analysis of software error cause-effect relationships. IEEE Trans. SW Eng. **17**, 830–838 (1991)
14. Bruning, S., Weissleder, S., Malek, M.: A fault taxonomy for service-oriented architecture. In: 10th IEEE High Assurance Systems Engineering Symposium (HASE'07), pp. 367–368 (2007). ISSN: 1530-2059
15. Chan, K.M., Bishop, J., Steyn, J., Baresi, L., Guinea, S.: A fault taxonomy for web service composition. In: Service-Oriented Computing, pp. 363–375 (2009)
16. Gunter, G., Janssen, C., Barbour, W., Stern, R.E., Work, D.B.: Model-based string stability of adaptive cruise control systems using field data. IEEE Trans. Intell. Veh. **5**(1), 90–99 (2020)
17. Stern, R.E. et al.: Dissipation of stop-and-go waves via control of autonomous vehicles: field experiments. Transp. Res. Part C **89**, 205–221 (2018)
18. Zave, P.: Secrets of call forwarding: a specification case study. In: Proc. Formal Description Techniques VIII, vol. 43, pp. 169–184, Chapman & Hall (1995)
19. Meyer, B.: Applying 'design by contract. Computer **25**(10), 40–51 (1992)
20. Derler, P., Lee, E.A., Tripakis, S., Törngren, M.: Cyber-physical system design contracts. In: 4th Intl. Conf. on Cyber-Physical Systems, pp. 109–118 (2013)
21. Delamaro, M.E., Maldonado, J.C., Mathur, A.P.: Integration testing using interface mutation. In: Proc. 7th ISSRE, pp. 112–121, IEEE (1996)
22. Delamaro, M., Maidonado, J., Mathur, A.: Interface Mutation: an approach for integration testing. IEEE Trans. SW Eng. **27**, 228–247 (2001)
23. Delamaro, M.E., Maldonado, J.C., Pasquini, A., Mathur, A.P.: Interface mutation test adequacy criterion: an empirical evaluation. Empir. Softw. Eng. **6**, 111–142 (2001)

On Using Ontologies in the Engineering of Intelligent Cyber-Physical Systems

Ingo Stierand[1]([envelope]) [ID], Lukas Westhofen[1] [ID], and Willem Hagemann[2] [ID]

[1] German Aerospace Center (DLR), Institute of Systems Engineering for Future
Mobility, Oldenburg, Germany
`{ingo.stierand,lukas.westhofen}@dlr.de`
[2] Fraunhofer Research Institution for Energy Infrastructures and Geothermal
Systems, Cottbus, Germany
`willem.hagemann@ieg.fraunhofer.de`

Abstract. The development of cyber-physical systems such as automated driving systems calls for proper engineering methods to ensure that the risk of causing harm to people is minimized. The increasing complexity of environment in which CPS have to function and the complexity of their interaction is becoming a potential source of risk. This becomes evident in recent standards such as ISO 21448, which specify respective requirements on considering the environment in the engineering process. Developing concepts and methods that effectively address these requirements is subject of ongoing research. This paper contributes to the research by putting ontologies into focus. It argues that making an ontology of the operational domain an explicit design artifact can greatly support the engineering process. To this end, the paper discusses its use in several key phases of an ISO 26262-compliant engineering process along a small running example, and looks into state of the art related to ontologies.

1 Introduction

Cyber-physical systems (CPS) are systems that interact with their physical environment [48]. They are often used to work in the presence of human users and thus must be engineered properly in order to minimize the risk of any harm caused to people interacting with them. Because of the importance of this safety aspect, there exist standards such as ISO 26262 [29] in the automotive domain, which provide guidance on how to address it in the development process. The standard mandates the identification of potential risks imposed to users and their environment. Safety goals and safety integrity levels must be derived as input for the definition of a concept to mitigate the identified risks, which is then implemented in the hardware/software (HW/SW) development phase of the system. The standard also mandates that all these steps must be covered by analyses

This work was partially funded by the *Federal Ministry of Education and Research* (BMBF) as part of *MANNHEIM-AutoDevSafeOps* (reference no. 01IS22087C).

to ensure that the system actually adheres to the safety goals, i.e., all requirements obtained along the design process can be traced back to the top-level requirements in a verifiable way to allow statements about their compliance.

Safety-critical system design is a well-established research area and provides solutions to many involved engineering tasks. The state of practice reflected in ISO 26262 however focuses on functional safety, i.e., on risks that are caused by faults in the system. More precisely, at the end of consideration always lies some hardware issue as root cause (which indeed may be the result of an environmental event such as lightning strike). On the other hand, CPS are becoming more intelligent and perform increasingly complex safety-critical tasks. Systems that interact unsupervised with complex environments require a common 'understanding' of the world in which they are functioning. An effective engineering and reasoning about safety requires considering the impacts of shortcomings of such understanding. Performance limitation of sensors and perception chains, insufficient processing of the perceived environment, and unsafe specifications of behavior in these environments become potential causes for hazards. These aspects have been coined as safety of the intended function (SOTIF) in the automotive domain, which is addressed in the ISO 21448 standard [30] as a complement to ISO 26262.

The operational design domain (ODD) of the system is a central element in SOTIF. The ODD is rather abstractly specified as "conditions under which a given [...] system is designed to function" [30]. It includes both the environment and the conditions of the system itself, such as the capabilities for sensing and perceiving the environment. ISO 21448 defines *triggering conditions* as those conditions in the ODD that can activate existing insufficiencies in the system, which in turn may lead to hazards. Consequently, the standard defines the identification of triggering conditions as one of the main activities. Although the standard gives examples for triggering conditions, the definition also remains quite abstract: It is a condition of a *scenario*, which constitutes a sequence of *scenes*.

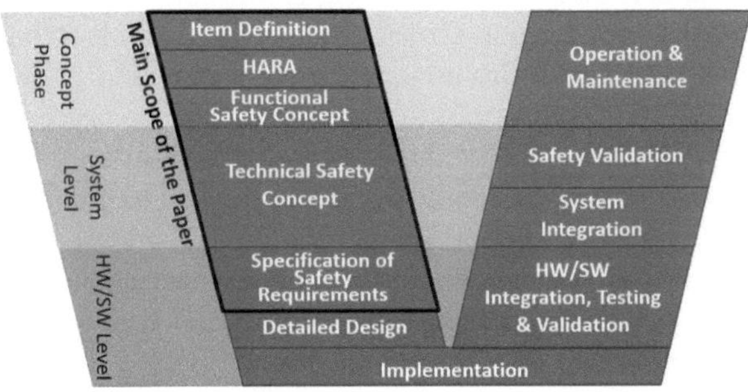

Fig. 1. Development process phases and main scope of the paper.

In order to apply SOTIF in a development process as effective as it is the case today for functional safety, there is a need for suitable means for specifying ODDs in terms of scenarios, scenes, and (triggering) conditions. These means should be consistently applicable throughout the whole development process, including architecture design. The development of solutions for this need is a vibrant research topic. The aim of this paper is to contribute to this research by putting the use of ontologies as a powerful specification means into focus. Similar to Yang et al. [68], we argue that integrating an ontology as a basic engineering artifact can be a key enabler for solving many of the challenges posed by the development of intelligent CPS. We specifically focus on the domain of auto-mated driving and therefore on an ISO 26262-compliant development process as depicted in Fig. 1. We discuss a selection of design phases and the benefits that can be taken from the application of ontologies. A contract-based system design is presented, where an ontology serves as the interpretation domain for the specification of interfaces, facilitating the connection between requirements and the architecture. In summary, our work provides arguments that ontologies can serve as a valuable tool for creating coherent and consistent design processes and managing design complexity.

The structure of the paper is as follows. Section 2 gives an account on the term ontology, and puts it into the context of development processes. It also introduces a running example that is used throughout the paper. The following sections are devoted to the development phases that are bold lined in Fig. 1. Each section starts with a short description of the role of the phase according to ISO 26262 and ISO 21448, develops the running example along this role, and refers to existing work. This is complemented by a discussion about requirements on the ontology in these phases. Section 7 concludes the paper.

2   Preliminaries

2.1   Ontologies as Foundation for Specifications

The term *ontology* can be translated as "logical discourse of that which is". This discourse goes back to the ancient Greek philosophers, and is striving millennia of philosophical and theological discussions. In computer and information science, an ontology is understood as a naming and definition of *concepts* and *relations* between these concepts in a particular *domain* of discourse. Wikipedia provides a good explanation of the rational: ontologies help "to limit complexity and organize data into information and knowledge."

Not surprisingly, one was searching for means to establish such 'understand-ing' on the machine level by machine-understandable knowledge representation and logic-based reasoning. Its roots from the artificial intelligence domain to model human cognition processes reach back to the 1960s [17]. The basic idea is to represent 'knowledge' by a number of *concepts* (such as *Human*) and their *relations* (*Parent*), and to derive new 'knowledge' from a set of logical rules (*Grand Parent = Parent of Parent*) [50]. While such ontology-based reasoning turned out to be over-simplified to mimic human-like reasoning, it persisted as a useful tool in many application domains [26,57]. Research in this area led to a common theoretical framework known as *description logic* (DL) [6]. Equipped with well-defined semantics, DL defines how concepts and *roles* (i.e., relations) can be constructed, and new 'knowledge' be derived by *inference* operations. Various DL variants with different expressiveness have been identified. Most variants can be embedded via a translation scheme into decidable fragments of first order predicate logic, such as the two-variable fragment or guarded fragments [52], and thus allow for fully automated reasoning about this kind of knowledge.

Constructing an ontology does not depend on DL. Both languages and methodologies for ontology construction vary greatly in different application domains. They are well-established in database design and software engineering in form of entity-relationship diagrams and class diagrams, respectively. The Object Management Group defines a mechanism for mapping between ontology languages [18] and a formal ontology meta-language itself [40]. Enterprise and business modeling makes use of ontologies [59,63]. Wehrstedt et al. [64] discuss ontology construction in Industry 4.0 applications for information exchange among smart production cells. In general, ontologies act as a foundation for specifications over the domain of discourse, e.g., for defining requirements in an ontology-based requirement language as we do here. Separating project-specific specifications, e.g., requirements, from the underlying ontology also fosters re-use and sharing of ontologies.

A key objective in the use of ontologies is to ensure that all parties have the same understanding of relevant concepts and relations. When developing an automated driving system (ADS), we desire to implement a valid interpretation of the relevant domain concepts, e.g., legal terms. However, how can we ensure that the implemented system correctly reflects the original interpretation? This is what we denote by the *congruence problem*. Ontologies have been found to help in addressing this issue [67]. Again, establishing congruence does not depend on the level of formalization. Getting a common understanding between domain experts and engineers requires thorough discussions and no DL. However, formalization of ontologies can help to pinpoint inconsistencies and semantic mismatches.

2.2 Running Example

We delineate our ideas along a running example based on an exemplary traffic rule, namely the safe distance rule of the German road traffic act, which was also used in prior work [67]. The following sections show how ontologies can serve as a fundamental concept when propagating traffic rules through a typical safety engineering process. For this, we assume to develop an Automated Lane Keeping System (ALKS) for German highways. Such a system "controls the lateral and longitudinal movement of the vehicle" [61]. For simplicity, we assume that it consists of two components for longitudinal and lateral control, respectively. The ALKS is a SAE Level 3 system, thus takes over the driving responsibility until the human is able to take back control. Hence, a safety engineering process has to a) ensure that the ALKS complies to the relevant traffic rules during its operation and b) demonstrate this convincingly to all stakeholders.

Our running example highlights the role of ontologies in this process. It presents a rudimentary excerpt of a limited number of safety engineering steps. In its core, an ontology is used and refined in each step of the example, starting with a coarse and informal ontology at the item and ODD definition up to a detailed formal version for e.g. requirement verification. As the next sections detail, this refinement concerns the degree of (in)formality, adding, removing, and changing concepts and roles, and updating their semantics. This process aims at resulting in a formalized ontology that is suitable for specifying requirements for the implementation of necessary safety mechanisms. For brevity, the full ontology of the example is presented in Fig. 2, combined for each step.

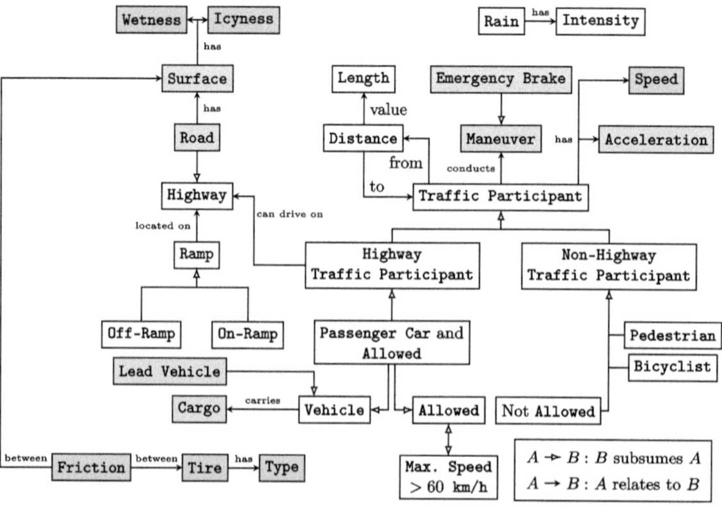

Fig. 2. An excerpt of an exemplary ontology to be used within a safety engineering process of a highway ALKS. White boxes are introduced in the item definition, gray boxes in the hazard analysis and risk assessment, and red boxes in the safety concept. (Color figure online)

The general procedure of safety engineering presented in the upcoming sections is based on the works by Graubohm et al. [25] and Stolte et al. [58]. Their method of deriving a functional safety concept and the involved steps [25, Sec. 4, Fig. 1 (a)] (such as the hazard analysis) thus function as a blueprint for this work.

3   Item Definition

According to ISO 26262, the item definition is part of the concept phase, with the objective "(a) to define and describe the item, its functionality, dependencies on, and interaction with the driver, the environment and other items at the vehicle level, and (b) to support an adequate understanding of the item [...]" [29]. There is no strict correspondence of the item definition to SOTIF, where it is part of the "Specification and Design" activity [30]. It however specifies a large number of resulting work products, including a description of the ODD, performance targets of sensors, potential performance insufficiencies and foreseeable misuses, but also countermeasures. ISO 21448 states this activity as part of an iterative process that includes the whole concept phase as per ISO 26262, but with a focus on SOTIF instead of functional safety. This section focuses on the ODD.

An ontology serves as a foundation for a set of constraints representing the ODD and thus for any of the ODD-related components of an item definition, e.g., the behavior at the vehicle level. Moreover, the ontology can be used for formally analyzing the ODD description, e.g., finding inconsistencies, empty ODDs, or applicable traffic rules. In this step, a first version of the ontology is created based on the use case of the product (e.g., the business model and target country).

3.1 Running Example

Recall that our example is concerned with the propagation of traffic rules through the safety engineering process. For the applicable norms, we find the UNECE R157 to be applicable to ALKSs. It states that "the activated system shall comply with traffic rules relating to the dynamic driving task in the country of operation" [61, Clause 5.1.2], i.e., the German road traffic act (StVO) in our example.

As to identify these rules during the item definition, the ODD needs to be specified. We exemplary assume the following constraints for the ALKS activation:

- It can only be active (a) on highways, and (b) if rain intensity is < 2 mm/h.
- Even if these conditions are met, it cannot be active (c) on off- and on-ramps, (d) on construction sites, (e) if there are vehicles present not allowed on highways, (f) if there are pedestrians present.

Such an ODD description can be seen as a set of constraints over a possibly pre-existing (e.g., legal) ontology. I.e., the underlying model specifies what entails a 'highway' and answers specifics, e.g., whether the driveway to a highway rest area is considered an off-ramp. An item definition thus references an initial ontology, e.g. in form of a diagram or natural language. An example is given in Fig. 2.

It is only due to the ontology-based ODD definition that legal experts can reasonably select the applicable StVO rules. Among others, we analyze the safe distance rule (§4.1.1 StVO): *'The distance to a preceding vehicle shall be, generally, large enough to allow stopping behind this vehicle even if it is suddenly braked.'*

We infer that this rule is applicable to the ODD by checking whether the concept 'vehicle' of the premise of §4.1.1 is non-empty w.r.t. the ODD. For this, we assume the intersection of passenger cars and vehicles with a designated maximum speed over 60 km/h to not be empty, as well as synonyms and homonyms to be reflected in the ontology. According to the given ontology, passenger cars are vehicles, passenger cars with a designated maximum speed over 60 km/h are highway traffic participants, and highway traffic participants can drive on highways. Therefore there exist things that can drive on highways and are vehicles (namely, allowed passenger cars) and §4.1.1 can be active within the ODD. Thus, the list of applicable traffic rules contains §4.1.1 of the StVO. Section 4.1 continues this and exemplarily shows a hazard analysis for the violation of this traffic rule.

3.2   Related Work

We specifically focus on traffic rules, as their adherence is mandatory for systems SAE Level 3 and upwards. There is a natural connection between ontologies and traffic rules: The legal expert requires accessing and storing a precise and consistent terminology [13,67]. Besides traffic rules, the running example highlights that an informal structure of the system's operating conditions is required, as e.g. presented by Czarrneky [19,20]. Similarly to this, Scholtes et al. give a unified structure for a high-level, informal taxonomy of traffic entities [54]. On a more formal note, ASAM OpenXOntology aim at providing a mechanism for the'extensibility of the underlying ontology' of ASAM standards, e.g., Open-ODD [2]. Our work proposes to incorporate such ideas into the item definition coherently.

When defining an ODD during item definition one has to rely on a modeling language, an aspect we did not expand on in our running example. Thus, let us shortly highlight existing approaches relying on ontologies: Irvine, Schwalb et al. [31,55] present an ODD language with a mechanism to import OWL ontologies and to connect natural language descriptions and formal semantics. Erz et al. give another example of ontological refinement by combining a UML-based ontology with Zwicky boxes for the definition of ODDs [24]. Early proposals for tools exist, e.g., by defining constraints on an underlying ontology [47].

3.3   Requirements on the Ontology Language

In this early phase of the system development, an ontology supports structuring of the entities present in the application domain. It is the product of an iterative process in which concepts and relations from different domains, such as ODD or traffic rules, are merged and unified. The ontology must support natural language definitions of concepts and may already offer tools to transform these natural language definitions into formal concept definitions. Empty concepts must be avoided. From the perspective of ODD-engineers it would be desirable to have tool support for graphical editing and interactive browsing of the ontology as well as the maintenance and import of existing ontologies from different domains.

An important role is played by the "is a"-relation ($\rightarrow$), which indicates that one concepts is a sub-concept of another. In order to determine whether relevant domain constraints or specifications have to be taken into account (e.g., as shown above along the distance rule), the "is a"-relation shall be specified as exhaustively as possible, because of the transitivity of this relation. Furthermore, it is advisable to make use of logical compositions of concepts whenever applicable, i.e., using unions (e.g., On-Ramps *or* Off-Ramps are Ramps), intersection (Passenger Car *and* Allowed), and even negation of (basic) concepts, to identify inconsistent concepts (which often indicate modeling errors) and to reduce complexity. The predominant rules of inference are classical syllogisms and have been intensively studied as term logics. It is advisable to investigate whether tool support for ontologies can be enriched by specialized solvers, as syllogisms are in the fragment of monadic first order logic and, hence, decidable [9,16].

Of the other relations ($\rightarrow$), the "has"-relation assigns certain properties to concepts and therefore has a functional character ("functional roles"). Important properties of relations such as functionality, reflexivity or transitivity ("part of" vs. "can drive on") should be annotated.

4 Hazard Analysis and Risk Assessment (HARA)

Based on the item definition, ISO 26262 mandates HARAs as to identify hazardous events (combinations of hazards with scenarios), where potential malfunctions may impact safety. These scenarios shall be consistent (they are not contradictory) and sufficiently complete (they encompass, in union, all safety-critical scenarios). Subsequently, the risk of each hazard is (qualitatively or quantitatively) assessed, and, if necessary, *safety goals* are derived. The objectives coincide with ISO 21448, where ISO 26262 is directly referred to. The main difference is the focus on system failures vs. functional insufficiencies that are activated by triggering conditions. The corresponding activity is specified in Chap. 7 "Identification and evaluation of potential functional insufficiencies and potential triggering conditions". The SOTIF standard explicitly specifies the concept phase as an iterative process. One reason is the fact that the standard defines the elimination of high-risk scenarios by restricting the ODD a valid safety measure, which in turn may require updating the HARA.

Using an ontology supports this step in three ways: Firstly, its discrete structure can be used in keyword-based brainstorming approach as to cover all possibilities systematically. Secondly, congruence is ensured, both during specification of hazards, scenarios, and safety goals, e.g., having a correspondence between concepts from traffic rules and safety goals, and during quantitative assessment of risk, e.g., mapping simulation data for controllability estimates to real-world data of exposures. Finally, formal methods become applicable: If we assume the ontology to be a valid model of the world, we can assess the consistency and completeness of a catalog of formal scenarios w.r.t. the ontology. For this, the step refines the initial ontology of the item definition by incorporating all concepts required to express hazards and scenarios into it.

4.1 Running Example

We left our running example in Sect. 3.1 with declaring the safe distance rule as relevant within the exemplary ODD. The subsequent step is a hazard analysis and risk assessment, based on the item definition. For our example, we consider the hazards induced by the potentially hazardous behavior (in the sense of Putze et al. [49]) of violating the safe distance rule (HB1). The goal is *to assess the risk associated with this potentially hazardous behavior*. Induced hazards can be:

– A distance to the lead vehicle inappropriate for timely braking (Hz1), or
– an inappropriate evasive maneuver (Hz2).

Subsequently, hazards are combined with scenarios: Hz1 can e.g. be present during emergency braking (Sc1) or cargo falling from a lead vehicle (Sc2). These events induce several harms, e.g. injuries to persons in the lead (H1) or ego vehicle (H2). To uncover low friction conditions (e.g. wet or icy surfaces) as a triggering condition (TC1) for HB1, we can use structured brainstorming based on an ontology of safety-critical factors (which may be different from the one of Fig. 2, in which case congruence has to be ensured). Figure 3 shows all artifacts.

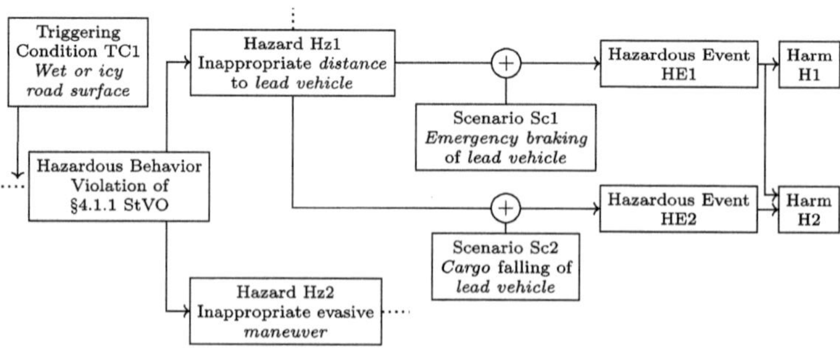

Fig. 3. An excerpt of a hazard analysis for the exemplary hazardous behavior'violation of the safe distance rule'. Employed ontological concepts are *italic*.

It is beneficial to describe all artifacts, i.e., hazardous behavior, hazards, scenarios, and triggering conditions, in terms of the ontology. To ensure their semantics to be consistent with the legal interpretation of hazardous behaviors, the use of an ontology that is congruent with the item definition ontology is required. We thus refine the initial ontology, as shown in Fig. 2 (gray boxes): Based on Hz1, we add **Lead Vehicle** and **Distance** as a property between two traffic participants. TC1 requires the addition of **Road** and **Surface**. Since hazard analyses are typically scenario-based, corresponding scenarios are also described using the ontology. For this, we add **Emergency Braking** and its defining concept **Acceleration**. In the concept phase, functional scenarios are often used [39], requiring the ontology to be a foundation for informal scenario descriptions.

For brevity, we omit a delineation of the subsequent risk assessment of HB1 at this point. We note, however, that various data sources may be employed for risk assessment, e.g. traffic simulation data for assessing the controllability, or real-world data for estimating the occurrence of triggering conditions. Before any quantity can be estimated on such data, they have to be mapped to the ontology used during hazard analysis. Otherwise, semantic mismatches can arise. For example, if the hazard analysis assumes a vehicle to be any type of transportation machine but the data model tags only passenger cars as such, the occurrence of lead vehicles and thus the overall risk of HB1 may be underestimated.

As a final step, we derive the safety goal *'Road surface condition must be considered in the safety distance to any lead vehicle'* (SG1) from TC1, described in terms of the ontology. Section 5.1 presents a safety concept for SG1.

4.2 Related Work

Performing HARAs for complex design domains calls for structured methods. As an input to a hazard analysis, a list of ontology-based criticality phenomena (abstract and largely system-independent classes of danger) can be used [44]. This is also prerequisite for the formalization of such phenomena [65]. Advancing to the system-dependent hazard analysis, Kramer et al. propose an approach for the identification of hazardous scenarios [36]. These scenarios are based on a specification of the considered ODD. Ollier et al. similarly propose to use an ODD for hazard analyses, and state requirements on the ontology language for doing so, e.g. to 'support hierarchical representation with inheritance' [47]. Our running example showed that scenarios specified in the HARA can be defined based on an ontology, which is supported by the literature [10,22]. Such scenario definitions must be congruently interpreted between diverse stakeholders, as highlighted by Zhang et al. [69]. Moreover, Erz et al. use the same ontology for the ODD as well as scenarios (e.g., those specified in a HARA) to 'consolidate and harmonize' different views along the engineering process [24]. Finally, the Automotive Global Ontology [62] has been proposed as a unified labeling mechanism for congruence between data sets, a potential issue in risk assessment over diverse data sources.

4.3 Requirements on the Ontology Language

While the ontology in the former step referred to abstract concepts and their relations only, scenario descriptions put new demands on ontologies. On the one hand, a scenario description refers to distinguished — often named — objects that are concrete instances of concepts and on the other hand, qualitative or quantifiable properties, like `Leading Car`, `Distance` or `Acceleration`, are added.

For the ontology the open world assumption (OWA) applies, as scenarios typically fix some states and individuals of the world only, while keeping other states or the existence of further individuals open. Hence, the OWA allows the combination of hazards (e.g. inappropriate distance) and scenarios (e.g. emergency braking of the leading vehicle), without having to specify triggering conditions (e.g. an icy road) in advance. On the other hand, however, in certain cases one wants to limit the number of related individuals, e.g., specify that at most two vehicles are on some road. From a formal perspective many DLs support the specification of situational knowledge by means of a finite set of assertional axioms on individuals ("ABox") [41] and can support number restrictions of relations [8].

An ontology must support assigning properties such as speed or acceleration to concepts. The ontology shall also allow to define composite concepts based on these properties, e.g., a stopped vehicle is a vehicle whose speed is zero. In addition, it has to enable temporal constraints to be defined over concepts, possibly in a separate but formally connected language, e.g., a parking vehicle is a vehicle that stops for longer than three minutes. This is difficult to realize from a formal view point. In $\mathcal{ALC(D)}$, properties can be mapped by (chains of) functional roles into concrete domains and concepts can be defined referring to domain-specific predicates. However, to remain decidable, satisfiability in the concrete domains has to be decidable and the set of predicate names in the domain must be closed under negation [7]. Some properties, e.g., distance, are defined over tuples of concepts. Hence, an ontology language shall allow for n-ary roles, however, this can also be expressed by reification [15].

Another requirement is that the ontology shall support engineers performing HARAs. In detail, this means that good visualization and explorability of the ontology is important, that new concepts can be easily integrated into the ontology, and that any difficulties stemming from formalization issues, as indicated above, should not interfere with the engineer's actual work. Moreover, there is the desire to link the ontology with other tools, to map simulation results in the context of risk assessment to the ontology or to support creativity methods for hazard identification (e.g. Zwicky boxes), which may require a machine-readable format.

5   Functional Safety Concept

The development of safety concepts aims at achieving the safety goals as identified in the previous phase. ISO 26262 specifies two phases for their development. The *functional* safety concept relates to the system level architecture, and the *technical* concept to its implementation at the HW/SW level. Functional safety focuses on faults as causes, and thus safety concepts are mainly about detection and mitigation of such faults. It subsumes the specification of a consistent set of safety requirements and their allocation to the architecture. An established way for implementing safety requirements is the integration of mitigation measures in terms of fault detection and safety mechanisms, such as the addition of redundancies. ISO 21448 splits the corresponding activity into five chapters. In the following we concentrate on Chap. 8 "Functional modifications addressing SOTIF-related risks". Chapters 9–12 focus on the V&V strategy, the evaluation of known and unknown scenarios, as well as the effectiveness of the applied measures. Different types of functional modifications are defined. The first addresses improvements of e.g. sensors, actuators and perception, and explicitly also the implemented algorithms, such as improvements by recognizing certain (triggering) conditions of the ODD. This is exemplified in Sect. 6. The second modification type is about raising the awareness of the system by other traffic participants, e.g. by enhancing visibility. Beside measures addressing foreseeable misuse, also functional restrictions to specific use cases are considered, e.g., ODD limitation, i.e., excluding potential triggering conditions s.t. the related hazards never occur.

When defining a safety concept, an ontology aids twofold: Firstly, it fosters congruence between safety requirements and the hazards allocated to their corresponding safety goals, ensuring the requirements address, in fact, the underlying hazards. Secondly, formal methods become again applicable. A common tool in this phase is consistency analysis of requirements, i.e., checking whether requirements area not implementable (a contradiction between a set or within a single requirement), or if (a part of) a requirement is always true (an unintentional tautology). For ontology-based requirement specification languages, consistency can be lifted to incorporate ontological aspects.

5.1   Running Example

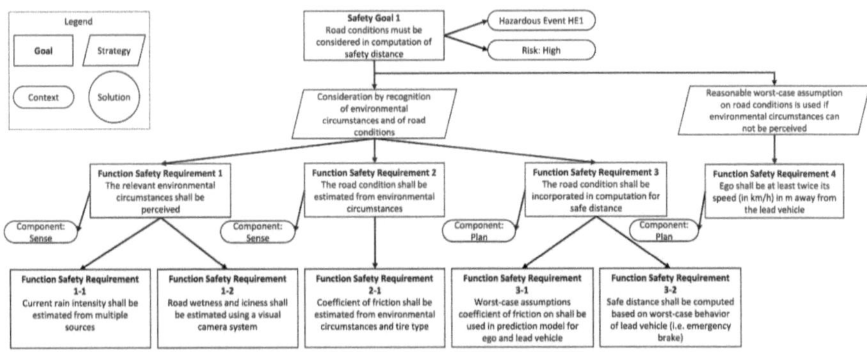

Fig. 4. Safety concept in Goal Structuring Notation [25,33] for the consideration of road conditions in the computation of safety distance (Safety Goal 1).

Figure 4 depicts the safety concept for our running example, which mainly addresses the first modification type. The top-level element Safety Goal 1 was obtained in the previous phase to address potential insufficiencies. We partially depict the safety requirements, which have been derived from this goal and specify different system modifications and perception improvements. Requirement 1 addresses potential insufficiencies of the perception, where Requirement 1–1 concerns the triggering condition *rain*, and Requirement 1–2 *water* or *ice* on the road surface. Requirement 2 specifies that these conditions must be used to derive the road condition using the *coefficient of friction*. Requirement 3 addresses a potential insufficiency of the planning algorithm and demands that the coefficient of friction must be considered in the calculation of a safe distance to a leading vehicle. As to increase reliability, an orthogonal strategy is introduced, which does not rely on road condition perception. Its requirement uses a common rule of thumb derived from the traffic rule of our example. It says that the distance to a leading vehicle must be at least twice the speed of the ego vehicle. Note that both strategies will later be incorporated in the system design using potentially different approaches.

The requirements introduced in the safety concept again induce a refinement of the ontology as depicted by the red boxes in Fig. 2. Here, the conditions of road surfaces, friction, information about vehicle tires, and speed were added.

5.2   Related Work

The development of methods for constructing SOTIF-related safety concepts is subject of ongoing research. We do not know of comprehensive tool-boxes as for functional safety. However, there exist ontology-based methods for the constructive specification of ADS. For example, the SOCA method uses an ontology as an explicit input to derive system level requirements [14]. In a similar fashion, Salem et al. propose to derive behavioral requirements from traffic rules and rely on ontologies to incorporate legal knowledge [51].

Appendix A of ISO 21448 provides an extensive argumentation template in terms of the goal structuring notation [33], where our example fits into (cf. [30], G-Dev2.1.1, Fig. A.3). There is considerable amount of work concerning safety concepts based on functional safety. See [25] for a comprehensive discussion.

An approach for the definition of mitigation strategies based on safety performance indicators has been popularized e.g. by Koopman [35], and is also part of the US standard UL 4600 [60]. Here, metrics are exploited to make individual safety aspects measurable, such as to detect insufficient risk mitigation through incident rates. Westhofen et al. review metrics for detecting such incidents [66].

5.3   Requirements on the Formalization

There is no definite answer to the question how far requirements should be formalized, because it highly depends on the methods and tools used in the development process. For this paper we seek far reaching automated tool support for development, verification and validation, which relies on sufficiently formalized requirements. As we will see in Sect. 6, this supports identifying observables that must be present in the system in order to implement the safety concept, and in guiding establishing them in the system architecture. Figure 5 depicts an approach for the specification of formal requirements, which has been developed to formalize traffic rules [67]. It consists of two layers. A formal ontology in terms of DL constitutes the base layer, in our case the ontology from Fig. 2. The upper layer is established by a definition of predicate and function terms, which take the ontology as their domain. Along this, a light-weight fragment of duration calculus has been defined in order to allow specifying temporal terms.

Figure 5 shows the terms `keeps-safe-distance`, which is a predicate, and `min-dist-in-front-of`, which is a function, with their definition. Arrows indicate dependencies in the respective definitions. The approach allows terms of the

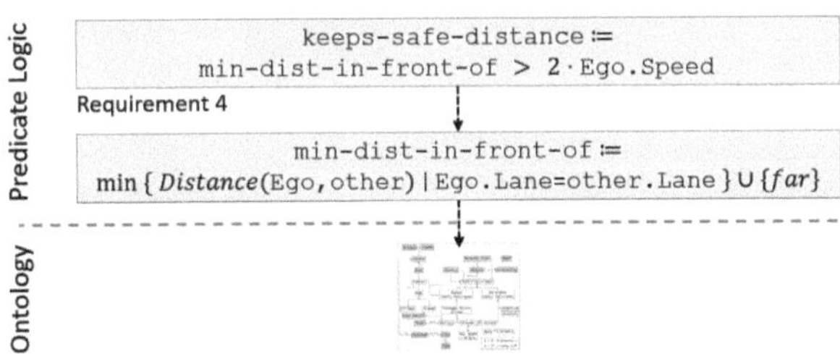

Fig. 5. Hierarchy of semantic properties, made of terms from ontology and predicates. (See Fig. 2 for an enlarged illustration of the ontology.)

ontology and predicate layer to be used in the same way. Moreover, it enables tool support for checking consistent use of definitions in formalized requirements.

6   System-Level and HW/SW-Level Design

The concept phase discussed in the previous sections is followed by product development. ISO 26262 divides it into system level and the HW/SW level design. At the system level, the realization of the functional safety concept is developed. To this end, a system-level architecture is created to which the safety requirements are allocated. This forms the base for constructing the technical realization. At the HW/SW level, the implementation takes place. These phases also include verification and integration testing, in reverse order from the technical components back to the system level, where the safety validation takes place. The SOTIF standard demands additional activities concerning the operation phase, which mainly involves the design (and later execution) of the monitoring process. Monitoring as to the standard serves two objectives: to evaluate (1) estimated occurrence probabilities of unknown hazardous scenarios, and (2) the general risk of context evolution during operation.

Using ontologies in the product development serves mainly a consistent interpretation of the requirements that are to be implemented. This not only concerns a congruent understanding between requirement and system engineers, but also the development—and integration—of different subsystems. It particularly helps avoiding issues at the "right-hand side" of the V-process, when verification and validation takes place. Refinements of the ontology in these phases, if any, are driven by the implementation, such as how distances shall be measured, e.g., in Frenet coordinates along the lane or as Euclidean distance, and the determination of measuring units (e.g., meter or yards).

6.1   Running Example

While the standards do not recommend any specific approach for system design, our running example employs a particular *model-based* methodology,

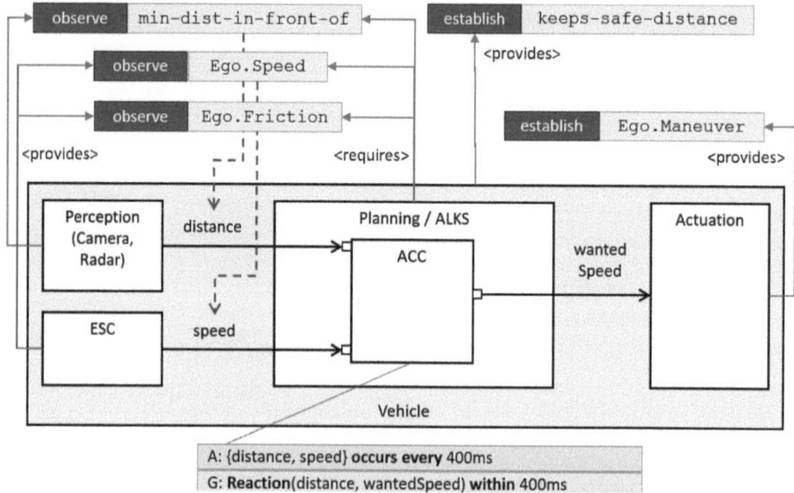

Fig. 6. System-level architecture with capabilities, and ACC component with contract. Safety Requirement 4 is allocated to the top-level vehicle component.

namely *contract-based design* (CBD). Model-based systems engineering combines domain-specific languages based on well-defined semantics with desired tool support to reduce design effort and to avoid mistakes. While CBD is language agnostic, it works well with models and tools for architecture design, such as SysML, Eclipse APP4MC [1], and AUTOSAR.

The fundamental concept in CBD is a component, and requirements in terms of contracts are assigned to components. It is the very nature of CBD to be inseparably connected to architectures. The context of any contract is a component and its local environment. The example employs so-called A/G-contracts, which consist of two parts: provided that a component is working in an environment specified by the *assumption* A, the component behaves as specified by the *guarantee* G. We refer to [12] for a comprehensive discussion on contract-based design. Details on the pattern language used for the example and its application can be found in [23].

The example proceeds by considering (a simplified version of) an adaptive cruise control (ACC) discussed in [11]. Figure 6 depicts an excerpt of the system-level architecture. The ACC component serves as a sub-component of the ALKS for longitudinal control. It has a contract assigned as part of the development process, and specifies its timing behavior. The assumption states that it gets updates on its input ports `distance` and `speed` every 400 ms. The ACC takes these inputs and calculates a `wantedSpeed` value for the actuation subsystem. The guarantee `Reaction(distance,wantedSpeed) within 400 ms` ensures that the calculation is finished before the next input arrives.

Our running example introduces *capabilities* in order to establish a consistent interpretation of requirements as a desired property stated in the beginning of

Sect. 6. A capability is any property as defined in Sect. 5.3 by (1) adding an obligation to it, and (2) assigning it in either a *provides* or *requires* role to one or more components. Two obligations have been defined: *observe* and *establish*. Figure 6 shows several capabilities at the top part.

The assignment of capabilities to components imposes responsibilities that are consistent with the contract viewpoint: The responsibility for a *provided* capability is with the assigned component, while *requiring* it moves the responsibility to the environment. An *establish*-capability expresses a (classical) requirement that must be provided by only one component. In our example, the top-level component `Vehicle` has assigned the capability to establish Safety Requirement 4 stated in the functional safety concept (cf. Fig. 5). An *observe*-capability can be provided by multiple components, depending on where it is needed in the system architecture.

Looking at Fig. 6 again, the ACC component certainly contributes to the `keeps-safe-distance` requirement: It gets a `distance` value from the perception chain in order to set a `wantedSpeed` value supposed to support establishing it. The contract however does not provide any mean to ensure that the `distance` is interpreted in the same way as in Safety Goal 1.

The allocation of capabilities to component interfaces establishes such common interpretation. Allocations are depicted as dashed arrows in Fig. 6. The meaning of an allocation is to make the corresponding property visible at the interface. For example, `min-dist-in-front-of` is allocated to the `distance` interface of the ACC component, hence values observed at this interface are meant to comply with the definition of the property. In turn, the contract can be rephrased as:

A: {`min-dist-in-front-of`, `Ego.Speed`} **occurs every** 400ms.

G: **Reaction**(`min-dist-in-front-of`, `wantedSpeed`) **within** 400ms.

This concludes our example, where ontologies serve as seamlessly applied specification mean in a development process, supporting congruent interpretation up the system architecture design. Note that Fig. 6 shows only an intermediate snapshot of the development: While the ESC is assigned as a provider of the capability to observe `Friction` of the ego vehicle, there is no corresponding interface. Its addition would be a required next design step for implementing Requirement 3-1 from Sect. 5 in the ACC as part of the safety concept.

6.2   Related Work

Capability modeling is not new and is for example an established tool in business modeling [56]. It also has emerged as engineering support, where systems and system components can be specified in terms of the capabilities they provide or require [21]. The application in the context of modeling technical systems is an ongoing research area and particularly focused on capability models with sound semantics. For example [53,64] both can be seen as approaches to formalize capabilities by a notion of contracts, and to utilize a concept of component services in order to establish online safety argumentation in dynamic systems. The presented example goes along the same line of attack. For related work with focus on implementation in the actual system, we refer to [3].

While it is sometimes considered difficult to integrate contracts with formal specifications into engineering processes, they are able to cover various functional, timing and safety aspects in industrially relevant applications [23,32,45,46]. Moreover, the application of contracts has been further extended along the development cycle, including automated test case generation [38] and also runtime, where monitors are generated and deployed to observe system behavior [43].

6.3 Requirements on the Formalization

An obvious recommendation is using a common underlying ontology in the concept phase and in system development (in our case for specifying capabilities, and in turn contracts), as this provides mean to reason about congruent interpretation — in this case between requirements engineers and architects. To a certain degree, this could be achieved with informal specifications. However, formalized specifications for ontologies, capabilities and contracts opens the way to higher degree of automation support. Examples include syntactical correctness such as completeness of role relations (required capabilities must be provided by adjacent components), and consistent use of property definitions. Finally, capabilities with precise semantics can (in many cases) be automatically translated into executable code. In combination with (also automatically generated) executable code for contracts, this would support engineers in the implementation of safety mechanisms, and of monitors thus also serving a request by ISO 21448.

7 Conclusion

Developing intelligent CPS like ADS introduces the need to consider aspects that are beyond *functional safety*. In the automotive domain, these aspects are coined as *safety of the intended function* (SOTIF). Here, not only system faults are considered as root causes of potential hazards, but also limitations to perceive the environment and to react accordingly. In SOTIF, situational awareness becomes a key driver of the safety concept: *Triggering conditions* in the environment take the role of faults, and thus their effects must be mitigated. This raises the question how to make these aspects visible in the system design. The paper pleads for the use of ontologies as a first-class citizen among specification means. It provides examples in key design phases of ISO 26262 and 21448 compliant processes in general and by (a tiny) example.

Employing ontologies in system design is no novelty, and the paper references a large body of existing work in this area. To the best of our knowledge, however, what is new is to consider integrating them consistently and coherently into an end-to-end development process, including incremental refinement and system architecture development, where they are seamlessly combined with contract-based design.

The presentation is far from showing a complete picture and leaves many methodological gaps, foremost because only some phases of the development process have been covered. Further relevant topics, among others, include scenario modeling and testing [2,10,34], system modeling, such as for maneuver planning in road traffic [4,28,70], and runtime monitoring [5,27,37,42].

A desirable next step would be to assess existing work on the requirements that have been identified along our walk through. We would emphasis two directions. First, the formalization of ontologies that meet the identified requirements, with a particular focus on the expressiveness and decidability. Second, exiting engineering support, such as tools for modeling and validation.

References

1. App4mc. https://www.eclipse.org/app4mc
2. Asam openxontology user guide 1.0.0. Tech. rep., Hoehenkirchen, Germany (2022). https://www.asam.net/standards/asam-openxontology/
3. Alvarez-Coello, D., Wilms, D., Bekan, A., Gómez, J.: Generic semantization of vehicle data streams (2021). https://doi.org/10.1109/ICSC50631.2021.00028
4. Armand, A., Filliat, D., Ibanez-Guzman, J.: Ontology-based context awareness for driving assistance systems, pp. 227–233 (2014). https://doi.org/10.1109/IVS.2014.6856509
5. Baader, F.: Ontology-based monitoring of dynamic systems. In: Fourteenth International Conference on the Principles of Knowledge Representation and Reasoning (2014)
6. Baader, F., Calvanese, D., McGuinness, D., Patel-Schneider, P., Nardi, D., et al.: The Description Logic Handbook: Theory, Implementation and Applications. Cambridge university press (2003)
7. Baader, F., Küster, R., Wolter, F.: Extensions to description logics. In: Baader, F., McGuinness, D.L., Nardi, D., Patel-Schneider, P.F. (eds.) The Description Logic Handbook: Theory, Implementation, and Applications, pp. 226–268. Cambridge university press (2003)
8. Baader, F., Nutt, W.: Basic description logics. In: Baader, F., McGuinness, D.L., Nardi, D., Patel-Schneider, P.F. (eds.) The Description Logic Handbook: Theory, Implementation, and Applications, pp. 45–100. Cambridge university press (2003)
9. Bachmair, L., Ganzinger, H., Waldmann, U.: Set constraints are the monadic class. In: Proceedings Eighth Annual IEEE Symposium on Logic in Computer Science, pp. 75–83. IEEE (1993)
10. Bagschik, G., Menzel, T., Maurer, M.: Ontology based scene creation for the development of automated vehicles. In: IEEE Intelligent Vehicles Symposium (IV), pp. 1813–1820 (2018). https://doi.org/10.1109/IVS.2018.8500632
11. Bebawy, Y., et al.: Incremental contract-based verification of software updates for safety-critical cyber-physical systems. In: The 2020 International Conference on Computational Science and Computational Intelligence (CSCI) (2020)

12. Benveniste, A., et al.: Contracts for system design. Foundations and Trends in Electronic Design Automation, p. 296 (2018). https://doi.org/10.1561/1000000053
13. Bhuiyan, H., Governatori, G., Rakotonirainy, A., Wong, M.W., Mahajan, A.: Traffic rule formalization for autonomous vehicle. In: Proceedings of the International Workshop on Methodologies for Translating Legal Norms into Formal Representations (LN2FR 2022) in association with the 35th International Conference on Legal Knowledge and Information Systems (JURIX 2022), pp. 22–35 (2022)
14. Butz, M., et al: SOCA: domain analysis for highly automated driving systems. In: 2020 IEEE 23rd International Conference on Intelligent Transportation Systems (ITSC), pp. 1–6. IEEE (2020)
15. Calvanese, D., De Giacomo, G.: Expressive description logics. In: Baader, F., McGuinness, D.L., Nardi, D., Patel-Schneider, P.F. (eds.) The Description Logic Handbook: Theory, Implementation, and Applications, pp. 184–225. Cambridge university press (2003)
16. Cellucci, C.: The decidability of syllogism. In: Search of a New Humanism: The Philosophy of Georg Henrik von Wright, pp. 171–174 (1999)
17. Colmerauer, A., Roussel, P.: The Birth of Prolog, pp. 331–367. Association for Computing Machinery, New York, NY, USA (1996). https://doi.org/10.1145/234286.1057820
18. Colomb, R., et al.: The object management group ontology definition metamodel. In: Calero, C., Ruiz, F., Piattini, M. (eds.) Ontologies for software engineering and software technology, pp. 217–247. Springer, Heidelberg (2006). https://doi.org/10.1007/3-540-34518-3_8
19. Czarnecki, K.: Operational world model ontology for automated driving systems - Part 1: road structure. Tech. rep. (2018). https://doi.org/10.13140/RG.2.2.15521.30568, issue: July
20. Czarnecki, K.: Operational world model ontology for automated driving systems - Part 2: road users, animals, other obstacles, and environmental conditions. Tech. rep. (2018). https://doi.org/10.13140/RG.2.2.11327.00165, issue: July
21. Daun, M., Brings, J., Obe, P., Wolf, S., Böhm, B., Unverdorben, S.: Using view-based architecture descriptions to aid in automated runtime planning for a smart factory, pp. 202–209 (2019). https://doi.org/10.1109/ICSA-C.2019.00043
22. De Gelder, E., et al.: Towards an ontology for scenario definition for the assessment of automated vehicles: an object-oriented framework. IEEE Trans. Intell. Veh. **7**, 300–314 (2022)
23. Ehmen, G., Grüttner, K., Koopmann, B., Poppen, F., Reinkemeier, P., Stierand, I.: Coherent treatment of time in the development of adas/ad systems: Design approach and demonstration. In: WCX World Congress Experience (2018), sAE Technical Paper 2018-01-0592 (2018)
24. Erz, J., Schütt, B., Braun, T., Guissouma, H., Sax, E.: Towards an ontology that reconciles the operational design domain, scenario-based testing, and automated vehicle architectures. In: 2022 IEEE International Systems Conference (SysCon), pp. 1–8. IEEE (2022)
25. Graubohm, R., Stolte, T., Bagschik, G., Steimle, M., Maurer, M.: Functional safety concept generation within the process of preliminary design of automated driving functions at the example of an unmanned protective vehicle. In: Proceedings of the Design Society: International Conference on Engineering Design, vol. 1, pp. 2863–2872. Cambridge University Press (2019)
26. Gruber, T.R.: A translation approach to portable ontology specifications. Knowl. Acquisition **5**(2), 199 – 220 (1993). https://doi.org/10.1006/knac.1993.1008, http://www.sciencedirect.com/science/article/pii/S1042814383710083

27. Grundt, D., Köhne, A., Saxena, I., Stemmer, R., Westphal, B., Möhlmann, E.: Towards runtime monitoring of complex system requirements for autonomous driving functions. Electron. Proc. Theor. Comput. Sci. **371**, 53–61 (2022). https://doi.org/10.4204/EPTCS.371.4
28. Hulsen, M., Zöllner, J., Weiss, C.: Traffic intersection situation description ontology for advanced driver assistance, pp. 993 – 999 (07 2011). https://doi.org/10.1109/IVS.2011.5940415
29. International Organization for Standardization: ISO 26262: Road vehicles – Functional safety. Standard (2018)
30. International Organization for Standardization: ISO 21448: Road vehicles – Safety of the intended functionality. Standard (2022)
31. Irvine, P., Zhang, X., Khastgir, S., Schwalb, E., Jennings, P.: A two-level abstraction odd definition language: Part i. In: 2021 IEEE International Conference on Systems, Man, and Cybernetics (SMC), pp. 2614–2621. IEEE (2021)
32. Kaiser, B., Weber, R., Oertel, M., Böde, E., Nejad, B.M., Zander, J.: Contract-based design of embedded systems integrating nominal behavior and safety. Complex Systems Inf. Modeling Q. **4**, 66–91 (2015)
33. Kelly, T., Weaver, R.: The goal structuring notation – A safety argument notation. In: Proceedings of the of Dependable Systems and Networks 2004 Workshop on Assurance Cases (2004)
34. Klueck, F., Li, Y., Nica, M., Tao, J., Wotawa, F.: Using ontologies for test suites generation for automated and autonomous driving functions. In: 2018 IEEE International Symposium on Software Reliability Engineering Workshops (ISSREW), pp. 118–123 (2018). https://doi.org/10.1109/ISSREW.2018.00-20
35. Koopman, P.: How Safe is Safe Enough?: Measuring and Predicting Autonomous Vehicle Safety. 3rd edn
36. Kramer, B., Neurohr, C., Büker, M., Böde, E., Fränzle, M., Damm, W.: Identification and quantification of hazardous scenarios for automated driving. In: Zeller, M., Höfig, K. (eds.) Model-Based Safety and Assessment, pp. 163–178. Springer International Publishing, Cham (2020)
37. Li, T., et al.: STSL: a novel spatio-temporal specification language for cyber-physical systems. In: 2020 IEEE 20th International Conference on Software Quality, Reliability and Security (QRS), pp. 309–319 (2020). https://doi.org/10.1109/QRS51102.2020.00048
38. MBAT Consortium: Specification of model-based test case generation & execution methods and tools, deliverable d_wp2.3_2 (2012)
39. Menzel, T., Bagschik, G., Maurer, M.: Scenarios for development, test and validation of automated vehicles. In: 2018 IEEE Intelligent Vehicles Symposium (IV), pp. 1821–1827. IEEE (2018)
40. Mossakowski, T., Codescu, M., Neuhaus, F., Kutz, O.: The distributed ontology, modeling and specification language – DOL. In: Koslow, A., Buchsbaum, A. (eds.) The Road to Universal Logic. SUL, pp. 489–520. Springer, Cham (2015). https://doi.org/10.1007/978-3-319-15368-1_21
41. Nardi, D., Brachman, R.J., et al.: An introduction to description logics. In: Baader, F., McGuinness, D.L., Nardi, D., Patel-Schneider, P.F. (eds.) The Description Logic Handbook: Theory, Implementation, and Applications, pp. 5–44. Cambridge university press (2003)
42. Nenzi, L., Bortolussi, L., Ciancia, V., Loreti, M., Massink, M.: Qualitative and quantitative monitoring of spatio-temporal properties. In: Bartocci, E., Majumdar, R. (eds.) Runtime Verification, pp. 21–37. Springer International Publishing, Cham (2015). https://doi.org/10.1007/978-3-319-23820-3_2

43. Neukirchner, M., Lampka, K., Quinton, S., Ernst, R.: Multi-mode monitoring for mixed-criticality real-time systems. In: 2013 International Conference on Hardware/Software CodeSign and System Synthesis (CODES+ISSS), pp. 1–10 (2013). https://doi.org/10.1109/CODES-ISSS.2013.6659021

44. Neurohr, C., Westhofen, L., Butz, M., Bollmann, M.H., Eberle, U., Galbas, R.: Criticality analysis for the verification and validation of automated vehicles. IEEE Access **9**, 18016–18041 (2021)

45. Nuzzo, P., Sangiovanni-Vincentelli, A.L.: Hierarchical system design with vertical contracts. In: Lohstroh, M., Derler, P., Sirjani, M. (eds.) Principles of Modeling. LNCS, vol. 10760, pp. 360–382. Springer, Cham (2018). https://doi.org/10.1007/978-3-319-95246-8_22

46. Nuzzo, P., et al.: A contract-based methodology for aircraft electric power system design. IEEE Access **2**, 1–25 (2013)

47. Ollier, G., Razafindrabe, D., Adedjouma, M., Gerasimou, S., Mraidha, C.: Using operational design domain in hazard identification for automated systems. In: 2022 18th European Dependable Computing Conference (EDCC), pp. 109–112. IEEE (2022)

48. Park, K.J., Zheng, R.L., Liu, X.: Cyber-physical systems: milestones and research challenges. Comput. Commun. **36**, 1–7 (2012)

49. Putze, L., Westhofen, L., Koopmann, T., Böde, E., Neurohr, C.: On quantification for SOTIF validation of automated driving systems. In: Proceedings of the 2023 IEEE Intelligent Vehicles Symposium (IV). IEEE (2023)

50. Robinson, J.A.: A machine-oriented logic based on the resolution principle. J. ACM **12**(1), 23–41 (1965). https://doi.org/10.1145/321250.321253

51. Salem, N.F., et al.: Ein beitrag zur durchgängigen, formalen verhaltensspezifikation automatisierter straßenfahrzeuge. arXiv preprint arXiv:2209.07204 (2022)

52. Sattler, U., Calvanese, D., Molitor, R.: Relationships with other formalisms. In: Baader, F., McGuinness, D.L., Nardi, D., Patel-Schneider, P.F. (eds.) The Description Logic Handbook: Theory, Implementation, and Applications, pp. 142–183. Cambridge university press (2003)

53. Schneider, D., Trapp, M.: Conditional safety certification of open adaptive systems. ACM Trans. Auton. Adapt. Syst. **8**(2) (2013). https://doi.org/10.1145/2491465.2491467

54. Scholtes, M., et al.: 6-layer model for a structured description and categorization of urban traffic and environment. IEEE Access **9**, 59131–59147 (2021)

55. Schwalb, E., Irvine, P., Zhang, X., Khastgir, S., Jennings, P.: A two-level abstraction odd definition language: Part II. In: 2021 IEEE International Conference on Systems, Man, and Cybernetics (SMC), pp. 1669–1676. IEEE (2021)

56. Scott, J.: Business capability MPAS: the missing link between business strategy and it action **5**

57. Stearns, M.Q., Price, C., Spackman, K.A., Wang, A.Y.: Snomed clinical terms: overview of the development process and project status. In: Proceedings. AMIA Symposium, pp. 662–666 (2001). https://pubmed.ncbi.nlm.nih.gov/11825268

58. Stolte, T., Bagschik, G., Reschka, A., Maurer, M.: Hazard analysis and risk assessment for an automated unmanned protective vehicle. In: 2017 IEEE Intelligent Vehicles Symposium (IV), pp. 1848–1855. IEEE (2017)

59. The Assurance Case Working Group (ACWG): Goal Structuring Notation Community Standard Version 2

60. Underwriter Laboratories: ANSI/UL 4600 - Standard for Evaluation of Autonomous Products. Standard (2022)

61. UNECE, Geneva, Switzerland: UN Regulation No. 157: Uniform provisions concerning the approval of vehicles with regard to ALKS. UN Regulation (2022)
62. Urbieta, I., Nieto, M., García, M., Otaegui, O.: Design and implementation of an ontology for semantic labeling and testing: automotive global ontology (ago). Appl. Sci. **11**(17), 7782 (2021)
63. von Rosing, M., Laurier, W.: An introduction to the business ontology. Int. J. Concept. Struct. Smart Appl. **3**(1), 20–41 (2015). https://doi.org/10.4018/IJCSSA.2015010102
64. Wehrstedt, J.C., Brings, J., Caesar, B., Daun, M., Feeken, L., Hildebrandt, C., Klein, W., Malik, V., Wirtz, B., Wolf, S.: Modeling and analyzing context-sensitive changes during runtime. In: Model-Based Engineering of Collaborative Embedded Systems, pp. 125–146. Springer, Cham (2021). https://doi.org/10.1007/978-3-030-62136-0_6
65. Westhofen, L., Neurohr, C., Butz, M., Scholtes, M., Schuldes, M.: Using ontologies for the formalization and recognition of criticality for automated driving. IEEE Open J. Intell. Transport. Syst. **3**, 519–538 (2022)
66. Westhofen, L., et al.: Criticality metrics for automated driving: a review and suitability analysis of the state of the art. Arch. Comput. Methods Eng. **30** (2022). https://doi.org/10.1007/s11831-022-09788-7
67. Westhofen, L., Stierand, I., Becker, J.S., Möhlmann, E., Hagemann, W.: Towards a congruent interpretation of traffic rules for automated driving-experiences and challenges. In: Proceedings of the International Workshop on Methodologies for Translating Legal Norms into Formal Representations (LN2FR 2022) in association with the 35th International Conference on Legal Knowledge and Information Systems (JURIX 2022), pp. 8–21 (2022)
68. Yang, L., Cormican, K., Yu, M.: Ontology-based systems engineering: a state-of-the-art review. Comput. Ind. **111**, 148–171 (2019)
69. Zhang, X., Khastgir, S., Jennings, P.: Scenario description language for automated driving systems: a two level abstraction approach. In: 2020 IEEE International Conference on Systems, Man, and Cybernetics (SMC), pp. 973–980. IEEE (2020)
70. Zhao, L., Ichise, R., Liu, Z., Mita, S., Sasaki, Y.: Ontology-based driving decision making: a feasibility study at uncontrolled intersections. IEICE Trans. Inf. Syst. **E100.D**, 1425–1439 (2017). https://doi.org/10.1587/transinf.2016EDP7337

Bluetooth Low Energy for Safety-Critical Real-Time Applications

Holger Hermanns[(✉)] [ID], Michaela Klauck [ID], and Govinda Sicheneder [ID]

Saarland University, Saarland Informatics Campus, Saarbrücken, Germany
`hermanns@cs.uni-saarland.de`

Abstract. Bluetooth low energy (BLE) is a low power wireless technology well established in consumer products. While BLE is designed for reliable data transfer, the worst-case transmission delay is unbounded. This is problematic especially for industrial real-time applications that hinge on predictability of upper bounds on transmission delays. This paper investigates probabilistic delay bounds for different configurations of BLE. It employs a model-centric approach together with state-of-the-art probabilistic model checking techniques. We start off with measuring BLE transmission errors and latency to understand basic BLE performance characteristics. These measurements are then integrated into a probabilistic timed automata model of BLE communication which is shown to faithfully extend beyond the measured configurations. In particular, it enables us to explore the novel Isochronous channel mechanisms which recently were introduced to the Bluetooth specification with version 5.2. We show that Isochronous channels, albeit targeting the consumer market, promise significant benefits also in industrial real-time applications compared to regular BLE. These findings are made concrete by wrapping them into a typical industrial use case. We verify that BLE is reliably able to transfer safety-critical control signals within a 50 ms deadline with a probability of at least 99.92%, and better if Isochronous channels are used.

1 Introduction

Bluetooth Low Energy (BLE) is a wireless low-power communication technology. It is well known for its usage in consumer products and available in almost all smartphones and laptop computers today. Besides end-user interaction, BLE has also been proposed as being capable of industrial usage, e.g., for internet of things (IoT) applications or industrial wireless sensor networks. A core feature of BLE is a strong specification and certification by the Bluetooth Special Interest Group (Bluetooth SIG) . The specification encompasses all layers in the OSI model, including the application layer, to ensure interoperability [5]. Using Bluetooth for real-time applications comes with distinguished challenges. Bluetooth operates on the license-free 2.4 GHz ISM radio band. These frequencies are

---

Authors are listed alphabetically.

© The Author(s), under exclusive license to Springer Nature Switzerland AG 2026
M. Fränzle et al. (Eds.): Werner Damm Festschrift, LNCS 15471, pp. 76–95, 2026.
https://doi.org/10.1007/978-3-031-97537-0_5

shared with several other technologies such as Wi-Fi. Bluetooth needs to coexist with these other technologies and other Bluetooth devices. To manage radio interference, Bluetooth uses an adaptive frequency hopping scheme (AFH), error detection, acknowledgement and retransmission of packets [5]. These mechanisms allow Bluetooth to deliver packets reliably even under difficult conditions [21,23]. The Bluetooth specification itself however does not provide transmission latency guarantees. For plain Bluetooth, the worst-case transmission delay can theoretically be infinite [20,23]. At the same time, BLE is designed for low power consumption. This is achieved predominately by transmitting data only at fixed time intervals, called connection events. The benefits of reliable packet delivery and low power consumption however come at the cost of increased latency [20].

Fig. 1. Ducktrain concept by DroidDrive GmbH [7].

The above mentioned properties of BLE make it a promising candidate for real-time applications under power constraints provided the latency can be kept in check. We are especially interested in safety-critical real-time applications powered by on-board batteries. In these contexts, low power consumption and guaranteed upper latency bounds are essential. To shed some light on this, we consider a visionary use case inspired by the *Ducktrain* [7] concept, depicted in Figure 1. In this, self-driving cargo trailers are envisioned to follow a bicycle. The trailers are supposed to keep the distance to the lead bike autonomously based on the information available. For emergency braking and other management reasons, a separate BLE connection between bike and trailer can be an essential asset. In particular, it can be considered as the backbone of a reliable fail-safe braking mechanism, where effectuation of the bike's brake lever is wirelessly signaled along the bike and trailers, making the trailers brake in synchrony with the bike ahead. This scenario induces very tight constraints on transmission delays, similar to what has been studied by Graf et al. [2] for a proprietary wireless technology and a simpler design. To get some insight into the timing constraints induced by this scenario, for a Ducktrain convoy traveling at (maximum) 30 km/h, a transmission delay of 50 ms translates into 0.4 meters of distance traveled.

The purpose of this paper is to study in how far BLE can indeed support such time-critical scenarios, what kind of guarantees can be provided, and what configurations will optimize the reactivity of the design. The available research on

BLE transmission delay for time-critical applications is limited . Existing studies investigated the impact of radio interference on BLE performance [22–24]. Some studies measured the latency impact caused by radio interference [23,24]. Other model-based investigations have been published regarding the suitability of BLE for time-critical applications [9,15,20]. However, these studies were limited to average round-trip time predictions with different bit error rates. To the best of our knowledge, no research has been published investigating specific latency time bounds for BLE. This paper fills this gap. We provide analytical techniques enabling the prediction of specific time bounds for BLE communication given an estimated error rate. These error rates are derived from real-life measurements and integrated into a carefully designed model of the protocol behavior. This model-based approach utilizes probabilistic timed automata networks [19] to faithfully represent the important behavioral details of BLE dynamics. It furthermore exploits state-of-the-art verification and validation technology, known as probabilistic model checking [1]. Altogether, this enables the derivation of probabilistic guarantees on successful message delivery within a predefined time bound. In other words, we provide a model-based and measurement-backed quantitative answer to the question in how far BLE can be used for real-rime systems. For the modeling and analysis we use MODEST [10], a modeling and analysis environment for stochastic timed systems, together with the MODEST TOOLSET [13].

Notably, Bluetooth 5.2 [4] has introduced Isochronous channels (ISO) to BLE connections, with the intention to thereby enhance consumer market penetration. Isochronous channels in BLE are designed for audio applications which require low latency transmissions. To our knowledge, no research has been published to date, investigating the effects of Isochronous channels on BLE latency. The paper therefore explores the potential of Isochronous channels to reduce latency and increase reliability for real-time applications. This is achieved by adjusting the MODEST model to comply with the low-level Bluetooth specification for Isochronous channels. This, in turn, enables us to investigate potential improvements over regular BLE for real-time applications.

Contributions. All in all, this paper makes the following original contributions:

- We develop an abstract MODEST model of BLE and BLE with Isochronous channels.
- We use probabilistic model checking to determine BLE latency and show that it is possible to meet a strict communication deadline with quantifiable high probability.
- We measure BLE message latency under various real-world conditions and use this data to validate the MODEST model.
- We show that Isochronous channels can improve BLE reliability significantly.

Outline of the paper. Section 2 discusses related work. Section 3 reviews the relevant parts of the Bluetooth specification. Section 4 reports on the BLE modeling using MODEST. In Section 5 we discuss BLE measurements under various conditions and in Section 6 we use these results in a model checking study validating the model. Section 7 discusses the findings and concludes the paper.

2   Related Work

Conceptually closest to our work is the one of Baró Graf et al. [2] who used the MODEST TOOLSET to verify a wireless bicycle brake system. In that paper a message delivery probability of more than 99.999% was obtained. The wireless channel was tailored to *MyriaNed* with a proprietary *Chess gMAC* protocol. Our model is instead tailored to the characteristics of Bluetooth, partly achieved by incorporating elements also found in the modeling of the Bounded Retransmission Protocol [12].

Spörk et al. [23] measured connection performance for BLE with different interference scenarios. The paper introduced a "number of connection events" (n_{CE}) which corresponds to counting the number of connection events needed to complete a round-trip transmission. The paper studied the impact of interference with Wi-Fi on n_{CE} which was found to be addressable by lowering the connection interval. n_{CE} serves as a performance indicator which is used to adjust the connection interval at run-time to optimize round-trip time (RTT) delay. The authors followed up [24] with investigations regarding different physical layer modes under Wi-Fi and BLE interference, proposing an AFH channel blacklisting scheme to improve packet delivery rate by up to 22%. We take this work as one conceptual inspiration, but are here focused on real-time applications with hard deadlines.

Töpsch [27] and Lüdersen [16] measured Wi-Fi interference on BLE performance. They proposed an AFH algorithm for Zephyr and showed improvement in BLE throughput and latency by blacklisting interfering Wi-Fi channels. Our measurements of packet error rates in Zephyr's link layer are inspired by this work.

Silva et al. [22] report on measurements of BLE signal quality in an anechoic chamber and with a Bluetooth device implanted into the human body. The experimental findings showed little impact for all conditions with small distances, while the absorption of the human body for distances above 3 meters led to connection loss. These results are in line with our findings for small distances.

Natarajan et al. [17] investigated coexistence in the 2.4 GHz band. The protocols under investigation were BLE, IEEE 802.15.4 (LR-WPAN) and IEEE 802.11b (Wi-Fi). The results showed BLE interference from Wi-Fi and IEEE 802.15.4. We only consider Wi-Fi interference in this paper.

Other works on BLE interference are based on analytical models or simulations with varying bit error rates [8,15,20,21,26]. Rondón et al. [20,21] developed a simulation model for BLE connections, used to study BLE worst-case transmission delay under different bit error rates, similar to Gomez et al. [9]. Both studies show an increase of latency for larger connection intervals and bit error rates. This coincides with our findings using MODEST.

3   BLE Background

The Bluetooth core specification 4.0 was released in 2010 and introduced "Bluetooth Low Energy" (BLE) as a new technology. BLE is targeted at ultra-low power applications in consumer electronics but also at industrial applications [3]. With the 5.2 version of the Bluetooth core specification, Isochronous channels (ISO) have been added in 2019, targeting audio applications, and branded as Bluetooth LE Audio [4]. It offers a replacement for Bluetooth Classic which still is in use for audio applications. The current Bluetooth core specification is 5.3 [5].

The Bluetooth specification ensures interoperability across different manufactures and products. The specification includes all layers of the OSI model, starting with the application layer down to the physical layer. In this paper we focus on dependable real-time communication and investigate the specification for BLE and ISO for such applications. To understand the low level communication, we introduce some key aspects from the BLE controller (Physical Layer (PHY) and the Link Layer (LL)).

The PHY uses the ISM radio frequencies which are shared by other communication technologies, most notably Wi-Fi. This may lead to interference, even if several mechanics are implemented in the controller to detect interference with other radio signals and to correct transmissions errors.

3.1   Link Layer

In a BLE connection both parties exchange data packets periodically. A periodic connection event starts at a so-called anchor point. The time between anchor points is defined by the connection interval (CI). The CI can be set by a developer in a range from 7.5 ms to 4 s with 1.25 ms increments [5]. Within a CI multiple packets can be exchanged. BLE modules can play either a central or a peripheral role. A central device scans for peripheral BLE devices to be connected so as to utilize the information hosted by them.

Bluetooth packets transmitted in a connection have the following format:

- Preamble - Used by the receiving radio to synchronize the radio module and the signal transmission.
- Access Address - Used by the radio to identify connections, filter frames and detect errors.
- PDU Header - Contains sequence number and Acknowledgement Bits.
- Protocol Data Unit (PDU) - Payload data processed by the Host part of the stack.
- Cyclic Redundancy Check (CRC) - Used by the link layer for error detection.

3.2   Error Handling and Reliability

Reliability of data delivery is important for most applications. BLE guaranties a successful delivery for data as long as the connection between devices is not lost.

To guarantee this, all BLE data packets have to be acknowledged (ACK) by the recipient. If a data packet ends up not being acknowledged, it will be resent by the link layer until an ACK is eventually received.

Error handling within BLE is essential for controlling the worst-case latency. It is specified as part of the Bluetooth specification [5]. In combination with the CI, it is possible to predict latency depending on the error rate [20].

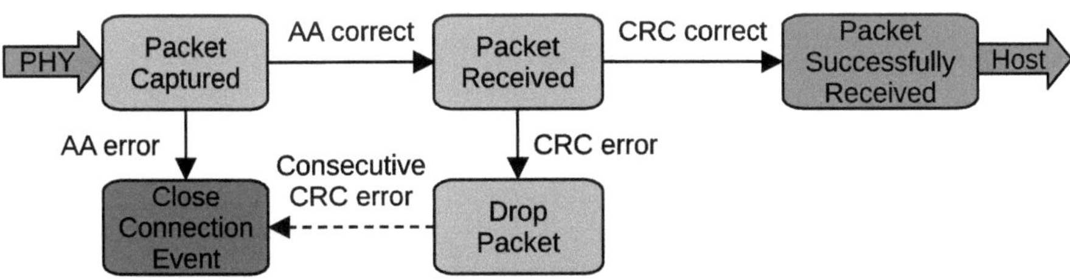

Fig. 2. BLE link layer error handling.

The Access Address (AA) is used to filter packets on the hardware level. After capturing any packet, the BLE PHY will immediately check the AA. Only if the AA matches the current connection, the packet is accepted as successfully captured and forwarded to the link layer. All other packets are ignored. If the link layer expected a packet but did not receive one, the connection event will be terminated early because the packet either was not send or it was lost due to a collision on the current radio channel which has corrupted the AA. When a packet was captured successfully the cyclic redundancy check (CRC) field will be checked next. Only after a successful CRC check the packet is accepted as successfully received and processed further. If the CRC check failed, the packet will be discarded and the link layer will explicitly reply with a Not-Acknowledge (NACK) message. If two consecutive packets fail due to CRC errors, the connection event will be terminated assuming the current channel has too much interference. This process is illustrated in Figure 2.

3.3   LE Audio and Isochronous Channels

Prior to BLE version 5.2, audio applications were not supported by BLE. However, Bluetooth audio devices such as wireless speakers or headphones are among the most popular consumer products, but thus far had to rely on Bluetooth Classic [4]. In comparison to Bluetooth Classic, BLE is a modern digital architecture and needs significantly less power. For the purpose of making the benefits of BLE accessible to audio applications, the Bluetooth SIG introduced a new audio specification called *LE Audio* for BLE. At the same time, new audio applications such as multiple synchronized audio streams are addressed by *LE Audio*. As the pivotal new mechanism to support audio streaming with low latency, the Bluetooth SIG added *Isochronous channels* to the BLE specification with version 5.2

[4]. These channels improve latency, maintain constant bandwidth and add time synchronization to BLE.

For two devices to utilize Isochronous channels, they form a connection via a so-called Connected Isochronous Stream (CIS)[5]. A CIS is similar to a default BLE connection but redefines some key aspects on the controller level. We will summarize these aspects in the next paragraph before we compare and analyze Isochronous channels further in Section 6.

The communication for a CIS is limited to discrete points in time. The duration and the interval is defined by an ISO interval which is similar to the CI for regular BLE. The ISO interval is the time interval between two consecutive ISO events and can be set by the application developer in a range from 5 ms to 4 s. An ISO interval is further split into subevents. Every ISO interval starts with one subevent where the central sends one packet, and the peripheral replays with one packet. If one device has more data to send or a sent packet had errors, it is possible to use additional subevents. Dependent on the maximum packet size, the controller can calculate an upper time-limit required for one packet exchange. This calculation results in the maximum duration of one subevent which in the sequel will be called `Sub_interval`.

The link layer uses CRC validation to check and acknowledge received packets. CIS packets can set a flush point after which they would not be retransmitted in case of errors. This can be used to enforce an upper limit on the packet transmission delay. However, this comes at the price of giving up reliable transmission of all data, albeit being useful for data which expires after a certain point in time, e.g., streamed audio data. As we are interested in maximizing communication reliability, we do not make use of flush points. The reliability in a CIS is further increased by faster frequency hopping for every subevent instead of every connection event. This enables the link layer to retransmit packets in the same connection event in case of errors, in contrast to the settings in plain BLE where two consecutive CRC errors induce closure of the connection event.

4   Modeling

This paper pursues a model-based approach to study the particularities of BLE and ISO with respect to transmission delays. It uses a modeling and description language for stochastic timed systems, MODEST [10]. The syntax of MODEST resembles the programming language C and the modeling language *Promela* [14]. On a semantic level MODEST describes the behavior of systems as networks of probabilistic timed automata (PTA) [19]. Conceptually, a PTA is an automaton model extended with (1) a notion of real-time in form of clock variables as in Timed Automata (TA) and (2) discrete probabilities as in Markov Decision Processes (MDP) or Markov chains. For the sake of conciseness, we here introduce the MODEST language constructs by example as per their use in the BLE and ISO models we develop. The reader is referred to [6,10,11] for more details on the language and its applicability across a large spectrum of contexts.

4.1   Modest Model

BLE communication is modeled as a network of three concurrent processes, (put together with nested parallel operators **par{}**). A first process models the application layer (not presented here). It generates symbolic data packets to be sent by a second process, the central process. This process models the link layer of a central device which can send data or empty protocol packets at each connection event. A third process models the link layer of a peripheral device which receives packets from the central device and sends responses back. We distinguish models for regular BLE communication and for Isochronous channels. Both models share the same process structure and differ only for the central process, in accordance to what is stipulated in Bluetooth specification 5.2.

Central Process — Common Section. The central process is divided into two alternatives for regular BLE communication and ISO communication. To measure the transmission time for one symbolic packet, the application process (running in parallel) will increase the queue by one packet at the start of its execution. The application data can only be processed at the start of a connection event. To consider the waiting time for the first packet in the transmission queue before it is processed by the central process, we add a random delay with the function **DiscreteUniform()** and a clock variable **c_start**.

```
clock c_start = 0;
real x = DiscreteUniform(0, Connection_Interval);
when urgent(c_start >= x) {==};
  par{
    :: urgent Central()
    :: urgent Peripheral() }
```

After this delay the central and peripheral processes are executed. The central process manages a global clock variable **ci** and the overall connection event timing. The actual transmission of data is delegated to one of two distinct processes, one for BLE and one for ISO. The central process is restarted after each connection event is finished and the **ci** clock is reset.

```
process Central(){
  {= ci=0 =};
  if(ISO)
    { ISO_Send() }
  else
    { BLE_Send() };
  invariant(ci <= Connection_Interval)
  when(ci >= Connection_Interval) Central() }
```

Central Process — BLE_send(). The BLE_send() process has two local variables to manage the state of the connection event.

- **count** - The number of successfully sent packets within the connection interval.

- `fail_counter` - The number of consecutive failed packets.

For one connection event the process can send multiple packets until the connection event is closed or the transmission queue is empty. At least one packet is sent in every connection event. In case the transmission queue from the application layer is empty, this will be an empty protocol packet. In case two consecutive CRC errors are detected or one Access Address error occurs, the connection event is closed immediately.

```
process BLE_send(){
  int count = 0;
  int fail_count(0..2) = 0;
  do{
    ::when((count == 0 || queue >= 1)
              && fail_count <= 1)
      urgent put {==};
      alt{
        ::get_ack
            {= fail_count = 0, count++,
            queue--, data_sent = true =}
        ::get_crc_error {= fail_count++ =}
        ::get_aa_error}  {= fail_count = 2 =}
    ::when(fail_count >=2
              ||(count >= 1 && queue == 0))
        urgent break }}
```

The transmission of a symbolic packet is abstracted by an action `put`. The action is synchronized with the peripheral process. The peripheral process introduces errors probabilistically and replies with one of three actions, together representing all possible transmission outcomes. The possible action alternatives (thus the `alt{}` statement) are:

- `get_ack` - Successful transmission, sent data acknowledged.
- `get_crc_error` - CRC error, data not acknowledged.
- `get_aa_error` - Packet lost during transmission.

Central Process — ISO_send(). Isochronous channels are redesigned based on BLE channels. The base functionality for both models is the same in case no errors occur.

The key differences are captured by the model in case of errors. Connection events are further divided into subevents. We model this by introducing an additional clock variable called `sub_c`. The number of subevents used in one connection event is limited by the connection interval and the `Sub_interval` length (set to 4 ms). The model can use as many subevents as needed until the end of the connection event is reached. In contrast to BLE, if an error occurs, the connection event will not be closed for Isochronous channels. Instead, the packet is retransmitted in the next subevent on a different radio channel.

```
process ISO_Send(){
  clock sub_c = 0;
  int count = 0;
  do{
    ::when(count == 0 || queue >= 1)
      {= sub_c = 0 =};
      do{
        urgent put {==};
        alt{
          ::get {= count++, queue--
                , data_sent = true =};
            urgent break
          ::get_crc_error {==}
          ::get_aa_error {==}; urgent break }};
      invariant(sub_c <= sub_interval)
      when(sub_c >= sub_interval) {==}
    ::when(ci >= Connection_Interval
            ||(count >= 1 && queue == 0))
      urgent break }}
```

Peripheral Process. Two global parameters (**Error_Rate** and **CRC_Rate**) are used in a **palt()** statement, representing probabilistic alternatives. The parameters define a probability distribution on possible responses to the synchronized action **put**. For our use case, we assume that a physical packet exchange requires 2 *ms* [18]. The model uses 2 *ms* as a fixed transmission delay TD.

```
process Peripheral(){
  clock c = 0;
  put {= c = 0 =};
  when urgent(c >= TD)
  palt{
    :100 - Error_Rate - CRC_Rate: {==}; urgent get
    :CRC_Rate : {==}; urgent get_crc_error
    :Error_Rate: urgent get_aa_error};
  Peripheral()}
```

We repeat that space constraints force us to refer to [6,10,11] for details on the MODEST language constructs.

4.2 Model checking

With the model at hand, we can use the probabilistic model checking engines included in the MODEST TOOLSET [11,13], to check relevant properties on the model. The property of interest in our context checks for the minimal probability of a successful transmission within a given deadline. For the tool, this property looks as follows

```
property sent_prob = Pmin(<>[T<=Deadline] data_sent);
```

and is to be read as: *What is the minimal probability to observe action* `data_sent` *within a time interval of length at most* `Deadline`*?*

To make these questions concrete, a number of parameters need to be fixed before submitting the model and property to the model checker. This is supported by the MODEST TOOLSET's option `-E` and pertains to the following constants in the model:

- `Error_Rate` - Parameter corresp. to packet error rate.
- `CRC_Rate` - Parameter corresp. to CRC error rate.
- `Connection_Interval` - Parameter controlling the length of a connection interval.
- `ISO` - Boolean flag to switch between BLE and Isochronous mode.
- `Deadline` - Deadline to wait for appearing in the `sent_prob` property.

Among these parameters, the last three are dependent on the configuration deployed and the concrete use case. The former two parameters however are of fundamental nature, and our intention is to work with estimates that are obtained from measurements taken under adverse conditions. This reflects the safety-critical nature of the application context we target. In a nutshell, we suppose that an error rate obtained empirically in a setting with large distances and several obstacles will not be lower than the one found in a carefully crafted industrial context.

5   Measurements

We conducted several experiments with Zephyr [25] to evaluate the capabilities of BLE for low-latency applications and to gather input data for parameters of the MODEST model. For this, we added functionality to Zephyr to measure raw packets in low-level sections of the BLE controller. We measured the BLE performance in various realistic scenarios and introduced additional Wi-Fi interference to the radio communication. The connection interval is the central parameter to influence BLE latency. We investigated different connection intervals from 7.5 *ms* to 100 *ms*.

5.1   Experimental Setup

We used two *nRF52 DK* boards as central and peripheral devices. The peripheral device had a simple GATT profile to receive payload data. The device accepted data and participated in link layer communication such as sending acknowledgements. The central device was able to generate test data and send data packets to the peripheral device using the GATT profile. The data packets contained one byte of payload. Zephyr 2.3 [25] was used as a real-time OS to implement the described experiments. The experiments were limited to regular BLE communication since Zephyr support for Bluetooth 5.2 and Isochronous channels was still in an experimental state at the time of writing this paper, making it impossible to carry out similar experiments with Isochronous channels using Zephyr.

We empirically compared several realistic communication scenarios with varying distances and with or without deliberate Wi-Fi interference. The experiments were carried out in a private apartment in a densely populated residential area, so background disturbances in the 2.4 GHz radio frequency band have been the norm during experimentation. Two nRF52 boards were placed in different locations in three rooms of the apartment. Figure 3 marks these locations as A, B, C, D, E and F, with the central device always positioned at location A. Locations D and E did not have a direct line of sight to location A.

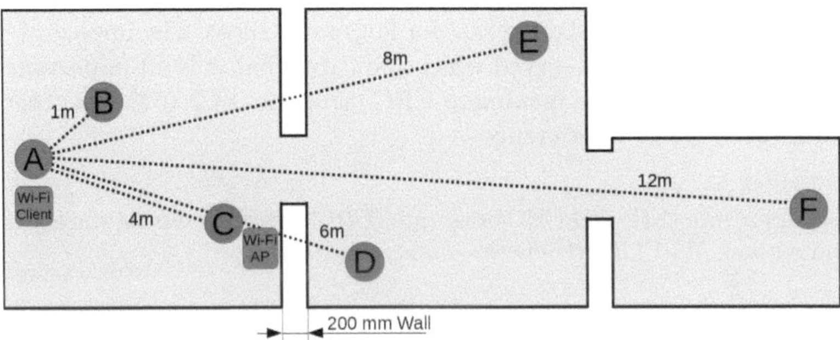

Fig. 3. Graphical representation of the locations and distances used in the experiments.

The central device was used to measure BLE performance metrics. Each experiment lasted for 1 minute and was repeated three times. The results of these three experiments formed a sub-experiment. In every experiment we sent one data packet every second to measure the RTT. We changed the connection interval for every sub-experiment to 7.5, 25, 50, 75 or 100 ms. All experiments were executed without Wi-Fi interference and with enabled Wi-Fi interference. The latter made use of a Wi-Fi access point (*AVM Fritzbox 3490* configured to use Wi-Fi channel 6 in the 2.4 GHz band) sending data packets by Ethernet to a mobile Wi-Fi client. The distance from the access point to the client was kept constant at 4 m, Figure 3 shows the overall spatial layout.

5.2   Error Rate

Lost packets are BLE packets that were sent, but either did not reach the other side or the acknowledgment was lost. Lost packets are not directly measurable within Zephyr. To measure packet loss, we implemented several function calls in the low level BLE controller, each calling a new data collection function, inspired by Töpsch [27]. This enabled counting the total number of BLE packets sent, the number of received and acknowledged packets, and the number of CRC errors. The number of lost packets is obtained by subtracting the number of acknowledged packets from the number of sent packets. To calculate the error rate, we divided the lost packets by the number of packets sent, and turned this into a percentage value, Packet Error Rate. The CRC Error Rate is calculated analogously, using the fact that CRC errors can directly be detected in the controller. Notably, packet errors include CRC errors, and so do the resulting percentages.

5.3   Results

To understand how distance and interference influence the BLE latency, we analyzed the packet error rate. Table 1 lists the measured error rates for each location as an average over the sets of experiments. From the results we observe a clear trend to an increased error rate for larger distances. The impact of "Wi-Fi interference" can also be observed which indicates that it is an important factor for BLE signal quality. The maximum CRC error rate of 2.76% was observed at location E with Wi-Fi interference.

Table 1. Error rate (ER) and CRC error rate (CRC) observed during the experiments with and without Wi-Fi interferences.

Location	no Wi-Fi		Wi-Fi	
	ER	CRC	ER	CRC
A	0.09%	0.00%	4.42%	0.28%
B	0.46%	0.03%	18.42%	1.02%
C	0.47%	0.04%	19.46%	0.25%
D	1.65%	0.12%	22.31%	0.17%
E	5.17%	0.33%	30.90%	2.76%
F	8.51%	0.52%	31.11%	2.60%

6   Validation and Evaluation

We now return to the safety-critical use case alluded to initially (Figure 1), where the brakes of a self-driving cargo trailer need to be effectuated from remote, namely as a result of the bike rider ahead of the trailer pressing the brake lever at the handlebar. For simplicity, we assume the same braking technology (brake shoe and effectuation mechanism) to be used both at the bike and at the trailer. We furthermore assume a (maximal) cruising speed of 30 km/h or 8 m/s. With this speed, the convoy travels a distance of 40 cm within 50 ms. This means that if the safety distance maintained between cargo trailer and bike during cruise is at least 40 cm, then a collision of bike and trailer is avoided if both effectuate their brakes near-synchronously, but with a time difference of at most 50 ms. So, all in all we need to show that BLE can reliably communicate with a maximum transmission delay of 50 ms. (The model checking is not at all bound to the value 50 ms, MODEST TOOLSET can work with arbitrary time bounds.)

This means that we will model check the property `sent_prob` discussed at the end of Section 4 with parameter `Deadline` set to 50 *ms*, unless otherwise stated. With respect to the other model parameters (`Error_Rate`, `CRC_Rate`, `Connection_Interval`, and `ISO`), we evaluate a variety of configurations. For link quality, we focus on scenarios that clearly over-approximate the true conditions. Concretely, we work with parameters as determined empirically for experiment location E. In this location, the line of sight is interrupted by a brick wall and the distance is 8 *m*. So, the experimental results in location E can be assumed to over-approximate any real-world condition.

With this, we proceed in several steps regarding the remaining parameters under study. We first compare model-based and measurement-based results, in order to validate that the MODEST model indeed accurately models the decisive value of interest, namely communication latency, encoded in property `sent_prob`. Then we study the behavior of the model for different settings of the other parameters.

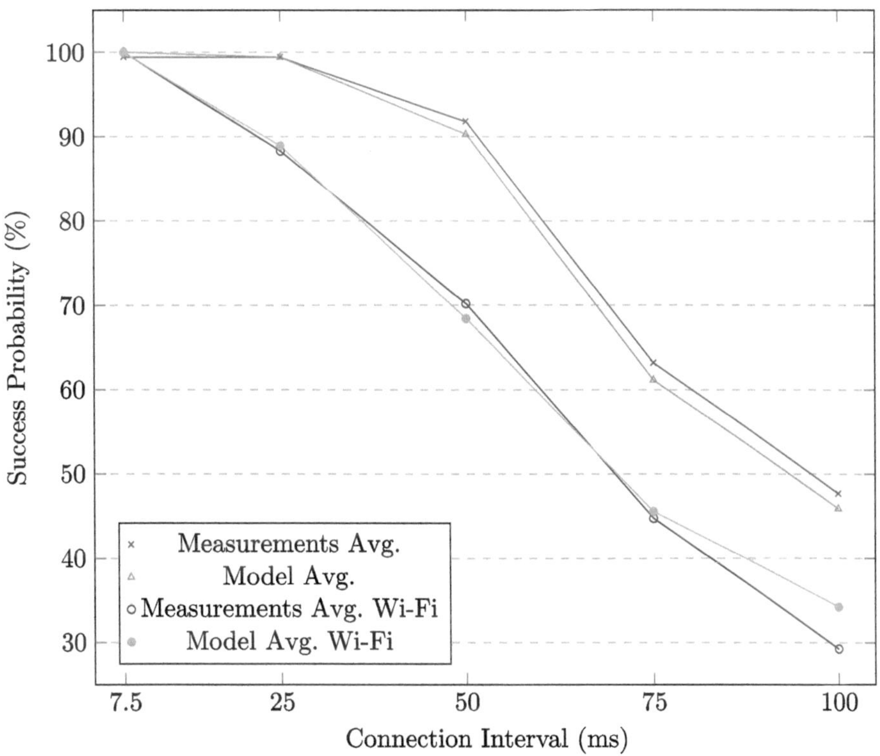

Fig. 4. Comparing the model results ("Model") to experimental results ("Measurements") for 50 *ms* deadline at location E with and without Wi-Fi interference.

6.1 Model Validation

To validate the model, we compare the model checking results to experimental results from Section 5 (for location E). The experimental data consists of three experiments per connection interval and interference condition. Each experiment measures the error rate over a one-minute interval. From these three experiments, we calculate the arithmetic mean (average) error rate. The resulting average error rate from the experiment is used as a parameter to run the model for each connection interval. In Figure 4 we compare the probability for packets to be successfully transferred within a 50 ms deadline (RTT). The red/orange graphs are without Wi-Fi interference with error rates around 5%. The blue/green graphs are with Wi-Fi interference and error rates around 30%. The results show, that our model is able to predict the BLE performance very accurately. The differences of the curves are within 2% with one exception. At the 100 ms connection interval the measured results for the Wi-Fi interference condition are 5% below the model prediction.

We observe from these investigations that our model predicts BLE connection performance accurately for connection intervals up to 75 ms. While the accuracy is uncertain for connection intervals from 100 ms, we consider the results as a strong indicator for latency performance. For time-bound predictions regarding the use case, we will only generalize the BLE model for connection intervals below 100 ms.

6.2 Evaluation

We now turn to purely model-based evaluations of BLE and Isochronous channels, obtained by probabilistic model checking of the various model instances. Especially, we study BLE in comparison with ISO. For the latter we have no way at hand to derive experimental results to compare with, due to lacking implementation support, as highlighted before. In other words, we are turning to temporal properties for which, at the time of writing of the paper, it has been impossible to study them by simply running experiments and measuring. This highlights the benefit of probabilistiv model checking tool support for our evaluation case.

Connection Interval. As the transmission delay is influenced by the error rate for both models, we investigate the influence of the length of the connection interval on the success probability. In Figure 5, the plots BLE and ISO depict the probability for a successful transmission within a 50 ms deadline assuming a constant error rate of 30% (displayed together with two other plots explained below). The connection interval is set to 7.5 ms, 10 ms, 20 ms up to 100 ms with 10 ms step size. For connection intervals greater than 50 ms, the connection interval exceeds the deadline. Therefore, the probability is influenced by the chance for the connection event to start after the 50 ms deadline was passed.

If we look at the results for connection intervals from 7.5 ms to 50 ms in Figure 5, we can observe that both curves start with 100% and the probability

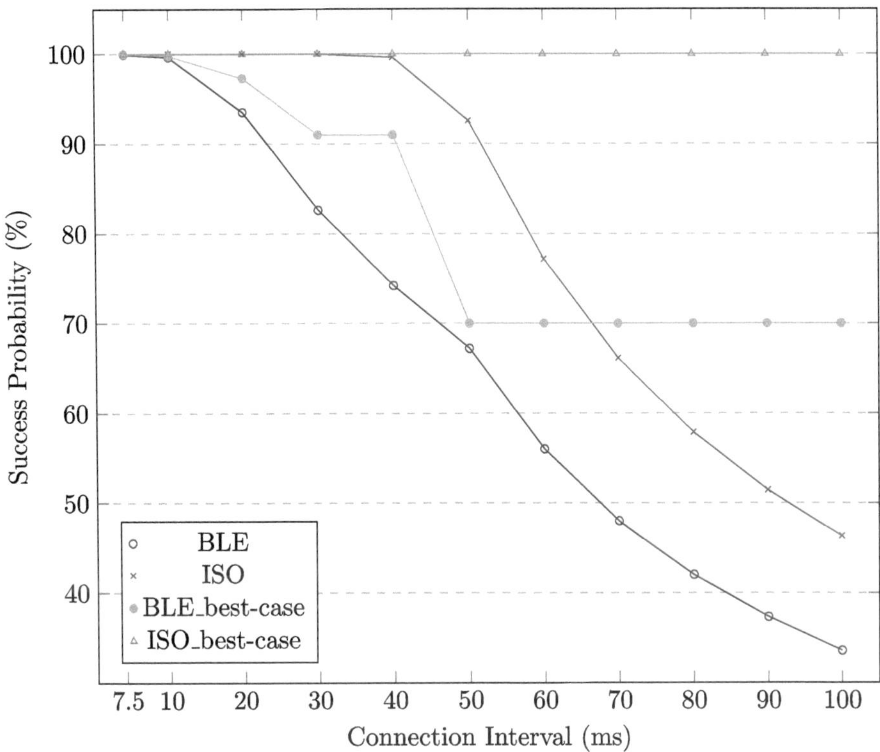

Fig. 5. The probability for one data packet to be transmitted and acknowledged within 50 *ms*. The error rate is set to 30%.

for BLE decreases for higher connection intervals. For Isochronous channels the probability only decreases only slightly for these low values. At 50 *ms* the probability has dropped to 93% due to an overlap of the deadline and the physical transmission delay of 2 *ms* for a packet exchange.

With connection intervals of more than 50 *ms*, both models continuously decrease in probability because of the deadline. For example, the 100 *ms* connection interval for Isochronous channels reaches 46%, which is rooted in a 50/50 chance of success whether the connection event starts within the 50 *ms* deadline plus influences of the physical transmission delay. Evidently, the connection interval should in general be chosen smaller than the expected deadline for latency sensitive applications.

To isolate the link layer performance from the application layer we report on another pair of experiments in Figure 5 (BLE_best-case/ISO_best-case). Here, the transmission delay is set to 0 *ms* while keeping the deadline 50 *ms*. This models a best-case timing scenario and thus can showcase the performance benefits of Isochronous channels under ideal conditions.

The plots demonstrate that the probability of a successful transmission for BLE within the deadline depends on the connection interval. We can see that for short connection intervals the probability is close to 100%, but decreases for larger values. Once the connection interval is equal or greater to the deadline, the probability is at 70%. This can be explained by the protocol design. Within a connection event the protocol can only tolerate one CRC error and no Access Address error. Whenever consecutive CRC errors or one AA error occurs, the event is closed. In these cases, the protocol has to wait for the next event to retry the transmission. For connection intervals greater than the deadline this results in only one transmission attempt, which has a success probability of 70% due to the packet error rate of 30% in this example. For the lower connection interval values the reasoning is equivalent, only the number of transmission attempts increases.

In contrast, we observe that the success rate stays near 100% for the Isochronous channel results. This echoes that there are improvements to the protocol, due to the newly added subevents. A subevent is one of several subevents of a connection event. If a transmission fails in one subevent, the protocol attempts a retransmission in the next subevent. The concrete number of possible transmission attempts depends on various parameters. For example, assuming a connection interval of 50 ms, and a 5 ms subinterval, this results in 10 possible transmission attempts. For BLE with the same 50 ms connection interval, this would result in one to three transmission attempts. It becomes evident that Isochronous channels therefore are much more resilient to errors on the underlying transmission medium.

Deadline. We now look into the interplay of time and probability when a packet will be transmitted and acknowledged under a given condition. To understand this, we plot the probability with a fixed error rate of 30% and a connection interval of 25 ms in Figure 6. The transmission buffer delay is enabled again with a range of 0 to the `connection_interval`.

With this data we can predict the performance for BLE and Isochronous channels for time bounded applications. We can see for BLE, that the probability increases slower. After 100 ms we reach a 99.04% probability for the data to be acknowledged under the chosen condition. On the other hand, for Isochronous channels (ISO) we reach 99.99% probability at 50 ms. Under these conditions, Isochronous channels are more reliable and can be used for real time application with a 25 ms connection interval. To achieve similar results with BLE, we plotted the reliability for BLE with a 7.5 ms connection interval. Here BLE reaches a 99.92% probability after 50 ms. For comparison, we plotted the probability for ISO for the same 7.5 ms connection interval which has a 99.99% probability after 40 ms. This shows that the difference in reliability decreases for smaller connection intervals but ISO still has an advantage over BLE.

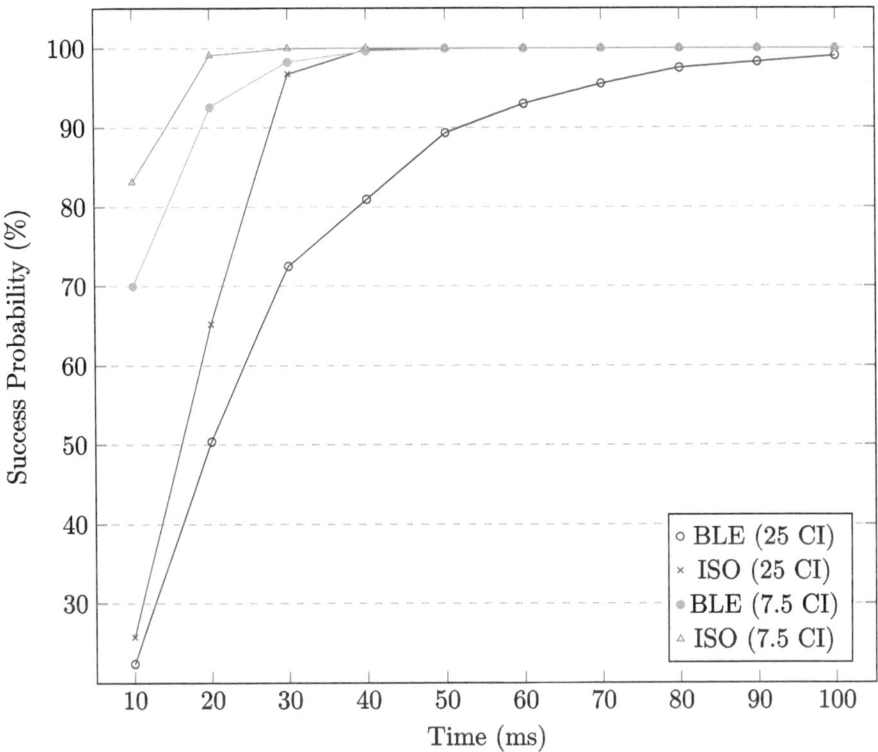

Fig. 6. Probability for a successful transmission depending on the time. The error rate is set to 30%.

7   Conclusion

This paper has investigated the adequacy of BLE with and without Isochronous channels in the context of safety-critical real-time use cases. In a nutshell, we studied exemplary scenarios which induced a 50 ms RTT bound to be met reliably. To study this problem we developed different behavioral models of the communication particularities in the modeling language MODEST. Several basic parameters were determined through measurement campaigns that are likely of interest in their own right. For the subsequent model-based evaluation, we chose the worst case parameters, so as to clearly over-approximate the risk in real behaviour.

We used probabilistic model checking with the MODEST TOOLSET for evaluating the model in various configurations. In these evaluations the BLE model was first validated by comparison with experimental results wherever possible. But due to lacking support for Bluetooth 5.2 in Zephyr, we extrapolated the behavior regarding Isochronous channels based on the model and its parameters as measured. So, probabilistic model checking was the crucial approach to evaluate the various configurations, because measurements were impossible.

All in all, the model checking results appear very promising for low latency and real-time applications, encompassing the Ducktrain use case scenario from Figure 1. There, a safety distance of 40 *cm* motivates a transmission delay of at most 50 *ms*. Indeed, the model checker has determined a 99.92% probability lower bound for a BLE packet to be acknowledged within 50 *ms* for a 7.5 *ms* connection interval and 30% packet error rate. This can be improved to 99.99% probability by instead using Isochonous channels with a connection interval set to 25 *ms*. It seems promising to improve these principal results even more by applying channel blacklisting as discussed by Spörk et al. [24]. Further analysis will be required to investigate the efficacy of channel blacklisting for dynamically moving applications, while in this paper we were able to address dynamic movement by a static worst-case over-approximation. Furthermore, the latest Zephyr protocol stacks by now do provide support for Isochronous channels in Bluetooth, and this opens the door for another experiment campaign to validate the model predictions in reality.

Acknowledgements. This work has received support by VolkswagenStiftung as part of Grant AZ 98514 – EIS, and by the DFG under grant No. 389792660 as part of TRR 248 – CPEC. This line of work got its initial trigger at CAV 1998 in Chicago, Illinois, USA, at the very first encounter of Werner Damm and the first author of the present paper, ending up in a deep discussion of the possibility of supporting hard real-time when communicating over wireless channels. These topics caught considerable momentum as part of Project S3 of the spectacular TRR 14 – AVACS where the Modest Toolset has been developed (cf. [10,12,13]) and, among others, applied to the verification of wireless technology in a safety-critical context first (cf. [2]).

References

1. Baier, C., de Alfaro, L., Forejt, V., Kwiatkowska, M.: Model checking probabilistic systems. In: Handbook of Model Checking, pp. 963–999. Springer (2018)
2. Baró Graf, H., Hermanns, H., Kulshrestha, J., Peter, J., Vahldiek, A., Vasudevan, A.: A verified wireless safety critical hard real-time design. In: 2011 IEEE WOWMOM Symposium. pp. 1–9. IEEE (2011)
3. Bluetooth SIG: Bluetooth Core Specification (6 2010), revision: v4.0
4. Bluetooth SIG: Bluetooth Core Specification (12 2019), revision: v5.2
5. Bluetooth SIG: Bluetooth Core Specification (7 2021), revision: v5.3
6. Butkova, Y., Hartmanns, A., Hermanns, H.: A Modest approach to Markov automata. ACM Trans. Model. Comput. Simul. **31**(3), 14:1–14:34 (2021). https://doi.org/10.1145/3449355
7. DroidDrive GmbH: Ducktrain. Website, https://ducktrain.io/, accessed on 2 July 2023
8. Dwijaksara, M.H., Jeon, W.S., Jeong, D.G.: A channel access scheme for bluetooth low energy to support delay-sensitive applications. In: 2016 IEEE PIMRC Symposium. pp. 1–6. IEEE (2016)
9. Gomez, C., Oller, J., Paradells, J.: Overview and evaluation of bluetooth low energy: An emerging low-power wireless technology. Sensors **12**(9), 11734–11753 (2012)

10. Hahn, E.M., Hartmanns, A., Hermanns, H., Katoen, J.: A compositional modelling and analysis framework for stochastic hybrid systems. Formal Methods Syst. Des. **43**(2), 191–232 (2013)
11. Hartmanns, A.: The Modest Toolset. Website, https://www.modestchecker.net, accessed on 2 July 2023
12. Hartmanns, A., Hermanns, H.: A modest approach to checking probabilistic timed automata. In: 2009 QEST Conference. pp. 187–196. IEEE Computer Society (2009)
13. Hartmanns, A., Hermanns, H.: The Modest toolset: An integrated environment for quantitative modelling and verification. In: 2014 TACAS Conference. Lecture Notes in Computer Science, vol. 8413, pp. 593–598. Springer (2014)
14. Holzmann, G.J.: The SPIN Model Checker - primer and reference manual. Addison-Wesley (2004)
15. Kindt, P., Yunge, D., Gopp, M., Chakraborty, S.: Adaptive online power-management for Bluetooth low energy. In: 2015 IEEE INFOCOM Conference. pp. 2695–2703. IEEE (2015)
16. Lüdersen, B.: Design and Implementation of Adaptive Frequency Hopping for Bluetooth Low Energy - Evaluation for Low Latency Applications. Bachelor's thesis, Christian-Albrechts-Universität zu Kiel (2020)
17. Natarajan, R., Zand, P., Nabi, M.: Analysis of coexistence between ieee 802.15. 4, ble and ieee 802.11 in the 2.4 ghz ism band. In: 2016 IEEE IECON Conference. pp. 6025–6032. IEEE (2016)
18. Nordic Semiconductors: Online power profiler for BLE. Website, https://devzone. nordicsemi.com/nordic/power/w/opp/2/online-power-profiler-for-ble, accessed on 2 July 2023
19. Norman, G., Parker, D., Sproston, J.: Model checking for probabilistic timed automata. Formal Methods Syst. Des. **43**(2), 164–190 (2013)
20. Rondón, R., Gidlund, M., Landernäs, K.: Evaluating bluetooth low energy suitability for time-critical industrial iot applications. International Journal of Wireless Information Networks **24**(3), 278–290 (2017)
21. Rondón, R., Landernäs, K., Gidlund, M.: An analytical model of the effective delay performance for bluetooth low energy. In: 2016 IEEE PIMRC Symposium. pp. 1–6. IEEE (2016)
22. Silva, S., Soares, S., Fernandes, T., Valente, A., Moreira, A.: Coexistence and interference tests on a bluetooth low energy front-end. In: 2014 Science and Information Conference. pp. 1014–1018. IEEE (2014)
23. Spörk, M., Boano, C.A., Römer, K.: Improving the timeliness of bluetooth low energy in noisy rf environments. In: EWSN 2019. pp. 23–34 (2019)
24. Spörk, M., Classen, J., Boano, C.A., Hollick, M., Römer, K.: Improving the reliability of bluetooth low energy connections. In: EWSN 2020. pp. 144–155 (2020)
25. The Linux Foundation: Zephyr Project. Website, https://zephyrproject.org/, accessed on 2 July 2023
26. Treurniet, J.J., Sarkar, C., Prasad, R.V., De Boer, W.: Energy consumption and latency in ble devices under mutual interference: An experimental study. In: 2015 IEEE FICLOUD Conference. pp. 333–340. IEEE (2015)
27. Töpsch, L.: Design and Implementation of Adaptive Frequency Hopping for Bluetooth Low Energy - Evaluation for High Throughput Applications. Bachelor's thesis, Christian-Albrechts-Universität zu Kiel (2020)

A Notion of Relevance for Safety-Critical Autonomous Systems

Astrid Rakow[(✉)]

German Aerospace Center, Institute of Systems Engineering for Future Mobility,
Oldenburg, Germany
`astrid.rakow@dlr.de`

Abstract. There are already manifold complex, highly autonomous systems that have build-in beliefs, perceive their environment and exchange information. These systems construct their respective world view and based on it they plan their future manoeuvres, i.e., they choose their actions in order to establish their goals based on their prediction of the possible futures. Usually these systems face an overwhelming flood of information provided by a variety of sources where by far not everything is relevant. This paper provides a formal characterisation of relevance for a safety-critical autonomous system at its mission.

Keywords: Relevance · Belief · Game Theory · Autonomous System

1 Introduction

The idea to formally characterise a notion of relevance for safety-critical systems originated as a by-product of a series of works with Prof. Dr. Werner Damm and my former colleagues that were centred around perception, communication and coordination of safety-critical cyberphysical systems.

Autonomous systems are operating in complex, dynamic and open environments, where they have to assess their situation and extrapolate possible futures, in order to decide on their current action. The question "What is relevant for an autonomous system S in its current situation?" comes naturally, underlining the importance of a theoretical characterisation of relevance. Full informedness is certainly not necessary for successful manoeuvres of autonomous systems. For instance, when an autonomous car approaches a pedestrian crossing, it has to decelerate, if pedestrians want to cross the road –irrespectively of their shirt colours or the exact number of pedestrians. Nevertheless, the number of pedestrians is relevant for the expectation, when the group will have crossed the road, influencing the decision whether to take a detour circumventing the crossing. For the latter case, the relevance of the group size results from the goal of minimising the travel time.

This research was supported by the German Research Council (DFG) in the PIRE Projects SD-SSCPS and ISCE-ACPS under grant no. DA 206/11-1.

M. Fränzle et al. (Eds.): Werner Damm Festschrift, LNCS 15471, pp. 96–112, 2026.
https://doi.org/10.1007/978-3-031-97537-0_6

The control of a system $\mathcal{S}$ can be considered as the implementation of a strategy that chooses actions (time-bounded services of the autonomous layer like "emergency braking") based on the currently agglomerated information. $\mathcal{S}$'s decision is based on the combination of its observations of the world and its insights into the world – e.g. $\mathcal{S}$ observed the speed limit and knows about the effect of acceleration. Since $\mathcal{S}$ has only limited sensing and communication capabilities and hence limited means to assess the situation, it faces uncertainties in determining its current situation. Several alternatives seem possible at a time, and it cannot tell which of them best reflects reality.

The question "What does a system $\mathcal{S}$ need to perceive and know and of what granularity must its beliefs be so that $\mathcal{S}$ can successfully perform an autonomous manoeuvre?" drives my research. I thereby strive to define a notion of relevance. To pave the way for a formal treatment of this question, I developed in [7] a formal framework that explicitly represents the beliefs of $\mathcal{S}$. A belief describes the internal representation of $\mathcal{S}$'s surroundings in terms of a set of the possible worlds [4]. Within this framework, I characterise an *autonomous-decisive system* $\mathcal{S}$ as a system that takes rational decisions based on its belief content and $\mathcal{S}$ is rational, when it chooses the actions it believes will lead to success. I consider an autonomous system to be successful, if it performs no worse in terms of its list of goals than the best possible system with access to the ground truth. In this paper I define *relevance* of knowledge, observations and a set of possible beliefs for an autonomous-decisive system $\mathcal{S}$ with goals ψ and an application domain W_D. Basically, a combination of knowledge $\mathcal{K}$, observations $\mathcal{O}$ and a set of possible beliefs $\mathbb{B}$ is relevant if we cannot omit anything while being equally successful. We present an algorithmical approach based on strategy synthesis to determine relevant combinations of knowledge, observations and sets of possible beliefs. Conceptually, the presented approach will be useful in the early design of $\mathcal{S}$, where simple and abstract models are considered.

I assume that a design-time world model W_D is given that characterises the application domain and captures test criteria of $\mathcal{S}$. Such a world model may be derived from scenario-databases and test catalogues. We envision that the starting point for this design step could be [2] of W. Damm and B. Finkbeiner to determine the optimal perimeter of a world model. We moreover assume that prior to our outlined analysis, the scope of beliefs $\mathbb{B}$ has been defined. So it has been defined which artefacts, objects, and interrelations will possibly be represented in $\mathcal{S}$'s beliefs. This contribution aims to guide the design of beliefs and the belief formation by highlighting the trade-off between (i) sensing capabilities, (ii) knowledge about the world and (iii) what can be expressed via beliefs, i.e. internally, to represent the surrounding world. Later design steps will have to generalise $\mathcal{S}$'s capabilities to deal with the known unknown aspects of the design-time world, taking into account that no world model will match the reality. An overview of the relevance framework is given in Fig. 1.

Outline. Section 2 presents related work. Section 3 and 4 are based on [7]. Section 3 briefly introduces the formal framework within which I define my notion of relevance. In Sect. 4, I characterize autonomous systems and summarize the main results of [7]. In Sect. 5, I present the notion of relevance and conclude in Sect. 6.

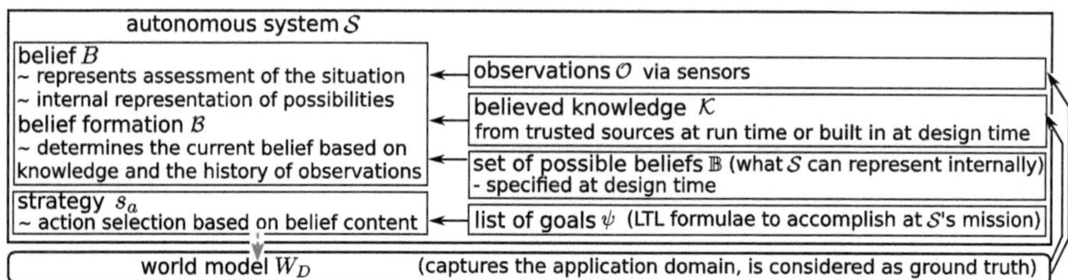

Fig. 1. Relevance in a nutshell: If "System S captures the relevant", then "S can autonomously achieve its goals as well as possible". The framework characterises autonomous systems: An autonomous system takes decision based on its beliefs. What S believes depends on (i) its observations. (ii) Its beliefs must satisfy the knowledge that is built-in and received at run time. (iii) S has only limited means to build its beliefs (i.e. to internally represent its surrounding). The set of possible beliefs $\mathbb{B}$ is hence finite. S chooses its actions to accomplish its prioritized list of goals ψ. The chosen actions influence the future evolution of the world. To judge whether S captures the relevant, the framework compares S's performance with the best performance that would be possible when knowing the ground truth, W_D.

2   Related Work

This work contributes to the ongoing conceptualisation of the notion of relevance. Although relevance is discussed in many fields such as philosophy, psychology or artificial intelligence, it is most prominently discussed in information science and information retrieval where it is considered to be a central notion [5,6,10,12]. Its importance for selection tasks in information retrieval has been recognised early on [5,10]. The first period of intensified research on this topic started about 1960 and the research is still ongoing [5,6]. While there is no consensus yet on the notion of relevance, it is agreed that relevance is a relationship. A document can be relevant to a problem. Moreover, it is agreed that relevance is multidimensional. In his influential paper [11], Saracevic developed a stratified system of relevance distinguishing *system relevance* (query ↔ information objects), *topical relevance* (topic expressed in the query ↔ topic covered by the retrieved text), *cognitive relevance* (state of knowledge/need of a user ↔ retrieved text), *situational relevance* (situation/task/problem ↔ retrieved text) and *motivational relevance* (user's intent/goal/motivation ↔ retrieved text). According to Saracevic the different strata dynamically interact and are interdependent.

While the early works aimed at a notion of relevance for retrieval systems such as those used in libraries, specific relevance notions have been developed for the more recent *mobile* information retrieval systems. In [3] De Sabbata et al. observe that for systems where the user might be moving (i) *the relation between space and time* and (ii) *a link to the real world* is an important factor for the relevance of the retrieved information. To emphasize the spatio-temporal nature, the authors introduce a new "space-time dimension". Furthermore, they introduce "world" as another new dimension, in order to capture the influence of different abstractions of reality. They argue that since the real information need is different from the query received by the system and the real world is different from the world perceived by the system, a relevance concept can be described as dealing with reality at the different abstraction levels: the real world; the documented world (recorded by the human and stored); the perceived world (perceived by the user); the system world (as known by the system).

My notion of relevance is in line with the above research, but it targets safety-critical autonomous systems. I characterize relevance in a doxastic, game-theoretic framework. Doxastic means that the beliefs of a system are explicitly represented. These beliefs represent the world as perceived (or rather constructed) by the autonomous system S. The framework distinguishes the formed beliefs from the ground truth world. Like e.g. [3], I consider mobile systems with a strong space-time dimension. The linkage of space-time is captured via the effects of S's actions on the ground truth world.

Already in 1966, Rees discussed the influence of prior knowledge on relevance [9]. Since an autonomous system S not only gets equipped with knowledge during the design, but –being part of a cyberphysical system– it can perceive knowledge updates while being on a mission, I more generally examine the influence of a knowledge base that can change over time.

The motivational dimension of relevance is captured by using a game-theoretic characterization, "Does a strategy exist that achieves S's goals?". Since I am interested in safety-critical systems, the other player, the environment including other agents, is adversarial.

In contrast to the above work my framework formally integrates several interdependent dimensions of relevance. It moreover allows us to effectively compute what is relevant and thereby takes the interdependencies into account. The resulting notion of relevance thus is subjective, dynamic, motivational, and cognitive.

In [2] Damm and Finkbeiner determine the optimal perimeter of a world model as the subset of a Kripke structure's propositions that is necessary to synthesize a winning strategy. I generalize their idea in order to define relevance for S. To this end, I distinguish between the ground truth, a.k.a. the design time model W_D, and the system's beliefs, based on which it takes decisions that in turn affect the ground truth.

For the works related to the doxastic framework, the reader is referred to [7]. More related work on the notion of relevance can be found in [8].

3   Ingredients of Our Doxastic Framework

In this section, the ingredients of the framework are introduced alongside a running example. *The two cars, Ego and Other, are on separate lanes heading towards each other. The left car, Ego, is our system. Its goals are avoiding collisions by all means and taking the left turn. From Ego's perspective, Other is uncontrolled. Ego and Other simultaneously choose actions. The actions of Ego are f, "moving one step forward if possible", t, "turn and move one step forward". Other is either a slow car or hasty. If it is hasty, it moves two positions forward from its initial position via action F.*

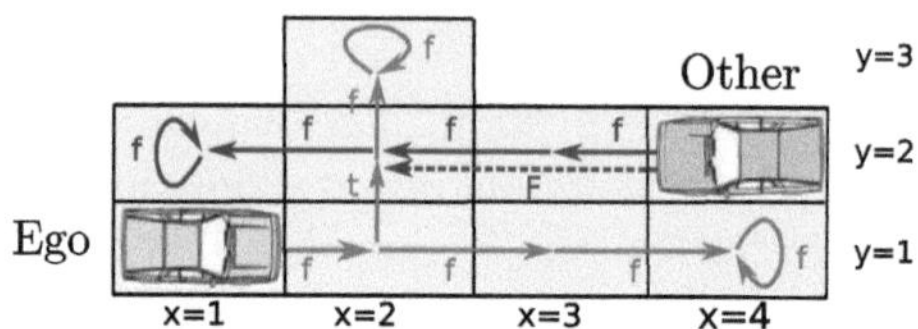

Fig. 2. Sketch of a simple world.

In the framework a *world* describes entities, their actions and how initial situations evolve in time and space.

Definition 1 (World). *Let $Act = Act_{Ego} \times Act_{env}$ be a finite set of actions and AP be a finite set of atomic propositions.*

A world $W = (S, Ed, L, I)$ is a labeled Kripke structure where

- *S is a set of states,*
- *$Ed \subseteq S \times S$ is a transition relation,*
- *$L : L_S \cup L_{Ed}$ where*
  - *$L_{Ed} : Ed \to 2^{Act}$ labels edges with a subset of Act and*
  - *$L_S : S \to 2^{AP}$ labels a state s with the set propositions $L_S(s)$ that are valid in s and*
- *I the initial states.*

The Kripke structure for Fig. 2 is presented in Fig. 3. The propositions $AP_{pos} = \{x_e = i | 1 \leq i \leq 4\} \cup \{y_e = i | 1 \leq i \leq 3\} \cup \{x_o = i | 1 \leq i \leq 4\}$ encode the positions of the two cars. Other's car type is either s (for slow) or h (for hasty). Let us assume, Ego cannot observe Other's car type but it can infer the type from Other's colour. For simplicity, let Other be either red or blue, and let red cars be hasty and blue cars be slow. Let us assume that Ego's colour sensor is faulty. While b (for blue) and r (for red) represent the true colour of Other, b_p (r_p) represent Ego's perceptions. For simplicity, we treat Other's color as boolean and hence write e.g. $\neg r$ for b. Let Ego's colour perception work correctly when Ego and Other are less than two tiles apart, otherwise the sensor switches colours. Ego, moreover, can observe its own position but not Other's. A label $abcdef \in \mathbb{N}^3 \times \{h, s\} \times \{b, r\} \times \{b_p, r_p\}$ encodes that $x_e = a, y_e = b, x_o = c$ are true and Other's car type is d, its colour is e whereas the perceived colour is f. Edges are labelled with sets of actions. I omit the sets' brackets for brevity. For example, the label ff denotes the set $\{$ff$\}$, that contains the one action ff, where Ego and Other simultaneously move one step forward, if possible. Actions that are not enabled lead to the sink state s_{undef} . The sink state, sink transitions, and the self-loops are neglected in the following.

In the framework, a so-called *design-time world* W_D is assumed to be given that captures the intended application domain and represents the *ground-truth*. For our running example W_D is the world of Fig. 3.

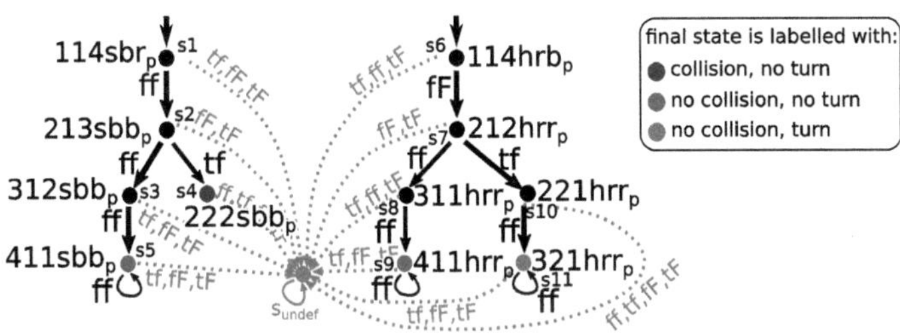

Fig. 3. Kripke structure of the setup sketched in Fig. 2

Our system Ego has to achieve a prioritized list of goals which are specified via linear-time temporal logic (LTL) formulae [1]. The temporal operator "globally" is denoted by $\square$, "eventually" by $\lozenge$, "next" by X and "until" by U. LTL is interpreted on traces, i.e. infinite sequences of valuations. For instance, collision freedom can be expressed as $\varphi_c = \square(x_e = 2 \wedge y_e = 2 \Rightarrow x_o \neq 2)$, and $\varphi_t = \lozenge(y_e = 3)$ expresses that Ego eventually does the turn. The priorities are given by $prio(\varphi_c) = 1, prio(\varphi_t) = 2$.

As an autonomous system, *Ego* will take decisions based on its beliefs about the world. These beliefs are influenced by its observations and the knowledge it believes to have. *Observations* represent e.g. sensor readings or received messages from other agents. They shed light on W_D, but they do not have to be truthful (cf. r_p, b_p). In our example *Ego*'s set of observations is $\mathcal{O} = \{x_e, y_e, \mathsf{r}_p\}$ (assuming $\mathsf{r}_p = \neg \mathsf{b}_p$). A *belief* describes what *Ego* currently thinks that the world is like. For instance, *Ego* may currently think that it saw a hasty vehicle, and that it will do the turn without causing a collision. Formally a belief is described by a set of (alternative) realities that represents *Ego*'s assessment of the world. A reality is a (possible) world together with a set of current states. The set of current states marks the present and thus separates the past from the future.

Definition 2 (belief, reality). *A* belief *B is a set of realities, $B = \{r_0, \dots, r_n\}$. A* reality *is a pair $r = (W, S_c)$ of a (possible) world $W = (S, Ed, L, I)$ and a set of believed current states $S_c \subseteq S$, where any current state is reachable from an initial state and every path has at most one current state.*

Since $\mathcal{S}$ has only finite resources, it can only represent finitely many beliefs. Therefore, it is assumed in the following that the set of *possible beliefs* $\mathbb{B}$ is finite, and it is assumed that the set of *possible worlds* $\mathbb{W} := \bigcup_{(S_c, W) \in \mathbb{B}} W$ is also finite. The choice of $\mathbb{W}$ and $\mathbb{B}$ constitutes an important design decision within the development process of a system, as it delimits the expressiveness of beliefs. For simplicity, it is assumed that $\mathbb{W}$ has the same atomic propositions and actions as W_D.

Let us assume, that Ego has been designed to handle a certain set of scenarios, for which it can evaluate what to do by extrapolating the future. These are sketched in Fig. 4(a)+(b): Other may be hasty or slow, and the road may be up to 5 tiles long. Let Ego's possible beliefs $\mathbb{B}$ be the beliefs that canonically evolve from these initial scenarios.

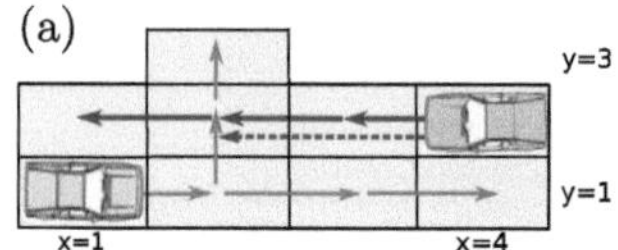

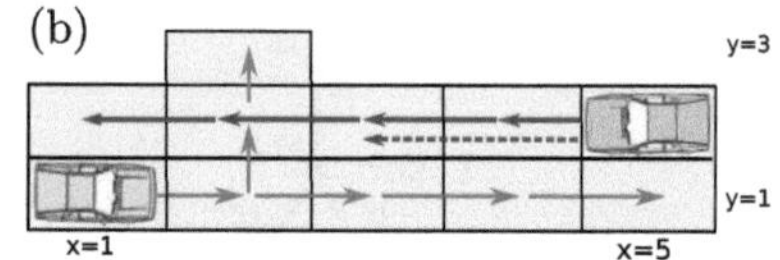

Fig. 4. Sketch of *Ego*'s possible beliefs. If *Other* is hasty, it uses the dashed arrow at first and then the solid arrows. If it is slow, it uses the solid arrows only.

Believed knowledge are statements that S believes in. An engineer can equip a system during its design with general statements about the application domain (e.g. *"Cars drive on the street."*). During S's mission, trusted sources (*like a traffic control system*) can provide additional information. These statements are expressed via an LTL variant, which I call *Belief-LTL (BLTL)*. A BLTL formula is defined via the following grammar: $\mathsf{K}\psi \mid \mathsf{K}^c\psi \mid \neg\varphi \mid \varphi \wedge \varphi'$, where ψ is an LTL formula and φ, φ' are BLTL formulae. A belief B satisfies $\mathsf{K}\psi$, i.e. $B \models \mathsf{K}\psi$, iff ψ is satisfied by all worlds of all alternative realities of B, i.e. ψ holds on all traces from any initial state. $B \models \mathsf{K}^c\psi$, iff ψ holds on all traces from any current state. A finite set $\mathcal{K}$ of BLTL formulae constitutes a *knowledge base $\mathcal{K}$*.

It is assumed that during the design an engineer determines the maximal set of statements, i.e. BLTL formulae, that a system S may get. This set is denoted as $\mathbb{K}$. The engineer examines under which circumstances certain statements can be provided. *For instance, global traffic flow information might only be available when Ego is able to receive notifications from the traffic control system; local announcements like "traffic jam ahead" might depend on whether trusted agents are close by.* Formally, the states of W_D get labelled by the knowledge base $\mathcal{K} = L_{\mathbb{K}}(s)$ that is available at state s. $B \models \mathcal{K}$, if $B \models \varphi$ holds for all $\varphi \in \mathcal{K}$.

Ego updates its beliefs e.g. when it gets new information from its sensors. The belief formation $\mathcal{B}$ captures formally how *Ego* builds its belief based on a history h of observations $\mathcal{O} \subseteq AP$ and the current knowledge base. In Definition 3 history h is a finite sequence of valuations of the observations $\mathcal{O}$, $\mathbb{H}_{\mathcal{O}}$ denotes the set of all such sequences and $last(\pi)$ denotes the last state of the path π.

Definition 3 (belief formation, knowledge-consistent). *The belief formation $\mathcal{B}$, $\mathcal{B} : \mathbb{H}_{\mathcal{O}} \times \mathbb{K} \to \mathbb{B}$, specifies the belief $B = \mathcal{B}(h, \mathcal{K})$ that S derives after perceiving a history h of observations $\mathcal{O} \subseteq AP$ and while believing in $\mathcal{K}$.*

A belief formation $\mathcal{B}$ is called knowledge-consistent, *if all formed beliefs satisfy the respective knowledge base $\mathcal{K}$, i.e., for all paths π of W_D holds, $\mathcal{B}(h, \mathcal{K}) \models \mathcal{K}$ where $h = L_S(\pi)|_{\mathcal{O}}$ and $\mathcal{K} = L_{\mathbb{K}}(last(\pi))$.*

Note, that $L_{\mathbb{K}}$ is a mean to anchor *Ego*'s beliefs in the ground truth. A state s can be labelled with a knowledge base $\mathcal{K} = L_{\mathbb{K}}(s)$ that reflects the ground truth, but $\mathcal{K}$ can also enforce wrong beliefs or distorted beliefs. In the following only knowledge-consistent belief-formations are considered.

Let us now consider an example of a knowledge-consistent belief formation, where Ego's possible beliefs $\mathbb{B}$ are as in Fig. 4. Let Ego have the knowledge base $\mathcal{K}$ of $\varphi_i =$"Other starts at $x = 4$", $\varphi_{ct} =$"The car type does not change" and $\varphi_{cs} =$"A *red car is hasty, while a blue car is slow." at all states. Then only realities arising from the scenario (a) of Fig. 4 are possible. Moreover, let Ego trust its sensors.* [1]*. Since Ego's colour perception is faulty, Ego is initially convinced that Other is hasty, while it is slow and vice versa. At the initial state, Ego hence thinks that it will do the turn. At the next time step, Ego moves one tile forward, while Other simultaneously moves either one or two tiles forward. In both cases, Ego then perceives Other's colour correctly and updates its belief. Figure 5 sketches the (knowledge-consistent) belief formation. The observed history ($11r_p$) is mapped to belief $B_{0,1}$ and ($11b_p$) $\mapsto B_{0,2}$, ($11r_p,21b_p$) $\mapsto B_{1,1}$ and ($11b_p,21r_p$) $\mapsto B_{1,2}$.*

Note in particular, that *Ego* can change its mind regarding the car type and can still be knowledge-consistent. φ_{ct} requires only that the car type does not change within a belief but the car type may change along a sequence of beliefs.

In this example, all beliefs are singletons, i.e. *Ego* believes only one reality is possible at a time. But if *Ego* is for instance unsure whether it is initially at position 1 or 2, it may believe in two alternative realities. In one reality it imagines itself at $x = 1$, in the other at $x = 2$.

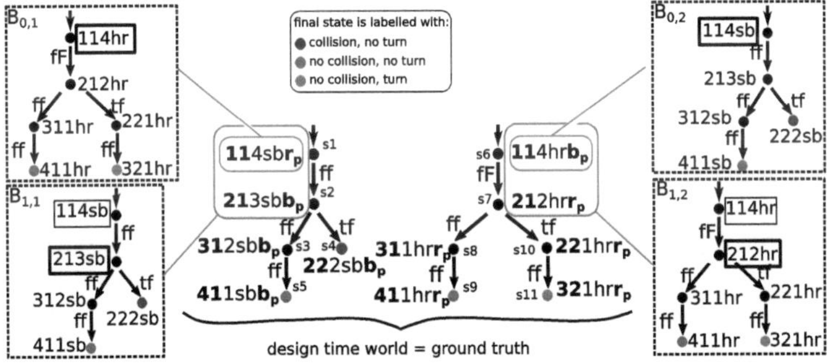

Fig. 5. Sketch of a belief formation. The W's observable propositions are in bold face. The history is within rounded boxes. The believed current state is within bold boxes.

By now, I introduced all components to model how an autonomous system builds its beliefs based on the history of observations and its knowledge. Let me summarize the components by defining the notion of *doxastic model*.

Definition 4 (doxastic model). *A doxastic model D of an autonomous system is given by a tuple ($W_D,\psi,L_\mathbb{K},\mathcal{O},\mathbb{B},\mathcal{B}$) of the design-time world W_D, the prioritized list of goals ψ, the knowledge labelling $L_\mathbb{K}$, a set of observations $\mathcal{O}$, the set of possible beliefs $\mathbb{B}$ and a belief formation $\mathcal{B}$.*

The world W_D is considered as ground truth during the design. The belief formation $\mathcal{B}$ describes how a system $\mathcal{S}$ maps its observations of W_D to beliefs that describe $\mathcal{S}$'s world view. The possible beliefs $\mathbb{B}$ can be regarded as the resources for the inner world view.

---

[1] This can be enforced via $L_\mathbb{K}(s) = \mathcal{K} \cup \{r\}$, if $r \in L(s)$ and $L_\mathbb{K}(s) = \mathcal{K} \cup \{b\}$ if $r \notin L(s)$.

Figure 6 gives an overview of the different components of the framework and an outlook of how it will be used to describe an autonomous system $\mathcal{S}$. In the next section we turn to the action selection and we will examine how the formed beliefs influence this selection. In order to define relevance, we will examine in Sect. 5 whether $\mathcal{S}$ can form beliefs about W_D such that it can operate successfully.

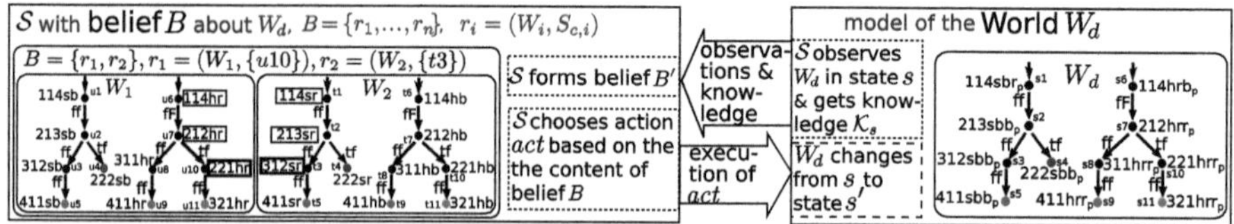

Fig. 6. Components of the framework and how they interact: $\mathcal{S}$ chooses actions in order to achieve its goals ψ on the "ground-truth" world W_D. It decides for an action act based on its beliefs. The execution of act influences the state of W_D. $\mathcal{S}$ continuously observes W_D and it gets believed knowledge (statements from trusted sources). Both influence how $\mathcal{S}$ forms beliefs. $\mathcal{S}$'s beliefs are a set of alternative realities. A reality is a world together with a set of current states. A reality describes what $\mathcal{S}$ thinks the world is like and at which state it is in this world.

4   Autonomous Decisions

In this section, I present a notion of *autonomous decision* and then characterize when a system exists that can take autonomous decisions to accomplish its goals. Notions of autonomy are discussed in various scientific fields [13]. Our formalization aims at capturing that (bel) an autonomous system must take decisions based on its internal world view (i.e. its belief) and that (rat) the system will take the choice alternative that promises the best outcome, i.e. the system is rational. We distinguish autonomous decisions from *automatic decisions*, that play out rule-determined choices that are based on the system's world view but are not its rational consequence.

In order to formalize when an *autonomous decision* can be taken successfully, I distinguish between *truth-observing strategies* (strategies that have access to ground-truth), *doxastic strategies* (strategies that can only access the formed beliefs) and *possible-world strategies* (strategies that "run" within a belief as emulation of the current action choice). The best truth-observing strategy represents what any system can possibly achieve. The best doxastic strategy represents what a system with a given belief formation can possibly achieve. If the best doxastic strategy performs as good as the best truth-observing strategy, then a doxastic system is successful. Rational systems choose what seems to be the best choice according to their belief content. The best possible-world strategy describes what a system believes to be the best strategy that can be applied to all alternative realities alike. An overview of the different strategies and the notion of *dominance* is given in Table 1.

Table 1. Strategy types & dominance in a nutshell

Strategy types:

– *truth-observing strategy* $s_t : (2^P)^+ \rightarrow Act_{Ego}$
observes the ground truth W_D via $P \subseteq AP_{W_D}$ and takes decisions based on the observed history; serves as comparative reference of what is achievable given W_D could be directly observed via P

– *doxastic strategy* $s_d : \mathbb{B}^+ \rightarrow Act_{Ego}$
observes the beliefs and takes decisions based on their history; represents the decision making of doxastic systems

– *possible-worlds strategy* $s_p : (2^{AP_B})^+ \rightarrow Act_{Ego}$
captures how $\mathcal{S}$ evaluates its action choice on all alternative realities of a belief B

A strategy s ψ*-dominates* s', $s' \leq s$, iff s' achieves the goal list ψ up to priority m' but s achieves ψ up to m with $m' \leq m$. s is ψ-dominant, if $s' \leq s$ for all strategies s'.

In order to motivate our formalization of "*Ego* chooses its actions based on the *content* of its beliefs", let us consider the following example, where *Ego* does not choose its actions based on its belief content. *Let us assume Ego's colour perception is severely broken, so it permanently switches red to blue and vice versa. In Fig. 7, the changed world model is given along with a belief formation $\mathcal{B}_p$ that relies on the colour perception, i.e., if the sensors say the other car is red (blue), then Ego believes it. $\mathcal{B}_p$ hence maps the observed history $(11r_p)$ to $B_{0,1}$ and $(11r_p, 21r_p) \mapsto B_{1,2}$, $(11b_p) \mapsto B_{0,2}$ and $(11b_p, 21b_p) \mapsto B_{1,1}$, ... Also sketched in Fig. 7 is the doxastic strategy with s_d' with $(B_{0,1}) \mapsto f$, $(B_{0,1}, B_{1,2}) \mapsto f$, ... and $(B_{0,2}) \mapsto f$, $(B_{0,2}, B_{1,1}) \mapsto t$, ... If Ego follows s_d', it takes a turn on W_D, if Other is hasty $((B_{0,2}, B_{1,1}) \mapsto t)$, and Ego drives straight on, if Other is slow $((B_{0,1}, B_{1,2}) \mapsto f)$. So s_d' achieves Ego's goals on the design time world W_D as well as possible and thus is a dominant strategy, but s_d' makes no sense from Ego's perspective. When Ego believes that Other is slow, it extrapolates that turning would cause a collision. But in this case, s_d' demands turning. Vice versa, s_d' chooses driving straight on, when Ego believes Other is hasty and taking the turn is fine.*

The above example motivates that I formalize "*Ego* decides based on the content of its belief" as "*Ego* always chooses an action, that a dominant strategy would choose at the believed current state". To capture this formally, I use the notion of *possible-worlds strategy*, s_p. A possible-worlds strategy (cf. Table 1) is applied to the alternative realities of *Ego*'s current belief B and results in believed traces which represent *Ego*'s simulations/extrapolations. $curAct(s_p, B)$ denotes the set of actions that are selected by s_p for the current-states of B. Formally $curAct(s_p, B)$, the current-state choices of s_p in B, is the set of actions with $act \in curAct(s_p, B) \Leftrightarrow \exists r = (W, S_c) \in B : \exists$ *path* π *in* W *from an initial state to a current state* $s_c \in S_c : s_p(L(\pi)) = act$.

Due to e.g. uncertainties, *Ego* might believe in several alternative realities. The best strategy might be to choose action act_1 in reality r_1 and act_2 in r_2. *For example, imagine that Ego wants to visit a filling station as soon as possible,*

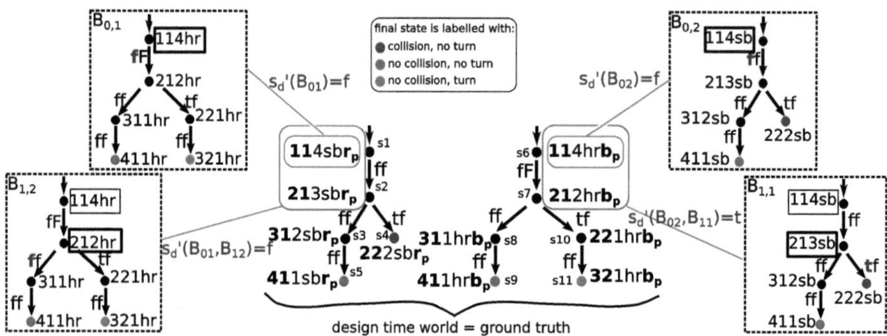

Fig. 7. Doxastic strategy s_d'; *Ego* believes in its broken sensor. Observable proposition are in bold black. The best choice for the believed current state is in bold blue. (Color figure online)

but does not know whether to turn left or right to get there. In this case, Ego imagines that in a reality r_1 the filling station is on the left and in a reality r_2 the filling station is on the right. The best strategy chooses to turn left in r_1 and to turn right in r_2. Since Ego does not know, whether it actually is r_1 or rather in r_2, it is indecisive what it should do best. We say that a possible-worlds strategy s_p is *current-state decisive in a belief* B, if $curAct(s_p, B)$ is a singleton.

Note that in the above example, there is no dominant current-state decisive strategy. But if Ego could ask for directions or if Ego just needs to find the filling station eventually, then dominant current-state decisive strategies would exist.

(P1) We can decide whether there is a current-state decisive possible-worlds strategy s_p achieving an LTL property ψ in a given belief B [8, p. 38], [7]. If it exists, we can synthesize such a strategy.

Because *Ego* strives for the best result, it will take the actions of the best strategy. The actions $curAct(s_p, B)$ of a possible-worlds strategy s_p are hence rationally justified, if s_p dominants all other possible-worlds strategies (cf. Table 1). We define the set of best choices for a belief:

Definition 5 (best choices in B, $bestAct(B)$). *Let a goal list ψ be given.*

Let $bestAct(B)$ be the set with $act \in bestAct(B) \Leftrightarrow \exists \psi - dominant$ possible-worlds strategy s_p in $B : act \in curAct(s_p, B)$. We call $bestAct(B)$ the best choices in B for ψ.

(P2) We can determine $bestAct(B)$ for a given ψ and B [8, p. 39], [7].

Note, that while all $curAct(s_p, B)$ should be a singleton sets in order to have a current-state decisive s_p, non-singleton sets $bestAct(B)$ are not a problem. *For example, imagine that Ego is supposed to go to an ice cream parlour. Let Ego believe in only one reality where one ice cream parlour is to its right and one is to its left. Ego can hence turn right (strategy s_{p_1}) and it can turn left (strategy s_{p_2}). Both strategies are current-state decisive and $Act(B)$ contains both actions,* **turn left** *and* **turn right***.*

Definition 4 introduced the notion of doxastic model $D = (W_D, \psi, L_\mathbb{K}, \mathcal{O}, \mathbb{B}, \mathcal{B})$. D captures how beliefs are formed, but not what actions the system will choose. A *doxastic system* $\mathcal{S}$ is a pair $\mathcal{S} = (D, \mathsf{s}_d)$ of a doxastic model D and a doxastic strategy s_d, $\mathsf{s}_d : \mathbb{B}^+ \to Act_{Ego}$ on W_D, i.e. $\mathcal{S}$ decides based on its beliefs. A doxastic system is considered to be autonomous, if its decisions are rational w.r.t. the content of its current belief, i.e. s_d must choose from $bestAct(B)$. In order to have an action in $bestAct\,(B)$, there has to be at least one current-state decisive strategy. I regard it is a design task to specify how an indecisive situation has to be resolved, since this work deals with safety-critical systems, where the final decision between two evils should be taken by an engineer.

Assumption 1: In the following the belief formation $\mathcal{B}$ forms only beliefs B in which a dominant current-state decisive strategy exists.

Definition 6 (autonomous decision & strategy) *The system $\mathcal{S}$ decides autonomously at a finite path π of W_D, iff it chooses an action $act \in bestAct(B)$,* $B = \mathcal{B}(L_S(\pi), L_\mathbb{K}(last(\pi)))$.

A doxastic strategy $\mathsf{s}_a : \mathbb{B}^+ \to Act$ is called an autonomous strategy *iff for all belief histories $\bar{B} \in \mathbb{B}^+$ it holds that $\mathsf{s}_a(\bar{B}) \in bestAct(last(\bar{B}))$.*

$\mathcal{S}$ autonomous-decisively achieves *the goal list ψ up to n, if it implements an autonomous strategy s_a, i.e. $\mathcal{S} = (D, \mathsf{s}_a)$, and s_a achieves ψ up to n.*

A system that is not autonomous-decisive chooses actions that are not rationally justified by the belief of the system.

An autonomous-decisive system $\mathcal{S}$ is *optimal*, if the autonomous strategy s_a is not ψ-dominated by any AP-truth-observing strategy[2]. $\mathcal{S}$'s belief formation $\mathcal{B}$ then builds beliefs, such that $\mathcal{S}$ is as successful as the best system with direct access to the complete ground truth, W_D.

We can decide for a given doxastic model without belief formation, $D^- = (W_D, \psi, L_\mathbb{K}, \mathcal{O}, \mathbb{B}, .)$, whether there is a knowledge-consistent belief formation $\mathcal{B}$ and an autonomous strategy s_a such that $\mathcal{S}$ is an optimal autonomous-decisive system, and we can synthesize the two (cf. Theorem 1).

Theorem 1 (Autonomous Decisiveness [8]). *Let $D^- = (W_D, \psi, L_\mathbb{K}, \mathcal{O}, \mathbb{B}, .)$ be a doxastic model without belief formation. We can decide whether there is a knowledge-consistent belief formation $\mathcal{B}$ and a doxastic strategy s_d such that $\mathcal{S} = (D^-, \mathcal{B}, \mathsf{s}_d)$ is an optimal autonomous-decisive system. If such $\mathcal{B}$ and s_d exist, we can synthesize them.*

Let me summarize the role of Theorem 1 for the design of $\mathcal{S}$. We can specify the application domain (W_D), the list of $\mathcal{S}$'s goals (ψ), the believed knowledge that $\mathcal{S}$ will get, what observations $\mathcal{S}$ can make ($\mathcal{O}$) and how its internal representation of the world is ($\mathbb{B}$). All that given, an engineer can determine whether it is at all possible to form beliefs ($\mathcal{B}$) such that $\mathcal{S}$ is able to autonomously-decide and succeed as if it knew the ground truth.[3] We can synthesize such a $\mathcal{B}$, if it exists,

---

[2] an AP-truth-observing strategy is based on perfect observations of W_D, cf. Table 1.

[3] Theorem 1 can easily be adjusted to compare $\mathcal{S}$ with systems having partial observations of W_D.

and also an autonomous strategy such that the resulting system is optimal. If there is no such $\mathcal{B}$, then an engineer can iteratively tweak ($L_{\mathbb{K}},\mathcal{O},\mathbb{B}$), i.e. provide more knowledge, observations or allow more detailed believes, until a $\mathcal{B}$ can be found.

The considered notion of autonomous system is very liberal regarding the belief formation. It is not enforced that a belief reflects what is observed and there are no constraints that beliefs evolve in a sensible way. So Theorem 1 is a necessary and not a sufficient condition for systems that have more constrained belief formations. If a belief formation $\mathcal{B}$ of Theorem 1 can be found, there is no guarantee that a sensible belief formation exists.

In the following, the focus is on *optimal* autonomous-decisive systems. For brevity, I usually just speak of *autonomous systems*.

5   Relevance

The main contribution of this paper is the notion of relevance for safety-critical autonomous systems, that is characterised as *what combination ($L_{\mathbb{K}}, \mathcal{O}, \mathbb{B}$) of knowledge labelling, observations and set of possible beliefs is relevant*. Basically, ($L_{\mathbb{K}},\mathcal{O},\mathbb{B}$) is relevant, if an autonomous system $\mathcal{S}$ can be optimal while neither of $L_{\mathbb{K}},\mathcal{O}$ or $\mathbb{B}$ can be reduced. We thus turn to questions like "Can $\mathcal{S}$ do with less observations?", "Can $\mathcal{S}$ do with less detailed beliefs?" or "Can we compensate that $\mathcal{S}$ misses observations by adding knowledge?". In the following we consider a system $\mathcal{S} = (D, \mathsf{s}_d)$ with $D = (W_D, \psi, L_{\mathbb{K}}, \mathcal{O}, \mathbb{B}, \mathcal{B})$ where $\mathcal{B}$ is a knowledge-consistent belief formation and s_d a doxastic strategy.

The combination ($L_{\mathbb{K}},\mathcal{O},\mathbb{B}$) is an important aspect in the design of an optimal autonomous-decisive system. Via $L_{\mathbb{K}}$ an engineer reflects $\mathcal{S}$'s prior knowledge and mechanisms for updating $\mathcal{S}$'s knowledge base during its mission. Via $\mathcal{O}$ she can reflect sensing capabilities of $\mathcal{S}$ and via $\mathbb{B}$, she captures what $\mathcal{S}$ can express internally, i.e. the resources for the internal world view.

In Sect. 5.1 I define when a belief formation $\mathcal{B}$ of a doxastic model $D = (W_D, \psi, L_{\mathbb{K}}, \mathcal{O}, \mathbb{B}, \mathcal{B})$ conserves the relevant. Intuitively, $\mathcal{B}$ conserves the relevant when $\mathcal{B}$ captures W_D sufficiently well for the goal list ψ, i.e. there is a s_d so that $\mathcal{S} = (D, \mathsf{s}_d)$ is optimal. In Sect. 5.2 I then define when ($L_{\mathbb{K}},\mathcal{O},\mathbb{B}$) is relevant, i.e. sufficient and minimal.

5.1   Conservation of the Relevant

To formally define when the belief formation $\mathcal{B}$ of a doxastic model D conserves the relevant, D's best possible autonomous-decisive performance (which is based on $\mathcal{B}$) is compared with the best possible performance that D could have with access to the ground-truth. $\mathcal{B}$ *conserves the relevant for autonomous-decisiveness*, if D's best possible autonomous-decisive performance is as good as the best possible ground-truth performance.

Definition 7 (relevance conservation for autonomous-decisiveness).
Let S_a be the set of autonomous strategies of a doxastic model $D = (W_D, \psi, L_{\mathbb{K}}, \mathcal{O}, \mathbb{B}, \mathcal{B})^4$.

The belief formation $\mathcal{B}$ of D conserves the relevant of the $\mathcal{O}$-observable design-time world W_D for autonomous-decisiveness, if all $s_a \in S_a$ are dominant w.r.t. AP-observing strategies s_t on W_D.

For an example of a belief formation that conserves the relevant, the reader is referred back to the example on p. 7. There a setting is sketched where the sensors are initially broken but when the decision has to be taken, the sensors provide the relevant information. The resulting belief formation allows a system to perform as well as when knowing the ground-truth. On p. 10 we saw an example of a belief formation that conserves the relevant for doxastic systems but not relevance for autonomous-decisiveness as defined above. In that example, Ego believed in its sensors that are permanently switching colours and Ego also believes that a red car is hasty and a slow car is blue. Therefore Ego autonomously chooses actions that are not the best in W_D.

5.2 Relevance of $(L_{\mathbb{K}}, \mathcal{O}, \mathbb{B})$

The above notion of *relevance conservation* characterises combinations of $(L_{\mathbb{K}}, \mathcal{O}, \mathbb{B})$ that are *sufficiently precise*, so that a system can be optimal. In the following, I define when $(L_{\mathbb{K}}, \mathcal{O}, \mathbb{B})$ is *relevant*. For relevance $(L_{\mathbb{K}}, \mathcal{O}, \mathbb{B})$ has to be "minimal" in addition to conservation of relevance. We hence define partial order relations on the set of knowledge labelling functions, the sets of observations and the sets of possible beliefs. We then infer a partial order to compare tuples $(L_{\mathbb{K}}, \mathcal{O}, \mathbb{B})$. The partial orders are defined as a mean to support the design decisions of an engineer:

(PO1) $\mathcal{O} \leq \mathcal{O}' :\Leftrightarrow \mathcal{O} \subseteq \mathcal{O}'$. We assume that a greater set of observations necessitates more sensors or more data processing. Hence, we want to determine the minimal set of observations.

(PO2) $\mathbb{B} \leq \mathbb{B}' :\Leftrightarrow \mathbb{B} \subseteq \mathbb{B}'$. The size of $\mathbb{B}$ determines the granularity of the beliefs and hence how much resources the system will need to encode them.

(PO3) $L_{\mathbb{K}} \leq L'_{\mathbb{K}} :\Leftrightarrow \forall s \in S : [L'_{\mathbb{K}}(s)] \subseteq [L_{\mathbb{K}}(s)]$, where $[\mathcal{K}]$ denotes the set of traces on all possible worlds $W \in \mathbb{W}$ that satisfy the believed knowledge $\mathcal{K} = L_{\mathbb{K}}(s)$. If $\mathcal{K}' \subseteq \mathcal{K}$, then $\mathcal{K}$ constrains the beliefs less and hence $\mathcal{S}$ knows less and represents more believed possibilities.

(PO4) $(L_{\mathbb{K}}, \mathcal{O}, \mathbb{B}) \leq (L'_{\mathbb{K}}, \mathcal{O}', \mathbb{B}') :\Leftrightarrow$ (PO1)-(PO3) hold.

Usually, several incomparable minima exist. $L_{\mathbb{K}}$, $\mathcal{O}$ and $\mathbb{B}$ are of course interrelated. Intuitively, knowledge ($L_{\mathbb{K}}$) about the world can replace observations that $\mathcal{S}$ needs otherwise. Having more resources for the representation of the inner world model ($\mathbb{B}$) allows $\mathcal{S}$ storing more of the made observations and allows it to make finer predictions. A tuple $(L_{\mathbb{K}}, \mathcal{O}, \mathbb{B})$ is weakly relevant, if no strictly smaller tuple $(L'_{\mathbb{K}}, \mathcal{O}', \mathbb{B}')$ can be found. This is called *weak*, since there can be

---

[4] By Assumption 1, a dominant current-state decisive strategy exists.

other tuples that are incomparable with $(L_{\mathbb{K}}, \mathcal{O}, \mathbb{B})$. "$X$ is relevant" (where X is either $L_{\mathbb{K}}$, $\mathcal{O}$ or $\mathbb{B}$), if all weakly relevant $(L_{\mathbb{K}}, \mathcal{O}, \mathbb{B})$ include X at the respective dimension.

Definition 8 (weak relevance). *Let a design-time world W_D and a prioritised list of goals ψ be given.*

$(L_{\mathbb{K}}, \mathcal{O}, \mathbb{B})$ is weakly relevant for (W_D, ψ), if

1. *there is a belief formation $\mathcal{B}$ of $D := (W_D, \psi, L_{\mathbb{K}}, \mathcal{O}, \mathbb{B}, \mathcal{B})$ that conserves the relevant for autonomous systems and*
2. *for all $(L'_{\mathbb{K}}, \mathcal{O}', \mathbb{B}') \neq (L_{\mathbb{K}}, \mathcal{O}, \mathbb{B})$ with $L'_{\mathbb{K}} \leq L_{\mathbb{K}}$, $\mathcal{O}' \leq \mathcal{O}$ and $\mathbb{B}' \leq \mathbb{B}$ there is no knowledge-consistent belief formation $\mathcal{B}'$ of $D' := (W_D, \psi, L'_{\mathbb{K}}, \mathcal{O}', \mathbb{B}', \mathcal{B}')$ that conserves the relevant for autonomous systems.*

$L_{\mathbb{K}}$ is weakly relevant if $(L_{\mathbb{K}}, \mathcal{O}, \mathbb{B})$ is weakly relevant with some $\mathcal{O}$ and $\mathbb{B}$ (analogously for $\mathcal{O}$ and $\mathbb{B}$).

The notion of weak relevance allows an engineer to determine, whether $\mathcal{S}$ can do with fewer observations/knowledge/belief (in exchange for increasing the other).

For an example consider the world model of Fig. 8 where in city R red cars are hasty while in city B blue cars are hasty. For readability we use r equals $\neg$b and R equals $\neg$B. Let the domain $\mathbb{K}$ of knowledge bases be the set $2^{\{\varphi_{cs}^R, \varphi_{cs}^B\}}$ with $\varphi_{cs}^R =$ "A red car is hasty, while a blue car is slow." and vice versa in φ_{cs}^B. In our example, when Ego is at the intersection and knows where Other is, it can determine whether it is best to turn. So one weakly relevant tuple is "Ego observes its own position and Other's position $(\mathcal{O}_1 = \{x_e, x_o\})$ while having no further knowledge". A second tuple is "$\mathcal{O}_2 = \{x_e, r\}$ while states s2" and s7 are labelled with φ_{cs}^R and s13 and s14 with φ_{cs}^B. In this case Ego can derive from its colour observation in combination with the speed-colour rules, how Other behaves. It is sufficient to provide the knowledge only at the intersection, but without the additional knowledge Ego cannot derive how Other will behave. A third tuple is $(\mathcal{O}_2 = \{x_e, r, R\})$ without further knowledge. If Ego can distinguish Other's colour and the city it is in, then Ego can precompute how Other will behave and hence it allows it to determine, whether to do the turn. In our simplistic example, two beliefs (e.g. $\mathbb{B} = \{B_{0,1}, B_{1,2}\}$ (cf. Fig. 5)) suffice; one encodes "ego should turn now" and one encodes "ego should drive straight on". The belief formation labels the states with the belief that represents the appropriate action choice. For an example where more detailed beliefs are necessary, let us assume that Ego additionally has to inform the user of its the planned maneuver, i.e. the goal list is extended by "Ego has to truthfully display its future moves.". In this case, Ego's beliefs have to distinctly represent the two maneuvers and their suffixes. The example illustrates that a system may reduce complex situations to small belief structures and illustrates the influence of a given goal. The following theorem assumes that an engineer specifies what observations $\mathcal{S}$ may make at most, what the maximal hardware resources for the beliefs are and what the maximal available knowledge is.

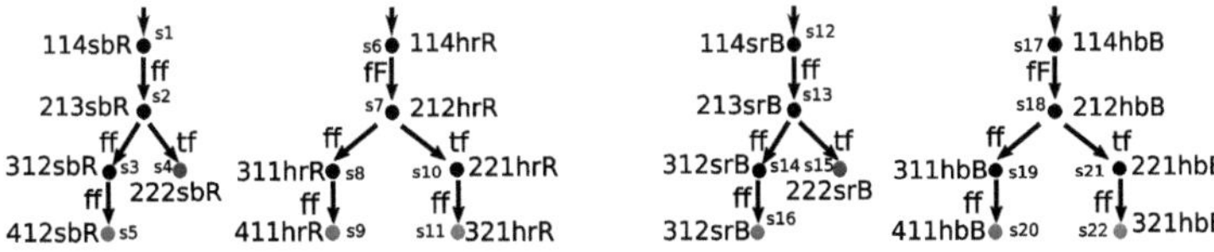

Fig. 8. W_D: In the city R red cars are hasty, while in city B blue cars are hasty.

Theorem 2. (Determine Weak Relevance). *Given W_D, ψ, the maximal available knowledge $L_{\mathbb{K}m}$, the maximal set of observations $\mathcal{O}_m$ that the system may make and the maximal set of possible beliefs, $\mathbb{B}_m$.*

We can determine all weakly relevant tuples for (W_D, ψ).

Proof (Theorem 2). Start with $L_{\mathbb{K}m}$, $\mathcal{O}_m$ and $\mathbb{B}_m$. Then build the lesser tuples and check whether there is a belief labelling $\mathcal{B}'$ that still conserves the relevant.

$\square$

Note that $\mathcal{O}_m$ can be at most AP. A sufficient candidate for maximal possible beliefs, $\mathbb{B}_m$, can be inferred from W_D.

The procedure sketched in the above proof is certainly not efficient. Strategy synthesis is expensive (2EXPTIME-complete for LTL) and the proof sketches a brute force approach. The choice of the maximal elements $L_{\mathbb{K}m}$, $\mathcal{O}_m$ and $\mathbb{B}_m$ will have a great influence on the overall runtime. For acceleration, the approach could easily consider constraints that restrict combinations $(L_{\mathbb{K}}, \mathcal{O}, \mathbb{B})$ to those an engineer is interested in. A further lever for acceleration might be to infer knowledge from previous synthesis attempts.

6 Conclusion

This paper is concerned with safety-critical autonomous systems, that assess their situation and decide what to do based on the formed beliefs. Since they operate in an overwhelmingly complex environments, the question arises "What is relevant for such a system $\mathcal{S}$?". To approach this question formally, I present a game-theoretic characterisation of relevance. The notion takes into account that $\mathcal{S}$ (i) takes decision based on its belief and it captures that (ii) the dimensions $(L_{\mathbb{K}}, \mathcal{O}, \mathbb{B})$ are interdependent -we might trade one for the other- and (ii) $(L_{\mathbb{K}}, \mathcal{O}, \mathbb{B})$ is influenced by the goal and context of $\mathcal{S}$. We reduce relevance to the question of whether there exists an autonomous system that can be successful. The present work considers a quite general notion of autonomous system with a very liberal belief formation. An interesting aspect for future research is hence the characterisation of salience properties of the belief formation $\mathcal{B}$, like notions of *belief stability* (how (drastically) may the beliefs evolve) or notions that enforce a *stable planning horizon* (how (drastically) may the future strategies change over time).

References

1. Baier, C., Katoen, J.P.: Principles of Model Checking, chap. 5, pp. 229–239. The MIT Press (2008)
2. Damm, W., Finkbeiner, B.: Does it pay to extend the perimeter of a world model? In: Butler, M., Schulte, W. (eds.) FM 2011. LNCS, vol. 6664, pp. 12–26. Springer, Heidelberg (2011). https://doi.org/10.1007/978-3-642-21437-0_4
3. De Sabbata, S., Mizzaro, S., Reichenbacher, T.: Geographic dimensions of relevance. J. Documentation **71**(4), 650–666 (2015)
4. Fagin, R., Halpern, J.Y., Moses, Y., Vardi, M.Y.: Reasoning About Knowledge. MIT Press (2003). https://doi.org/10.7551/mitpress/5803.001.0001
5. Huang, X., Soergel, D.: Relevance: an improved framework for explicating the notion. J. Am. Soc. Inf. Sci. Technol. **64**(1), 18–35 (2013)
6. Mizzaro, S.: Relevance: the whole history. J. Am. Soc. Inf. Sci. **48**(9), 810–832 (1997)
7. Rakow, A.: A doxastic characterisation of autonomous decisive systems. In: Luckcuck, M., Farrell, M. (eds.) Proceedings Fourth International Workshop on Formal Methods for Autonomous Systems (FMAS) and Fourth International Workshop on Automated and verifiable Software sYstem DEvelopment (ASYDE), Berlin, Germany, 26th and 27th of September 2022. Electronic Proceedings in Theoretical Computer Science, vol. 371, pp. 103–119. Open Publishing Association (2022). https://doi.org/10.4204/EPTCS.371.8
8. Rakow, A.: Framing relevance for safety-critical autonomous systems (2023). https://arxiv.org/abs/2307.14355
9. Rees, A.M., Saracevic, T.: The measurability of relevance. In: Proceedings of the American Documentation Institute, pp. 225–234. American Documentation Institute, Washington, DC (1966)
10. Saracevic, T.: Relevance: a review of and a framework for the thinking on the notion in information science. J. Am. Soc. Inf. Sci. **26**(6), 321–343 (1975)
11. Saracevic, T.: Modeling interaction in information retrieval (IR): a review and proposal. In: Proceedings of the 59th Annual Meeting of the American Society for Information Science, vol. 33 (01 1996)
12. Schamber, L., Eisenberg, M., Nilan, M.: A re-examination of relevance: toward a dynamic, situational definition. Inf. Process. Manage. **26**, 755–776 (1990). https://doi.org/10.1016/0306-4573(90)90050-C
13. Vagia, M., Transeth, A.A., Fjerdingen, S.A.: A literature review on the levels of automation during the years. what are the different taxonomies that have been proposed? Appl. Ergon. **53**, 190–202 (2016). https://doi.org/10.1016/j.apergo.2015.09.013, https://www.sciencedirect.com/science/article/pii/S0003687015300855

Optimization Problems in Cyber-Physical Energy Systems

Sebastian Lehnhoff and Jörg Bremer$^{(\boxtimes)}$

University of Oldenburg, Oldenburg 26129, Germany
`sebastian.lehnhoff@uni-oldenburg.de`, `joerg.bremer@uol.de`
`https://uol.de/ei`

Abstract. The current upheaval in the energy supply towards a more climate-friendly system is accompanied by a technological renewal, which allows completely new control and optimization approaches for the operation of the physical infrastructure through its digitalization. The energy supply system is thus becoming a cyber-physical energy system (CPES) with complex interactions between millions of distributed generation and consumption units. This paper provides insights into the characteristics of the resulting dynamic optimization tasks and how they can be attacked with distributed heuristics.

Keywords: Cyber-Physical Energy Systems · Multi-Agent Systems · Optimization

1 Introduction

Digitalized energy systems are characterized by a high degree of complexity in monitoring and controlling a large number of energy consumers and generators through a dynamically optimized infrastructure synchronized with societal behavior and controlled by internationally operated markets. From a technical point of view, digitalized energy systems can be described as cyber-physical systems (CPS) that have emerged from the interconnection of electromechanical components with IT-interfaces and IT-functionalities that are closely connected to the surrounding physical world and its ongoing processes, while providing and using data access and processing services available through the Internet. In other words, CPS can be broadly characterized as physical and technical systems whose operations are monitored, controlled, coordinated, and integrated by a computing and communicating core. The term "cyber-physical system" was created in response to the need for a new theoretical foundation for the study and development of large, distributed, complex systems. In this sense, cyber-physical energy systems have special characteristics that distinguish them from other CPSs:

- They are regarded as a critical infrastructure (CRITIS) and indispensable lifelines of modern societies.

M. Fränzle et al. (Eds.): Werner Damm Festschrift, LNCS 15471, pp. 113–126, 2026.
https://doi.org/10.1007/978-3-031-97537-0_7

- They have cross-continental size, e.g., in Europe from North Africa to Scandinavia, from Ireland to Asia.
- They show instantaneous propagation velocity of dynamic phenomena, e.g. imbalances or instabilities.
- Energy systems include a distinct market perspective, i.e. system-wide orchestration and resource provisioning through economic incentives and specialized market models.
- Their actors exhibit omnipresent conflicts of objectives, e.g. monetary, technical, environmental and (national/international) political interests.
- They are undergoing rapid and fundamental change due to the energy transition and digitalization.

Following this reasoning, smart grid research, development and engineering requires and specializes the need for a theoretical basis for interdisciplinary education, research and development of digitalized energy systems – significantly extending the scope of CPS.

Competences necessary for mastering this novel field are not only in the general area of "system intelligence" – algorithms for adaptive control and continuous dynamic optimization of the complex and expansive international power supply system – and the knowledge of methods to create and orchestrate "overall system competence" – complexity control through decomposition and abstraction, identification of and focus on generalizable principles, detection of decoupling points for effective governance, and avoidance of bottlenecks.

The outline of this paper is as follows. We start by introducing key optimization challenges in CPES and detail the problem characteristics that make them particularly hard to solve. Subsequently, we introduces COHDA – a versatile multi-agent-based heuristic that has been demonstrated to achieve good solutions for a variety of these challenges. We finish with a brief outlook on the feasibility of recently discussed optimization tasks in CPES.

2   Optimization Problems in Cyber-Physical Energy Systems and Their Characteristics

Many of the optimization and coordination problems that frequently occur in cyber-physical energy systems (CPES) are hard to solve for several reasons [27, 37,40].

Most problems can be classified as constrained, black-box, non-linear, and multi-objective [26]. This paper aims at providing an overview over common optimization problems in CPES as well as heuristic solution strategies. For sake of simplicity and because each multi-objective problem can be reduced to a single-objective problem by using a scalarization approach [24], in the following we will restrict our explanations to single objective problems.

2.1   Problem Size

In Germany alone, in the year 2022, the amount of electric energy produced by renewable energy resources grew to roundabout 254 TW [19]. Renewable energy is mostly generated by small, distributed and – due to its dependence on environmental factors – hardly controllable and predictable energy resource. This growth in the share of renewable energy leads to a growing complexity of the energy system and thus to a growth of the problem size of optimization problems within the system. In addition, a paradigm shift in energy management can be observed. The traditional paradigm "generation follows consumption" – meaning that generation is scheduled according to predicted consumption (which was treated as non controllable) – will be replaced by a scenario that incorporates controllable consumption in one or the other way [28,43]. Integrating controllable consumption on the other hand further increases the problem size for control tasks. The most promising solution for coping with problem size is distributed soling by self-organization or agent-based approaches [37].

2.2 NP-Hardness

Problems in energy management are of often times of discrete and combinatorial nature and categorized as NP-hard. NP-hardness occurs on two different levels. One example for a problem on a global level is given by the problem of coalition structure formation [12] which is known to be NP-hard [35] and can efficiently only be solved approximately [1]. The formation of suitable coalitions is a necessary preliminary stage that groups complementary energy resources to virtual power plants (VPP) as a controllable entity in the smart grid that can substitute obsolete larger power plants for ancillary services – complementarity is desired to increase robustness of a VPP. Different energy resources like photovoltaic plants, co-generation plants, batteries, and so forth, are literally drawn together by a communication network that enables interaction and thus new, self-organized orchestration and optimization solutions. Seen from the outside, one can abstract from the different energy resources as they coordinate themselves to jointly act like a single, large power plant. Finding optimal production schemes within such a VPP, often leads to a bi-level optimization when solved in a distributed way. The inner part can be NP-hard itself. This has for example been proven for the case of finding optimal operation schedules for a co-generation plant [11]; even for a single one. In the local case, complexity often grows exponentially with the dimensionality of the optimization problem whereas in the global case the problem size is responsible for complexity. Sometimes, local sub-problems are even harder, e.g., the #P-complete calculation of Shapley values for fair surplus distribution in VPPs [6].

2.3 Landscape Characteristics

Apart from size and growth of the search space, other characteristics render most problems hard to solve. In fact, as soon as one relaxes the discrete predictive scheduling to the continuous case, the problem is no longer NP-hard. Such relaxation can always be achieved by moving from the power level to the on-time domain in steady state, i.e., being switched on for a certain percentage of the duration of the considered time interval and off for the rest can be interpreted as operating the whole duration with the respective mean power level. Neverthe-less, most of the problems show fitness landscape characteristics that demand the use of specific heuristics for efficient solving. Fitness landscape analysis has been conducted on different problems within CPES. An examples is given in [11].

In [11], the issue of complexity in ensuring the feasibility of a solution was considered. The guaranteed generation of solutions that can be implemented in practice, especially within the framework of self-organized solution strategies, contributes significantly to the trust in such procedures. One of the essential findings is the fact that the guaranteed compliance with constraints is not inferior in complexity to the actual solution finding and must be considered equally in the algorithm design.

2.4   Uncertainties

Almost all problems in the real word are subject to uncertainties. Within the energy domain, these uncertainties mainly result from two sources: Weather and human interaction – we deliberately omit failures and faults since in CPES they are either correlated with weather-dependent influences or have a magnitude less frequent occurrence. Weather conditions inflict imprecision on predictions rendering plans unfeasible. Human interactions may lead to unpredictable dis-ruptions of the schedules operation of the system. Dynamic scheduling techniques are thus needed to achieve robust and flexible operation. Meanwhile, multiple approaches have been proposed to handle the unpredictable disruptions during operation, either by predictively developing a robust and stable schedule or by designing an efficient reactive policy to assign resources to tasks [39].

2.5   Real-Time Constraints

Finally, an additional challenge for many problems are induced by real-time constraints with hard deadlines. According to the German Transmission Code [2] ancillary services comprise frequency control, voltage control, re-establishment of power supply after a supply disruption or blackout, and power system operation and management [41] with different deadlines ranging from seconds to minutes dictating the completion time of necessary underlying optimizations.

An example for an agent-based energy management system in which soft-ware agents, representing consumers and producers, negotiate supply configura-tions (on-line checked during operation for compliance with permissible network operation limits) and apply any necessary corrective or stabilizing measures in a decentralized manner can be found in [45].

But, even problems with longer deadlines like day-ahead planning with steady-state models may meanwhile cause timing problems due to the ever grow-ing problem size.

2.6   Constrained Black-Box Models

For most problems in CPES, the individual flexibility – and thus the set of different operational capabilities – of many different, technically versatile energy resources [3] has to be taken into account. Flexibility modeling can be understood as the task of modeling constraints for energy units. Apart from global constraints (e.g. regarding the grid), constraints often appear within single energy components and affect local decision-making.

For optimization approaches in smart grid scenarios, black-box models capable of abstracting from the intrinsic model have proven useful [14], [15]. Such machine learning-based flexibility models are trained with a set of feasible energy generation profiles specific to the capabilities of the energy resource at hand – learning an abstract model of the decision boundary that separates feasible and infeasible generation profiles.

In order to integrate a flexibility model into an algorithm, decoders [13] have been introduced. In general, a decoder maps between genotype and phenotype solutions and in case of flexibility modeling maps a unconstrained genotype solution onto a nearby, feasible phenotype solution. For this mapping, the information from the flexibility model is used.

2.7   Self-organisation

Another important aspect of huge, distributed systems with individually acting components that are supposed to show a certain degree of autonomy as a whole is the concept of self-organization [16]. Self-organization is typically defined as the evolution of a system into an organized form of the components without any external pressures or control [34]. A broader definition given by Haken [20] who stated that a system was self-organizing if it acquires a spatial, temporal or functional structure without specific interference from the outside.

In a self-organizing system, global system behavior arises through local interaction between the individual components. However, this also means that no component has access to all global information and thus no component alone can create a complete system image for its own decision-making. On the one hand, this shows the strength of self-organizing systems (for example, in the ability to react appropriately to unforeseen situations [5]), but on the other hand, it creates problems, especially in systematic system and algorithm design.

There are four main reasons that prevent a straightforward implementation of self-organizing systems in industry. First, there is still a lack for systematic design of the numerous interactions that are necessary to establish self-organization with desired outcome [21]. Second, these numerous interactions often cause a communication overhead [42]. Third and probably most severe, self-organization results in nondeterministic outcomes. This is at the same time a strength as far-from-equilibrium and symmetry-breaking dynamics may lead to stable patterns that can and should be exploited, but also a weakness when it comes to adoption by industry; nondeterminism requires an appropriate verification, and verification methodologies are still on the research agenda. Finally, a self-organizing system cannot be developed with incremental advancements that are accepted by industry [34].

A possible remedy is the concept of controlled self-organization as proposed in [30], in which an additional system of observers and controllers monitors the actual system in its behavior and performance and, if necessary, intervenes to regulate it; for example, through suitable reparameterization. An example of the energy domain can be found in [9].

3   COHDA as a Versatile Solver

As a versatile tool for decentralized optimization in CPES, the Combinatorial Optimization Heuristics for Distributed Agents (COHDA) [22] has been introduced. The general concept of this heuristic is closely related to cooperative co-evolution [33]. The key concept of COHDA is an asynchronous iterative approximate best-response behavior. Each agent is responsible for just one dimension of an algorithmic-level problem decomposition. In this way, an agent in COHDA does not work on a complete (global) solution, but is responsible only for a local part. Each participating agent reacts to updated information from other agents by adapting the own previous decision on a possible (local) solution candidate.

COHDA defines a negotiation protocol that leads to repeated sending of local solution candidates to neighboring agents that lead to adaption of local solutions due to changed prerequisites and thus to a revision of previous agent decisions, if possible. The intermediate solutions of other agents (represented by published decisions) are regarded as temporarily fixed. Thus, each agent only searches along a low-dimensional cross-section of the search space and thus has to solve merely a simplified sub-problem. Nevertheless, for evaluation of the solution, the full objective function is used after aggregation of all agent's contributions.

In this way, the approach achieves an asynchronous coordinate descent with the additional ability to escape local minima by parallel searching different regions of the search space; and because former decisions can be revised if newer information becomes available. This approach is especially suitable for large-scale problems [8].

The agents make local decisions solely based on this information. The general protocol works in three repeatedly execute stages: perception, decide, and act (cf. [31]).

Perception Phase: During this first phase, the agent prepares for local decision-making. When a message is received from a neighboring agent that precedes in the directed communication topology, the data from the message is merged into the own knowledge-base. A small-world topology [44] has proven useful and is mostly used for communication. The data from the message usually consists of the updated local decision of the sender and the transient information on decisions of other agents that led to the updated decision. The agent first updates the own local knowledge base to gain an updated belief for the follow-up decision phase. COHDA is a heuristic and might thus get stuck in a local minimum. To better circumvent this, agents may postpone a decision until more information has been collected for better decisions.

Decision Phase: In this second phase, the agent is supposed to conduct a local optimization to yield the best decision for an own local action that puts the coalition forward as best as possible. Each agent is capable of evaluating the joint solution of all agents to do this. For finding the optimal local solution part, a low-dimensional part of the problem is solved. The term dimension can also refer to a sub-manifold containing low-dimensional local solutions as fraction of a much higher-dimensional global search space. As an example from the energy domain, a local contribution to a joint energy generation profile of many distributed energy resources could be a many-dimensional real-valued vector describing amount of generated energy per time interval for one single device. In general, other agents have made local decisions before. Based on collected information about these (intermediate) local decisions, a new solution candidate from the local (constrained) search space is sought that optimizes the global problem.

Act Phase: In the last stage, the agent compares the fitness of the best found global solution with the previous solution. For comparison, the global objective function is used. If the new solution has a better fitness (or lower error; depending on the specific problem setting), the agent finally broadcasts a message containing its new local solution contribution together with everything it has learned from other agents and their current local solution contributions (the decision base) to its immediate neighbors in the communication topology. Receiving agents then execute these three phases from scratch leading to potentially revised local solution contributions and thus to further improved overall solutions.

Figure 1 gives an overview on these three stages of an agent embedded in a coalition of an multi-agent system. If an agent cannot find any local contribution that improves the global solution, no message is sent. If no agent can find any improvement anymore, the process has reached an at least local optimum and eventually ceases because no more messages are sent. After the system has produced a series of intermediate solutions, the heuristic eventually terminates in a state where all agents know an identical solution. This one is taken as the final solution of the heuristic. Properties like guaranteed convergence and local optimality have been formally analyzed and proven [22]. Moreover, after a short setup time, COHDA possesses the anytime-property. Thus, the agent-protocol could be stopped at any time with a valid (sub-optimal) solution, if necessary.

3.1 Predictive Scheduling with COHDA

In order to make it more tangible, this section presents a COHDA-based solution to predictive scheduling, which is also known as offline scheduling [39]. In a two stage approach, suitable and timed allocations of n jobs to m machines are first sought in a job-shop problem in the manufacturing domain [18]. In energy management, related problems are common in VPP (refer to Sect. 2.2). A frequent task in VPP is solving the scheduling problem that assigns an operation schedule to each energy resource taking into account a bunch of objectives like accurate resemblance of the desired load profile, robustness of the schedule, costs, maximizing remaining flexibility for subsequent planning periods, and more. A

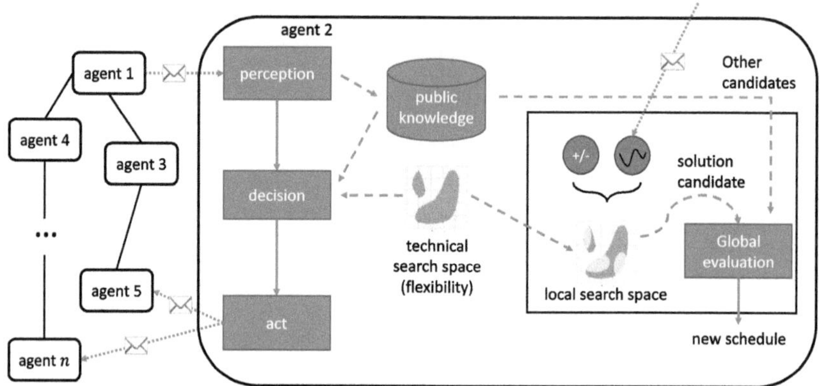

Fig. 1. General COHDA procedure for the case of predictive scheduling.

schedule in this context is a real-valued vector with each element denoting the amount of energy (or mean active power) generated or consumed during the respective time interval for a given future discrete planning horizon. A predictive scheduling algorithm tries to assign a schedule to each energy resource such that the sum of all schedules resembles a desired aggregated schedule as close as possible. Such aggregated schedule might for example be given as a result of some market bidding and the VPP is now supposed to jointly operate this schedule. COHDA has initially been developed as a decentralized solver specifically to this problem in VPP [23].

For predictive scheduling with high penetration of renewable energy resources, COHDA in combination with a classifier-based decoder for modeling individual flexibilities has demonstrated good performance. On the other hand, such decoder-based algorithms are designed for single entities and not able to cope with ensembles of energy resources. Combining training sets sampled from individually modeled energy units, results in folded distributions of power levels with unfavorable properties for machine learning-based approaches. Nevertheless, this happens to be quite a frequent use case, e.g., when commercial buildings with an ensemble of co-generation, heat pump, solar power, and controllable consumers take part in decentralized predictive scheduling. Apart from approaches to circumvent the folding problem already apparent at the sampling stage, [7] presented a hybrid method that equips COHDA with an internal solver that solves the problem in a bi-level optimization approach. To achieve this, COHDA may in general use any optimization approach during the decision phase to find the optimal schedule for the local ensemble that is then offered to the other agents.

3.2 Fair Surplus Distribution

Temporary collaboration, i.e., "teamwork" of individually owned and operated distributed energy resources demands for a proper and fair surplus distribution after product delivery. Distributing the surplus merely based on the absolute (load) contribution does not take into account that smaller units may provide the means for fine grained control as they are able to modify their load on a smaller scale. Additionally, the wear and tear of smaller (less efficient) systems might be higher compared to bulk installations.

A surplus distribution scheme based on Shapley values for fairness and k-majority voting games for incorporating several criteria has been proposed in [6]. In order to efficiently estimate the marginal contributions of all agents, different COHDA instances are run in parallel.

In this way, it becomes possible to determine the distribution key also in a distributed manner without the need for an external entity to centrally specify a distribution key.

3.3 Learning to Solve Unknown Problems

In order to do justice to the idea of autonomy in cyber-physical systems algo-rithmically, it is necessary to prepare the system components for an independent elaboration of solution strategies for unknown problems. In distributed systems, however, this also means that subsystems must cooperate with other (possibly unknown) subsystems in order to jointly develop a solution strategy in which the individual capabilities are suitably combined.

A targeted design of a specific emergent behavior is difficult to achieve, espe-cially with general purpose programming languages [32]. Consequently, [4] pro-posed to rather learn how to solve new problems in self-organized systems in a decentralized way. As a first approach to learn self-organized behavior which brings in individual capabilities, Cartesian Genetic Programming [29] has been used with COHDA as solver. As a result, agents are able to independently and cooperatively develop a solution strategy that makes exclusive use of various functions provided by the agents.

By using an approach from Genetic Programming, the different function-alities (performing computations, actuators to control the power grid, sensors, communication, ...) that an agent can offer can be interconnected by means of data links. Finding the optimal connections (in the sense that the resulting pro-gram solves the new problem well) is an optimization problem that can be solved in a distributed way with COHDA.

3.4 Privacy Preservation

Negotiation among agents that are controlling and orchestrating a set of dis-tributed processes relies on frequent data exchange in order to converge to a joint solution. Within multi-agent systems, trust has been on the research agenda for years [36]. Trust is also a major issue when humans have to interact with a sys-tem. Many CEOs identify trust and reputation as the key driver of their action [24].

Solving decentralized coordination problems within agent coalitions that exchange messages and information to build up agents' beliefs for problem solving, inevitably allows insight into other agents' options. As an example, in the predictive scheduling use case in VPPs frequently operable schedules are exchanged as proposal for own choice of action [21]. Each schedule contains data of the possible portion of energy that may be generated (or consumed) during a given time period. With each negotiation round, a new possible operable schedule is sent to several other agents together with transient information on other agents' schedules. This is the widely used gossiping principle that comes into play into many decentralized coordination algorithms [25]. Such information can be collected, aggregated, and exploited by agents. Examples can be found in [17], where detailed information on internal processes like heating profiles, and thus working hours, machinery load factors in case of internal consumption optimization with batteries, or current capacity utilization were derived. In [14] individual time of use tariffs were reconstructed from aggregated, collected schedules.

Parties potentially interested in participating in distributed energy management will refrain from engaging in agent-based computations if the economic risk of revealing data is larger than the benefit from participating. In order to achieve broad acceptance for the use of such algorithms, especially among commercial users, it is essential to handle the disclosure of data sparingly or to otherwise ensure that it cannot be misused. One possible solution for the energy domain, as discussed in [10] and [38], among others, is to integrate encryption approaches into the COHDA protocol. The most promising for this at present seems to be the concept of secret sharing [38].

4   Conclusion

Cyber-physical energy systems are particulary complex and hard to operate CPS that require unique solution methods and tactics. Most combinatorial problems in CPES imply optimization techniques that require exponential computation time. In this paper we have given an insight into CPES' special problem characteristics as well as a versatile and established go-to-method for solving diverse optimization tasks present here. It is intuitively plausible to opt for heuristic strategies that compute fast but yield sub-optimal solutions instead. However, computation performance has grown exponentially in the last decades (i.e. Moore's law) and there have been significant advances in on-line optimization [15]. However, nowadays there are numerous real-world examples of optimization problem instances in CPES that are of a size, which might be efficiently optimized on modern parallelized computer architectures (e.g. 10 devices in a household, 20 EVs below a low-voltage feeder, 100 households within a district etc.). Ultimately, it seems reasonable to revisit heuristic approaches to CPES problems in the literature on a regular basis that have been hitherto treatable only suboptimally.

Acknowledgements. The authors thank Werner Damm for many years of inspiration in treating energy systems as CPS with special characteristics. This has led to many fruitful discussions and insights into system model boundaries as well as solution properties when designing our heuristics.

References

1. Beer, S.: A formal model for agent-based coalition formation in electricity markets. In: IEEE PES ISGT Europe 2013, pp. 1–5 (2013). https://doi.org/10.1109/ISGTEurope.2013.6695442
2. Berndt, H., Hermann, M., Kreye, H.D., Kreye, R., Scherer, U., Vanzetta, J.: Transmissioncode 2007: Netz- und systemregeln der deutschen Übertragungsnetzbetreiber (2007). https://www.vde.com/resource/blob/937758/14f1b92ea821e9e19ee13fc798c1ee0e/transmissioncode-2007--netz--und-systemregeln-der-deutschen-uebertragungsnetzbetreiber-data.pdf
3. Bremer, J., Rapp, B., Sonnenschein, M.: Encoding distributed search spaces for virtual power plants. In: Computational Intelligence Applications In Smart Grid (CIASG), 2011 IEEE Symposium Series on Computational Intelligence (SSCI). Paris, France (2011). https://doi.org/10.1109/CIASG.2011.5953329
4. Bremer, J., Lehnhoff, S.: Towards evolutionary emergence. In: Ganzha, M., Maciaszek, L., Paprzycki, M., Slezak, D. (eds.) Position and Communication Papers of the 16th Conference on Computer Science and Intelligence Systems, pp. 55–60. Polish Information Processing Society, Polskie Towarzystwo Informatyczne, Warszawa, Poland, Online (2021).https://doi.org/10.15439/2021F111
5. Bremer, J.: Learning to Optimize, pp. 1–19. Springer International Publishing, Cham (2022). https://doi.org/10.1007/978-3-031-06839-3_1
6. Bremer, J., Lehnhoff, S.: Decentralized surplus distribution estimation with weighted k-majority voting games. In: Bajo, J., Vale, Z., Hallenborg, K., Rocha, A.P., Mathieu, P., Pawlewski, P., Del Val, E., Novais, P., Lopes, F., Duque Méndez, N.D., Julián, V., Holmgren, J. (eds.) Highlights of Practical Applications of Cyber-Physical Multi-Agent Systems, pp. 327–339. Springer International Publishing, Cham (2017). https://doi.org/10.1007/978-3-319-60285-1_28
7. Bremer, J., Lehnhoff, S.: Hybrid multi-ensemble scheduling. In: Squillero, G., Sim, K. (eds.) EvoApplications 2017. LNCS, vol. 10199, pp. 342–358. Springer, Cham (2017). https://doi.org/10.1007/978-3-319-55849-3_23
8. Bremer, J., Lehnhoff, S.: The effect of laziness on agents for large scale global optimization. In: van den Herik, J., Rocha, A.P., Steels, L. (eds.) ICAART 2019. LNCS (LNAI), vol. 11978, pp. 317–337. Springer, Cham (2019). https://doi.org/10.1007/978-3-030-37494-5_16
9. Bremer, J., Lehnhoff, S.: Controlled self-organization for steering local multi-objective optimization in virtual power plants. In: De La Prieta, F., et al. (eds.) PAAMS 2020. CCIS, vol. 1233, pp. 314–325. Springer, Cham (2020). https://doi.org/10.1007/978-3-030-51999-5_26
10. Bremer, J., Lehnhoff, S.: Encrypted decentralized optimization for data masking in energy scheduling. In: Proceedings of the 3rd International Conference on Big Data Research. p. 103–109. ICBDR 2019, Association for Computing Machinery, New York, NY, USA (2020). https://doi.org/10.1145/3372454.3372487
11. Bremer, J., Lehnhoff, S.: Ant colony optimization for feasible scheduling of step-controlled smart grid generation. Swarm Intell. **15**(4), 403–425 (2021). https://doi.org/10.1007/s11721-021-00204-7

12. Bremer, J., Sonnenschein, M.: Estimating shapley values for fair profit distribution in power planning smart grid coalitions. In: Klusch, M., Thimm, M., Paprzycki, M. (eds.) MATES 2013. LNCS (LNAI), vol. 8076, pp. 208–221. Springer, Heidelberg (2013). https://doi.org/10.1007/978-3-642-40776-5_19

13. Bremer, J., Sonnenschein, M.: Constraint-handling with support vector decoders. In: Filipe, J., Fred, A. (eds.) ICAART 2013. CCIS, vol. 449, pp. 228–244. Springer, Heidelberg (2014). https://doi.org/10.1007/978-3-662-44440-5_14

14. Bremer, J., Lehnhoff, S.: Information disclosure in vpp - information disclosure by decentralized coordination in virtual power plants and district energy systems. In: Wohlgemuth, V., Naumann, S., Arndt, H.K., Behrens, G., Höb, M. (eds.) EnviroInfo 2022. p. 133. Gesellschaft für Informatik e.V., Bonn (2022)

15. Chen, N., Goel, G., Wierman, A.: Smoothed online convex optimization in high dimensions via online balanced descent. In: Bubeck, S., Perchet, V., Rigollet, P. (eds.) Proceedings of the 31st Conference on Learning Theory. Proceedings of Machine Learning Research, vol. 75, pp. 1574–1594. PMLR (2018). https://proceedings.mlr.press/v75/chen18b.html

16. Collier, J.: Fundamental properties of self-organization. In: Causality, emergence, self-organisation, pp. 287–302 (2003)

17. Dabrock, K.: Privacy in der automatisierten prädiktiven Einsatzplanung von Energieanlagen im Smart Grid. Master's thesis, University of Oldenburg, Oldenburg, Germany (2018)

18. Dauzère-Pérès, S., Ding, J., Shen, L., Tamssaouet, K.: The flexible job shop scheduling problem: a review. Eur. J. Oper. Res. **314**, 409–432 (2023)

19. Energien-Statistik, A.E.: Erneuerbaren energien in deutschland daten zur entwicklung im jahr 2022. Umweltbundesamt, Dessau-Roßlau (2022)

20. Haken, H.: Information and Self-Organization: A Macroscopic Approach to Complex Systems. Springer Science & Business Media (2006). https://doi.org/10.1007/3-540-33023-2

21. Hinrichs, C., Sonnenschein, M.: Design, analysis and evaluation of control algorithms for applications in smart grids. In: Gómez, J.M., Sonnenschein, M., Vogel, U., Winter, A., Rapp, B., Giesen, N. (eds.) Advances and New Trends in Environmental and Energy Informatics. PI, pp. 135–155. Springer, Cham (2016). https://doi.org/10.1007/978-3-319-23455-7_8

22. Hinrichs, C., Sonnenschein, M.: A distributed combinatorial optimisation heuristic for the scheduling of energy resources represented by self-interested agents. Int. J. Bio-Inspired Comput. **10**(2), 69–78 (2017)

23. Hinrichs, C., Sonnenschein, M., Lehnhoff, S.: Evaluation of a self-organizing heuristic for interdependent distributed search spaces. In: Filipe, J., Fred, A.L.N. (eds.) International Conference on Agents and Artificial Intelligence (ICAART 2013), vol. Volume 1 – Agents, pp. 25–34. SciTePress (2013). https://doi.org/10.5220/0004227000250034

24. Kasimbeyli, R., Ozturk, Z.K., Kasimbeyli, N., Yalcin, G.D., Erdem, B.I.: Comparison of some scalarization methods in multiobjective optimization: comparison of scalarization methods. Bull. Malays. Math. Sci. Soc. **42**, 1875–1905 (2019)

25. Kermarrec, A.M., Van Steen, M.: Gossiping in distributed systems. ACM SIGOPS Oper. Syst. Rev. **41**(5), 2–7 (2007)

26. Lehnhoff, S., et al.: Towards fully decentralized multi-objective energy scheduling. In: 2019 Federated Conference on Computer Science and Information Systems (FedCSIS), pp. 193–201. IEEE (2019)

27. Macana, C.A., Quijano, N., Mojica-Nava, E.: A survey on cyber physical energy systems and their applications on smart grids. In: 2011 IEEE PES Conference on Innovative Smart Grid Technologies Latin America (ISGT LA), pp. 1–7. IEEE (2011)

28. Manfren, M., Caputo, P., Costa, G.: Paradigm shift in urban energy systems through distributed generation: methods and models. Appl. Energy **88**(4), 1032–1048 (2011)

29. Miller, J.F.: Cartesian Genetic Programming, pp. 17–34. Springer, Heidelberg (2011). https://doi.org/10.1007/978-3-642-17310-3_2

30. Müller-Schloer, C., Schmeck, H., Ungerer, T.: Organic Computing—A Paradigm Shift for Complex Systems. Springer Science & Business Media (2011). https://doi.org/10.1007/978-3-0348-0130-0

31. Nieße, A., Beer, S., Bremer, J., Hinrichs, C., Lünsdorf, O., Sonnenschein, M.: Conjoint dynamic aggregation and scheduling methods for dynamic virtual power plants. In: 2014 Federated Conference on Computer Science and Information Systems, pp. 1505–1514. IEEE (2014)

32. Parzyjegla, H., et al Weis, T.: Model-driven development of self-organising control applications. In: Organic Computing—A Paradigm Shift for Complex Systems, pp. 131–144 (2011)

33. Potter, M.A., Jong, K.A.D.: Cooperative coevolution: an architecture for evolving coadapted subcomponents. Evol. Comput. **8**(1), 1–29 (2000)

34. Prokopenko, M.: Design vs. self-organization. In: Advances in Applied Self-Organizing Systems, pp. 3–17 (2008)

35. Rahwan, T., Michalak, T.P., Wooldridge, M., Jennings, N.R.: Coalition structure generation: a survey. Artificial Intelligence **229**, 139–174 (2015). https://doi.org/10.1016/j.artint.2015.08.004, https://www.sciencedirect.com/science/article/pii/S0004370215001198

36. Ramchurn, S.D., Huynh, D., Jennings, N.R.: Trust in multi-agent systems. Knowl. Eng. Rev. **19**(1), 1–25 (2004)

37. Ramchurn, S.D., Vytelingum, P., Rogers, A., Jennings, N.R.: Putting the'smarts' into the smart grid: a grand challenge for artificial intelligence. Commun. ACM **55**(4), 86–97 (2012)

38. Rapp, B., Bremer, J.: Masking sensitive data in self-organized smart region orchestration. In: Proceedings of the ICIEI 2022: The 7th International Conference on Information and Education Innovations Manchester April 14 - 16, 2023. Association for Computing Machinery, New York, NY, USA, (in press) (2023)

39. Renke, L., Piplani, R., Toro, C.: A review of dynamic scheduling: context, techniques and prospects, pp. 229–258. Springer International Publishing, Cham (2021). https://doi.org/10.1007/978-3-030-67270-6_9

40. Schwarz, J.S., Steinbrink, C., Lehnhoff, S.: Towards an assisted simulation planning for co-simulation of cyber-physical energy systems. In: 2019 7th Workshop on Modeling and Simulation of Cyber-Physical Energy Systems (MSCPES), pp. 1–6. IEEE (2019)

41. Seidl, H., Mischinger, S., Rehtanz, C., Greve, M., et al.: dena-studie systemdienstleistungen 2030: Sicherheit und zuverlassigkeit einer stromver-sorgung mit hohem anteil erneuerbarer energien. In: Deutsche Energie-Agentur GmbH (dena), Energiesysteme und Energiedienstleistungen (2014)

42. Sneyd, J., Theraula, G., Bonabeau, E., Deneubourg, J.L., Franks, N.R.: Self-Organization in Biological Systems. Princeton university press (2001)

43. Sonnenschein, M., Rapp, B., Bremer, J.: Demand side management und demand response (Neufassung). In: Beck, H.P., Buddenberg, J., Meller, E., Salander, C. (eds.) Handbuch Energiemanagement, vol. 31. VWEW (Verlag), Frankfurt a. M. (2011)
44. Watts, D., Strogatz, S.: Collective dynamics of "small-world" networks. Nature **393**(6684), 440–442 (1998)
45. Wedde, H.F., Lehnhoff, S., Handschin, E., Krause, O.: Dezentrale vernetzte energiebewirtschaftung (dezent) im netz der zukunft. Wirtschaftsinformatik **49**(5), 361–369 (2007)

Dynamic Risk Management in Cyber Physical Systems

Daniel Schneider, Jan Reich, Rasmus Adler[(⊠)], and Peter Liggesmeyer

Fraunhofer IESE, Fraunhofer-Platz 1, 67663 Kaiserslautern, Germany
```
{daniel.schneider,jan.reich,rasmus.adler,
    peter.liggesmeyer}@iese.fraunhofer.de
```


Abstract. Cyber Physical Systems (CPS) enable new kinds of applications as well as significant improvements of existing ones in numerous different application domains. A major trait of upcoming CPS is an increasing degree of automation up to the point of autonomy, as there is a huge potential for economic success as well as for ecologic and societal improvements. However, to unlock the full potential of such (cooperative and automated) CPS, we first need to overcome several significant engineering challenges, where safety assurance is a particularly important one. Unfortunately, established safety assurance methods and standards do not live up to this task, as they have been designed with closed and less complex systems in mind. This paper structures safety assurance challenges of cooperative automated CPS, provides an overview on our vision of dynamic risk management and describes already existing building blocks.

Keywords: Safety · Cyber-Physical Systems · Runtime Safety · Dynamic Risk Management

1 Introduction

Over the last decades we witnessed a very strong trend of computerization and digitization. As a consequence, people's daily lives and ways of working have been undergoing significant transformations. For the future we are anticipating these trends to continue, maybe even further accelerate, and thus digital systems to be even more ubiquitous in everybody's daily lives. We expect them to permeate every facet of it, to support humans in any way imaginable, be it with respect to information, health, mobility or work. At the same time, work life will undergo further transformation as systems are increasingly assuming tasks that today are exclusively addressed by humans. Corresponding visions are being developed and pursued for many years now and a pretty impressive range of corresponding umbrella terms have been coined along the way, such as Ubiquitous Computing, Pervasive Computing, Ambient Intelligence, Internet of Things or Cyber-Physical Systems. In addition, there are domain specific umbrella terms such as Industry 4.0, Autonomous Driving, Connected Automated Mobility, Operating Room of the Future, Ambient Assisted Living, and so on. All these visions of future systems

M. Fränzle et al. (Eds.): Werner Damm Festschrift, LNCS 15471, pp. 127–146, 2026.
https://doi.org/10.1007/978-3-031-97537-0_8

are essentially enabled by a trinity of technological characteristics – **automation, interconnection, and AI,** where automation is the end and interconnection and AI are the indispensable means.

The potential is huge. The embedded systems industry, for example, is a significant job-creator in Europe, contributing a 34% share to the world production of embedded systems with particular strengths in the automotive sector, aerospace and health [1]. In the Electronic Components and Systems Strategic Research and Innovation Agenda 2023 (ECS-SRIA), the EU puts specific emphasis on the SoS topic, but also use of AI and pushing automation feature very strongly [2]. Further, AI in general has been identified as strategic technology and AI strategies have been devised by the key economic powers around the globe including Germany [3], Europe [4], US [5] and China [6]. As we previously elaborated in [7], aside from creating opportunities to improve and optimize existing businesses, digitalization also supports new forms of business models. In that respect, data-driven technologies are one central diver, paving the path for data-driven business models [8]. The key modification from traditional business models is that data-driven business models exploit data as the key value proposition [9]. As dependable, collaborative and autonomous systems rely on advanced sensor technologies and create huge amounts of data, data-driven business models are strongly appealing. Another driver are completely new types of applications which are enabled due to automation. Take mobility as a service as an example, which might revolutionize the way mobility is realized in our societies.

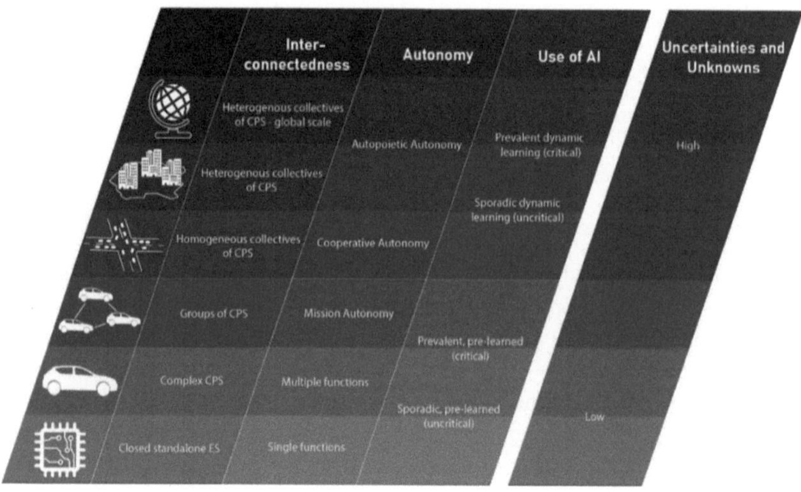

Fig. 1. Key trends as per [10]

But there are still significant challenges ahead which need to be tackled before this potential can be fully unlocked. These challenges can roughly be structured into three categories: *technological, structural/frame conditions* (e.g. laws, standards, interoperability, societal, education) and *assurance.* Technology seems to be relatively straight forward. Despite there always being open issues and needs for improvement, the currently available technology already goes a long way to actually enable future system visions to a significant extent. Considering structural challenges, i.e. the question as to how to establish the frame conditions that are required, we think there is no hard blocking point per se. On the other hand, this does not mean this class of challenges is easy or even straight forward to address. The problem here typically lies in questions such as who is responsible for driving what topics, what new structures are required, and so on. Last but not least, assurance related challenges seem to be the most urging ones. The technology capabilities to build systems with high degrees of automation have basically outgrown the established assurance capabilities, so that we momentarily have a mismatch severely hindering progress and innovation. Assurance challenges thus require special attention and additional research, Sect. 2 sets out to give a more detailed overview in this regard.

On an abstract level, assurance challenges result from unknowns and uncertainties which cannot be resolved during development time due to several reasons. At runtime, however, many of these unknowns and uncertainties could be resolved and resulting implications could then be treated. Thus, a general solution idea is to shift parts of the assurance process into runtime. Figuratively speaking, one could say such approaches are about making systems safety aware much in the way human navigate risk when operating safety-critical systems manually.

Enabling systems in this sense implies equipping them with two key capabilities: 1.) to be aware of the current risk as well as the risk related to intended behaviors, and, 2.) to be aware of own safety related capabilities (i.e., in automotive terms, those related to controllability or reducing severity). The technical means for realizing these capabilities can range from easy checks to sophisticated and intelligent reasoning. It is expected that, in parallel to increasing degrees of adoption of the aforementioned key traits (automation, interconnectedness, AI), systems will need to be empowered to assume increasingly comprehensive runtime assurance responsibilities [10].

Over the last years, runtime safety assurance approaches are increasingly accepted as a promising means by the safety engineering community and corresponding approaches, discussions and considerations can increasingly be found in publications, conferences, working groups and even in standardization groups, such as Digital Dependability Identities [11] being referred to in [12]. At our institute we investigated and developed different building blocks for enabling runtime assurance and devised a superordinate vision called Dynamic Risk Management (DRM). In Sect. 3 we introduce this vision, a conceptual solution architecture and describe several approaches for the core building blocks of that architecture in more detail.

2 Safety Engineering Challenges and Related Work

The trends shown in Fig. 1 directly translate in specific safety engineering challenges. High levels of automation lead to high levels of system complexity as well as, sometimes even more importantly, context complexity. Interconnectedness and distribution lead to challenges related to communication and interoperability, but particularly also to challenges due to unknowns with respect to the safety-related properties of connected and collaborating systems. Finally, utilization of ML/DL components for e.g. perception tasks inevitably leads to uncertainties, which again constitute a significant challenge for safety assurance. In this Section, we further structure and elaborate corresponding safety engineering challenges.

2.1 Assuring Autonomy – How to Deal with the Complexity?

Increasing automation levels imply increasingly transferring responsibilities from human operators to systems. This naturally translates into increasing complexity of systems, e.g., considering autonomous driving, due to now also assuming tasks related to perception and reasoning. At the same time, the complexity of the physical context of systems takes its toll, as context information is a fundamental input to be provided via perception and reasoned about by the system. Relatedly, with respect to assurance, systems need to be enabled to deal with internal and external risk factors [13]. An elaboration with respect to both of these dimensions follows below.

There is a widely recognized classification of automation levels in the automotive domain standardized in SAE J3016:2021. Automation levels 1 and 2 are purposed to support the driver, whereas the driver is still fully responsible to supervise the system continuously. Starting from level 3, the driver is allowed to take the eyes off the road and to be, at least for a given amount of time, out of the operating loop. However, in level 3, the driver must still be able to step in when the system issues a transition demand with sufficient time to reestablish situational awareness. On paper this seems to be a straightforward fallback mechanism, but the actual practicality is often questioned due to a mismatch between the time required to bring the driver back into the loop and the lookahead capabilities of today's perception stacks [14]. Thus, taking the driver completely out of the loop as foreseen for levels 4 and 5 seems to be the cleaner concept.

Popular definitions of autonomous systems refer to situation-specific behavior. Kagermann et al. [15] define that a system can be described as autonomous if it is capable of independently achieving a predefined goal "in accordance with the demands of the current situation…". A derived requirement is that the system is aware of the risks in the current environmental situation and that it behaves so that the risks will not become unacceptable. In contrast to conventional safety functions, this risk management function requires reasoning about the current environmental situation and potential harm scenarios [16], i.e. the **situational handling of external risk factors**. To this end, safety supervisors are used to monitor the nominal behavior (cf. German application rule VDE-AR-E 2842–6 part 3 clause 12.4), and evaluate the risks of potential system behaviors on different time horizons such as operational, tactical and strategic. Physics-based approaches like Time-To-Collision (TTC) and Responsibility-Sensitive Safety Model [17] are commonly used to formalize risk criteria. However, these physics-based approaches are not sufficient for developing human-like risk assessment capabilities, because they do not take into account knowledge about agent interactions. Thus, expert systems that imitate human risk reasoning must be developed, e.g. with real-time inferrable Bayesian networks [18]. Ultimately, as AI-based perception is often necessary for situational handling of external risk factors due to the required level of situation understanding, the relationship between uncertain AI-based perception and external risk factor estimation needs to be explored. Unsolved challenges in this research area are the identification of relevant risk factors and the definition of a risk criterion combining these factors to express the border between safe and unsafe situations on different time horizons.

Apart from the capability to determine the impact of complex environments on dynamic risk, another important aspect auf autonomous systems is their **self-sufficiency**, autonomicity or capability to maintain operation [19], i.e. achieving a goal without human control [15]. The degree of human independence is promoted in problem structuring frameworks like SAE J3016 or ALFUS [20]. To achieve self-sufficiency, systems need to dynamically handle *internal* risk factors, i.e. all kinds of component failures including failures due to functional insufficiencies or limitations like limited sensor range. **Situational handling of internal risk factors** enhances error detection and handling approaches by considering the risks of detected errors, which vary in different operational situations. Challenges in this research area emerge from the complexity in detecting all relevant errors and handling them so that the best possible functionality is provided. The approach in [21] proposes how to identify component degradation variants and considers the reconfiguration sequences for transitioning from one system configuration to another [22]. Trapp et al. [23] enhances this approach with respect to the risk-oriented handling of errors and proposes to shift parts of a functional safety engineering lifecycle to runtime. Further challenges emerge from components based on Machine Learning (ML), because their output is subject to uncertainty, which we understand as the likelihood that the actual output deviates in a particular way from the correct output. This aspect is further detailed in Sect. 2.3.

2.2   Assuring Interconnectedness – How to Modularize Safety?

Interconnectedness, which enables information sharing, cooperation and collaboration, is a key enabler for future CPS. The potential often seems to be underestimated though. Consider, for example, the field of mobility, where corresponding developments have been going rather slow over the last two decades. Cars (e.g.) are still understood and designed as mostly closed systems, where connectivity is primarily used for infotainment and convenience purposes (and not for optimizing the key functionalities and safety).

Considering safety challenges associated with interconnectedness, safety-related requirements regarding the communication system directly come to mind. But these are often not the most challenging ones, as there usually are ways to forge a sound safety concept around known shortcomings of a communication channel (e.g. black channel concept). A more significant obstacle to overcome are application-level challenges, i.e. assuring the safety of applications/functions that are rendered through a collaboration of different (maybe even dynamically) interconnected systems of different manufacturers. The key issue here is that the safety-related properties of the collaboration partners are not known and that there are no means to determine the actual safety-related guarantees of such a collaboration. Essentially, adequate means for safety modularization are required.

There is a broad range of safety modularization approaches, often contract-based, in the state of the art, mostly focusing on supporting safety engineering for single systems [24–31]. At the same time, there are several approaches taking more of a supply chain perspective, also considering model exchange and tooling, such as the work from the

European projects DEIS (the Open Dependability Exchange (ODE) metamodel and its instances called Digital Dependability Identities (DDI) [11]) and AMASS [32].

Utilizing such approaches at runtime was not a topic until relatively recently, John Rushby was one of the first researchers to propose an approach utilizing safety contracts for runtime monitoring to enable "just-in-time certification" [33]. This being still focused on single systems, the next step is to enable CPSoS to dynamically assess their safety-related properties at runtime. The ConSerts approach [34] utilize modular runtime safety models describing safety-related properties for a (constituent) system and a protocol for the dynamic assessment of these properties across composition hierarchies. Further details are provided in Sect. 3.1. There are further related approaches using runtime models for the assurance of adaptive systems [35, 36], distributed service-based systems [37] as well as considerations how a large scale adoption of runtime models might play out to manage heterogenous SoS ecosystems [38] as they are foreseen for the future [10].

Last, it is also noteworthy that safety interfaces and modularization play an important role for shared platforms, e.g. domain controllers, supporting co-existence of different mixed-criticality applications (was e.g. a topic in the European EMC2 project [39]) as well as dynamic updates (e.g. considered in the Up2date project [40]).

2.3 Assuring Machine Learning – How to Deal with Uncertainties?

In addition to interconnectedness, Machine Learning is clearly set to become the second key enabler for future (highly automated) CPS. Perception and behavior planning in autonomous systems are areas where applying ML is attractive and where we have been seeing very good results over recent years. The reason for this is that it is hard to engineer such functions by conventional means, thus it is appealing to let an artificial neural network (NN) learn the wanted behavior based on training data. An example for this is camera-based object detection, where it is difficult to manually specify how to detect e.g. a human. Training an ANN by means of (vast amounts of) training data ideally leads to a strong classification performance with just the right amount of generalization, but realistically this is also no small feat to achieve. Assuring safety is particularly challenging, as there is neither a sound and complete requirements specification nor adequate means to analyze and assure the learned model. There is a range of approaches that can be utilized to attain indications wrt. The performance of an ANN, but there is no silver bullet for generating hard evidence regarding the ANN performing as intended in any situation. As a consequence, assuring systems with ML components typically requires safety concepts which take safety responsibility away from the ML-based channel. Figure 2 exemplifies the different elements/channels such a safety concept might encompass, color coding implying a range from "black box" over different shades of grey to "white box". The ML component itself is a black box, but could be transformed into a somewhat greyish box by means of specialized ANN approaches (i.e. assuring the ML component itself), e.g. heatmapping from the field of XAI approaches. Moreover, parallel supervision channels might be established. This might be channels specialized for ANN, e.g. such that the "neuro"-channel is augmented by a "symbolic"-channel. Such approaches are also called hybrid AI approaches and a specific one is briefly introduced in Sect. 3.3. Apart from that, there is always the possibility to establish AI-agnostic supervisor channels.

Those again might include very simple and traditionally realized channels (i.e. easy to assure) that might serve as a last layer of defense, but there also might be more complex elements which again might even include AI or ANN themselves. The idea here would again be that the safety weight is put on the last layer of defense, but the outer layers shall ideally be good enough to ensure that the last layer is never actually needed.

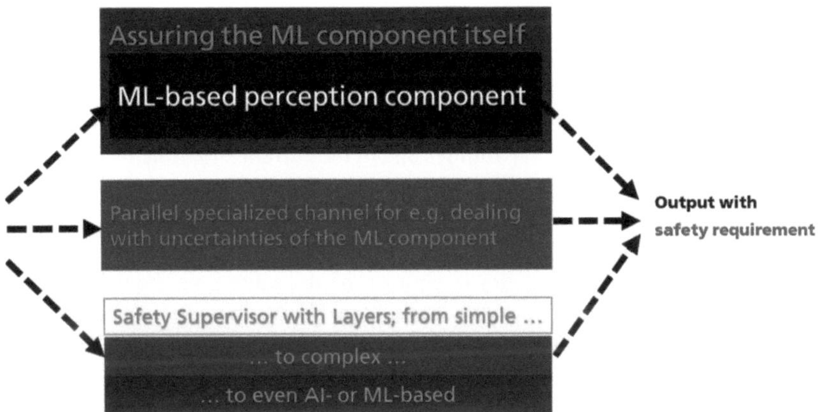

Fig. 2. Channels for ML assurance – from "black box" to "white box"

2.4 Safety Assurance Frame Conditions

A prerequisite for market introduction of AI systems and autonomous systems is that they are safe. Europe has traditionally very high constraints for market introduction of safety-critical products and is also leading with the introduction of laws for AI systems and autonomous systems. Existing vertical safety regulation is currently enhanced with respect to autonomy. For instance, the new Machinery Directive [41] will replace the Machinery Directive and address autonomous mobile machinery. Regulation for road vehicles, which are not in the scope of the Machinery Directive, was enhanced with respect to automated driving.

In addition to such vertical enhancements of regulation, AI systems are regulated horizontally by the AI Act. The regulatory safety requirements can be quite abstract in order to provide enough stability considering that technology is evolving very fast. Technical safety standards, which can be updated more easily, can provide more concrete requirements. If the safety standards are harmonized, then they can be used to evoke the presumption of conformity to the respective regulations. The development of the harmonized safety standards to the AI Act is currently under development following the standardization request draft [42].

The reason why current safety regulation and standardization is not sufficient is simply because systems were mostly closed systems and had low degree of automation when the standards have been developed. Accordingly, current safety standards provide insufficient guidance for developing cooperative automated systems. In the following, describing our vision of DRM, we illustrate which new topics and questions come up when we focus on high automation level and dynamic cooperation. These topics and questions need be considered when developing novel safety standards.

3   Dynamic Risk Management in CPS

Since several years we structure our research on assuring the safety of highly automated CPS according to a scheme we call Dynamic Risk Management (DRM). The main goal of DRM is to improve the performance of automated CPS by means of giving them a better safety awareness. I.e., instead of relying on worst case assumptions, systems shall be aware of their safety-related capabilities and of the current situation to make informed decisions regarding their safe behavior. On an abstract level DRM thus functions similar to how a human operates in critical applications:

Continuously monitor the current situation. Extrapolate and assess risks, related to both, environment and own behavior intentions. At the same time, be aware of safety-related capabilities of the involved systems. Plan how to operate safely under these circumstances while trying to optimize performance. Implement the plan.

In DRM this process is realized as a MAPE-MART / MAPE-K [43][44] cycle which is illustrated in Fig. 3. Monitoring needs to attain the required information for the analysis step in the cycle, i.e. for adequately assessing risk and safety-related capabilities. This comprises 1.) context information for dynamic risk assessment (DRA) via own sensors and/or information from external systems (e.g. edge/infrastructure, cloud), and, 2.) context information for dynamic capability assessment (DCA) by means of internal system monitors as well as information (wrt. Their safety-related capabilities) from external systems. Based on this information, the DCA and DRA analyses are conducted, each of which requiring dedicated approaches including runtime models to capture the necessary knowledge (as in MAPE-K). DCA and DRA will be described in more detail in Sects. 3.1 and 3.2, respectively. Based on the results of the analysis activities a plan for action is to be devised. We denote this as dynamic risk control (DRC) where the goal is to plan dynamic management of safety-related properties and to optimize the performance of the overall system (while guaranteeing safety).

It is important to note that we are considering a CPSoS scenario, thus there are different collaborating systems from different manufacturers that jointly render complex services or applications. Each of these constituent systems is required to provide (standardized and hence interoperable) MART fragments to adequately support DRA, DCA, DRC and potentially also "safe" sensor-based perception. The latter is mostly about explicitly dealing with perception uncertainties and is detailed in Sect. 3.3.

Figure 4 shows a logical runtime architecture for DRM. On the one hand, there are the DRM components (DCA, DRA and DRC) and on the other hand there is the nominal function of the system. In between there is the perception, which provides context information regarding the physical environment for both channels.

It is important to note that using DRM does not mean that anything safety-related is handled by DRM and that the nominal behavior channel is safety agnostic. Rather, DRM provides an additional layer of "safety intelligence", to dynamically adjust safety related parameters (which otherwise would have been fixed at development time) and to

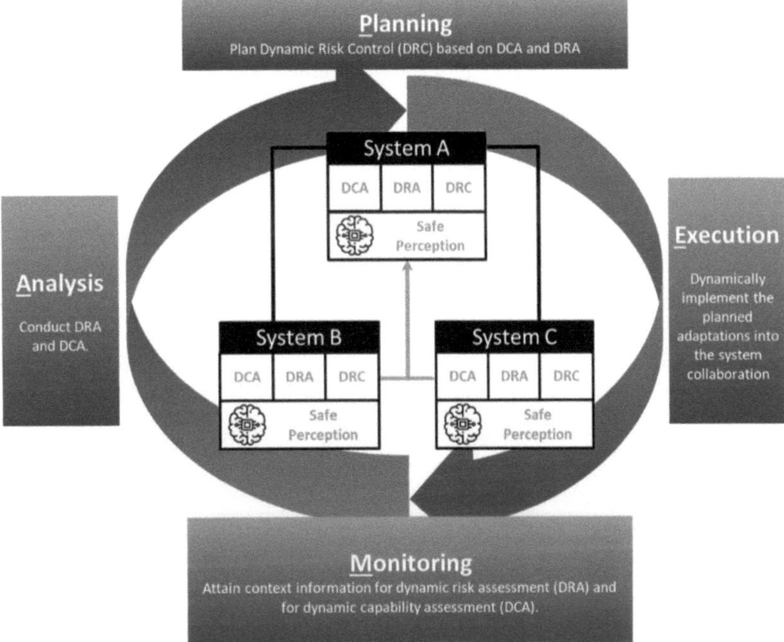

Fig. 3. DRM MAPE-MART/K scheme

continuously optimize performance. As an example, consider that the nominal function channel might implement RSS [17]. RSS is a rule-based safety approach, e.g. considering a safe lateral distance between cars, which is parameterized at development time and rather static (i.e. not context aware) at runtime. As a consequence, RSS might lead to overly cautious driving maneuvers (which does not necessarily improve safety, maybe even to the contrary), because the RSS rules are violated quite often in common human-operated driving [45]. DRM could improve this issue by dynamically adjusting the RSS parameters based on the current DRA and DCA results.

It is further conceivable that parts of the DRM might not be realized with safety integrity. Safety integrity is quantified in current functional safety standards with levels that define which safety measures should be applied to deal with hardware and software-related issues causing incorrect system behavior. For this limited scope, the safety measures are very effective and low target failure rates can be achieved. DRM has a broader scope. By estimating and controlling risks of the current situation, it targets the definition of safe behavior that has to be implemented correctly. For this implementation, it considers AI-based models and it is an open research question if one can find safety measures for such models that are as effective as those for conventional software [69]. DRA in particular might utilize AI-based models, to e.g. realize risk prediction, on top of simpler rule-based layers, thus implementing a layers of protection type of architecture. Ideally, the outer ("low integrity") layers of such an architecture work well enough so that the traditionally forged "last layer of defense" mechanisms are hardly ever required in the field. Still, they of course need to be there to bear the safety responsibility. DCA, on the other hand, can be realized with a high "safety-grade" integrity without too much hassle. ConSerts, for instance, can be developed based on common safety engineering approaches. The only addition being the consideration of variants of assumptions and corresponding guarantees and the integration of rather simple evaluation functions (contracts with variants modeled by means of Boolean logic) into the actual systems.

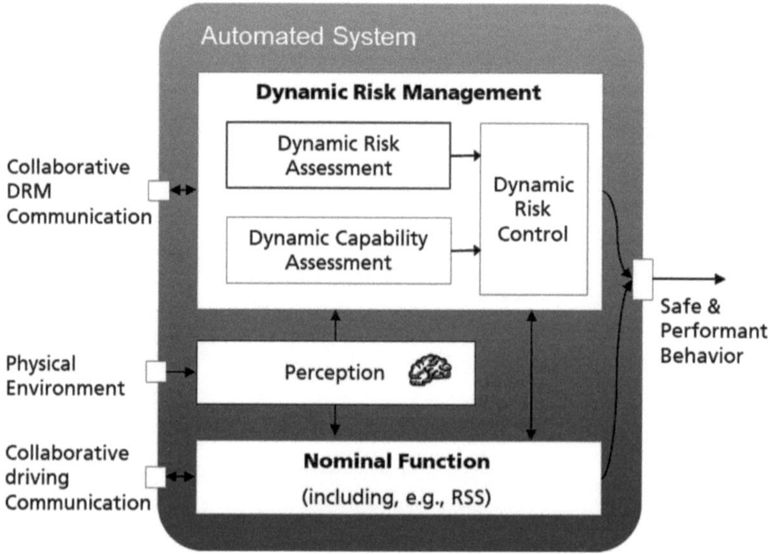

Fig. 4. Logical DRM architecture

3.1 Dynamic Capability Assessment (DCA)

DCA is about dynamic and continuous assessment of a systems safety-related capabilities. It is the counter-part to dynamic risk assessment, which is introduced in Sect. 3.2 and which essentially is about a (partially) dynamic hazard and risk analysis, or, in other words, a dynamic (safety) requirements analysis.

On the one hand, a systems capabilities might be dynamically influenced by internal aspects such as system health and resource availability aspects. On the other hand, there are external influences, such as the physical environment or the digital interface to functions/services of other systems. The internal aspects might be directly monitored and their impact on system level functions might be assessed. The external aspects can be monitored by means of corresponding perception capabilities (i.e. aspects that are associated with the physical environment such as weather conditions) and by communication. The latter comprises sharing information between systems (e.g. shared perception) as well as information regarding the safety-related properties of functions involved in a system cooperation. Regarding autonomous systems, one important aspect related external context is the operational design domain (ODD) of the system, which ideally has been defined explicitly and thus can be monitored as well.

Sharing this different kind of information across systems requires corresponding information formats/models and protocols, which today certainly is one significant hindering factor in exploiting the huge potential there is. Without being able to attain sufficient information (with explicit quality guarantees) regarding the environment, worst case assumptions need to be taken, which in turn severely constrain performance.

A key aspect to consider for a DCA approach is how safety responsibility is distributed in a dynamic SoS setting. It is common knowledge that safety is not compositional in a straight forward way, i.e. if you compose several "safe" systems into a SoS, the resulting SoS is not automatically "safe" as well. The issue here is that safety invariably is a property of the overall system in its context and the application/function it realizes. The constituent systems of a SoS cooperation lack this perspective, they can only assume responsibility for rendering their constituent function given that corresponding context assumptions are fulfilled. A solution approach for this aspect is to dynamically form system hierarchies. This implies that cooperative applications do not emerge "by chance", but that there is a system actively orchestrating other so render such an application. This orchestrator would then be, for the duration of that cooperation, be the root of a composition hierarchy and also assume the safety responsibility for the overall cooperation. It would have certain demands wrt. The safety-related properties of the others system functions and it would know what variations in their guarantees would imply regarding the safety of the overall cooperation. Overall, in context of e.g. autonomous driving scenarios, a heterarchical structure would be formed, where e.g. platoons, overtaking maneuvers and so on are rendered based on temporal concurrent application hierarchies.

A DCA approach should provide means for all the above aspects. Conditional Safety Certificates (ConSerts) [34] is such an approach which we have been investigating and applying for over 10 years now. ConSerts operate on the level of safety requirements. They are specified at development time based on a sound and comprehensive safety argumentation (e.g. an assurance case). They conditionally certify that the associated system will provide specific safety guarantees. Conditions are related to the fulfillment of specific demands (i.e. assumptions) regarding the environment. ConSerts comprise different potential guarantees which are bound to different variations of demands. In the same way as "static" certificates, ConSerts shall be issued by safety experts, independent organizations, or authorized bodies (depending on the respective application domain) after a stringent manual check of the safety argument. To this end, it is mandatory to prove all claims regarding the fulfillment of provided safety guarantees by means of suitable evidence and to provide adequate documentation of the overall argument – particularly including the sets of demands and their implications.

Let us briefly illustrate ConSerts based on the example used in [46]. In the agricultural domain, tractor implement management (TIM) enables implements to assume control over the tractor functions, such as setting the vehicle speed or the steering angle. To do this in the best possible way, the implement might consume sensor information from the tractor and from auxiliary third party sensors, such as a swath scanner or a GPS. Consequently, TIM scenarios are scenarios of cooperative CPS, realized by cooperation of different systems of different manufacturers.

For illustrating the engineering of ConSerts in this example the role of the implement manufacturer shall be assumed. The goal of the manufacturer is to develop a round baler with TIM support. From a functional point of view, it is clear (due to existing AEF [47] guidelines) how the interfaces between the potential participants look like and how they are to be used. However, the implement manufacturer does not know about the safety properties of these functions.

From a safety point of view, the engineering of the baling application starts top-down with an application-level hazard and risk analysis. Assume that the agricultural manufacturers agreed by convention that during the operation of a TIM application, the application (and thus the application manufacturer) has the responsibility for the overall cooperation, thus dynamically forming a cooperation hierarchy as motivated above. Therefore, the safety engineering goal is to ensure adequate safety not only for the TIM baling application or for the implement, but for the whole cooperation of systems that will be rendering the cooperative application at runtime. Application-level hazards of the TIM baling application could correspondingly comprise the tractor having an unwanted acceleration or steering during TIM baling. Causes might be located in the TIM baling application itself or in the tractor or in other cooperating systems (e.g. a third-party sensor). Causes in the TIM baling application and the implement are tackled by traditional safety engineering. Causes outside the system under development are translated into ConSert demands and runtime evidences that are to be evaluated at runtime. Thanks to the ConSert-based modularization it is thereby sufficient to only consider the direct dependencies of the system under development on its environment. The runtime evaluation can be done bottom up, i.e. the system at the leaves of the cooperation hierarchy determine their guarantees and propagate them up so that the root (here: the TIM baling application) can determine its guarantees. Based on these guarantees, the cooperation might be parameterized (e.g. set maximum speed) to ensure safety but also maximize performance.

Enabling dynamic management of system performance while always ensuring safety is a key benefit of DCA in general. From changes in system guarantees due to wear and tear to changing weather conditions, anything can potentially be considered. Thus, it is no longer necessary to work based on worst case assumptions (because you cannot know the actual conditions during operation), with DCA systems become aware regarding the safety-relevant context conditions.

Of course, these benefits come at a certain price because additional engineering is required, where some technical aspects might also be automated as investigated in [48]. Overall, ConSerts are a relatively lightweight DCA approach and they are well aligned with traditional safety engineering. The main difference being that unknown context is structured into a series of foreseen variants, which are then specified in terms of a relatively simple modular runtime model. While this already provides significant gains in terms of flexibility and realizable system performance (compared to a completely static approach), there is still further potential. More powerful DCA evaluation logic could be investigated beyond simple contracts with variants, e.g. with the capability to calculate with probabilities and with fuzziness.

3.2 Dynamic Risk Assessment (DRA)

DCA monitors aspects affecting the CPS capabilities to react dependably in a critical situation. However, whether a particular error or failure is safety-critical and poses an actual risk depends on the current operational situation the system finds itself in during runtime. Hazardous events and their associated risk, i.e. the likelihood of a transition from a system hazard to an accident as well as the potential severity of this accident, are always conditioned on the operational situation. Since the operational environment in the envisioned CPS use cases is highly dynamic, determining the impact of the current situation on the risk parameters at runtime can increase system performance, as the CPS may continue operation in situations, where it would have stopped in conventional worst-case assumed environmental situations. Consequentially, if we monitor the presence of risky/non-risky operational situations, we can increase CPS performance by a) being able to tolerate certain faults and failures if their associated risk is low in the current situation and b) actively reducing particular risk parameters with tactical decisions, where severity, controllability or the operational situation itself is changed. Dynamic risk assessment (DRA) techniques thus treat the CPS or a constituent system as a black box and provide means to analyze the consequences of CPS behavior deviations or present system hazards on the risk in the current operational situation.

For autonomous systems, in particular in the automotive domain, DRA techniques have been applied to address the trade-off between performance and safety risks. The existing approaches can be classified broadly into three categories. 1. DRA is incorporated into motion and trajectory prediction frameworks by specifying behavioral or kinematic constraints that are input to the trajectory planner [49, 50]. 2. DRA is performed during the online verification of an already planned yet potentially unsafe trajectory [17, 51]. 3. DRA is not performed in direct relation to a planned trajectory, but instead monitors risk parameters that enable distinction between the presence/absence of a hazardous event and is used to adapt the parameters of risk metrics used in trajectory planners [52–54].

Although different architectures for incorporating DRA into autonomous systems exist, all of them need to decide which particular risk-influencing situation features are to be observed in the present, project the evolution of these features into the future by using a set of assumptions, and rate the risk of the projected future situation. For each of these DRA sub-tasks, respective design time engineering activities are required. Detailed information on situation prediction techniques and concrete risk metrics can be found in relevant literature reviews (see [55–57]). To enable machines to perform DRA, modelling formalisms are required to technically capture the relationship between situation features and risks. For this purpose, different classes of models can be used, which are model-based (e.g. structural equation models), rule-based (Boolean models), generative (Gaussian Mixture Models, dynamic Bayesian networks, Hidden Markov Models), discriminative (Decision Tree, Random Forest, Support Vector Machines) or deep learning-based (Multi-Layer Perceptron, Convolution Neural Network – CNN, Recurrent Neural Network – RNN, Long-Short-Term-Memory Network – LSTM). Some of them are developed purely based on expert knowledge, and some of them can be adapted to new operational domains with machine learning techniques. Depending on the concrete modelling formalism selected, support for deterministic or probabilistic assumptions may be given as well as the consideration of uncertainties during feature perception is possible.

At Fraunhofer IESE, [55] introduced a concrete DRA approach for autonomous systems. The work contributed a conceptual taxonomy of DRA and provided a concrete application using concrete instances of controllability and severity metrics to rate collision risks for automated driving. In order to be applicable in many different operational situations, risk metrics were required to be situation-agnostic. Situation-specific features such as road structure, lighting and weather conditions, traffic rules, and actor interactions were consequentially not considered. To account for the fact that risk is influenced by exactly those situation-specific features, the work in [55] was extended to the Situation-Aware Dynamic Risk Assessment (SINADRA) framework [58]. The approach uses Bayesian networks as a modelling formalism and explicitly considers the mentioned situation-specific features as risk influences. A proof-of-concept and tool implementation of the approach is presented in [59]. The design-time method for building Bayesian situation prediction models with machine learning techniques is presented in [60]. More recently, SINADRA runtime models have been formally traced to model-based design-time safety engineering artifacts and applied in a distributed smart logistics mobility system [61]. An important aspect of a DRA model is its connection to a model-based representation of the operational design domain (ODD). First steps in this direction have been explored in [62].

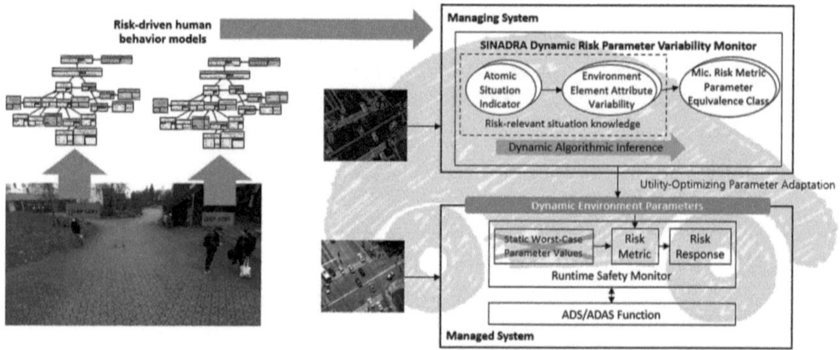

Fig. 5. Logical DRA architecture

The conceptual components of the SINADRA framework are shown in Fig. 5. The core artefact of the framework is the DRA Monitor Runtime Component. It is an executable software component that is deployed to a CPS along with the nominal functionality. Functionally, it contains a model that captures the variability of risk parameters with respect to risk-relevant atomic situation indicators. Thus, the output is a dynamic risk classification of particular (unsafe) behavior intentions of interest. Based on the runtime perception of relevant situation features and the inference determining the effect of the dynamic input feature vector on risk, the basis for a risk-informed parameter adaptation is given.

3.3 ML-Powered Perception

Adequate perception capabilities are a key ingredient for safe autonomy and also important inputs for DCA and DRA. As pointed out in Sect. 2.3, ML/DL based perception is in the limelight and there presently is a lot of attention being spend in research to get a grip wrt. The uncertainties associated with the output of ML/DL based components.

The idea is to measure uncertainty during operation and to use the explicit uncertainty information in DCA and DRA. A consequence of a high uncertainty value could thus be that the current sensor capabilities are assessed to be low (DCA) and that this propagates through to an ongoing cooperation scenario. As a consequence, some parameters of the cooperation (e.g. min distance in platooning) are adjusted to ensure safety. Likewise, the current risk might be assessed conservatively (i.e. as being high) due to perception uncertainties (DRA) and again this leads to adaptions to maintain safety. At this point, there might be an overlapping between DCA and DRA which needs to be partitioned carefully.

The general scheme has some likeness to the "classical" error detection and error handling. Considering traditional software errors, the error will always occur under some specific conditions. This holds also true for software that is developed by means of ML/DL, but the conditions can be hard to grasp in this case. Consider e.g. the well-known examples of adversarial attacks on traffic sign detection [63]. Still, general environmental conditions can be considered such as noise, e.g. weather/fog for a camera-based object classification. Depending on such conditions there is a likelihood that the current output of an ANN deviates in a particular way from the correct output. One approach to estimate this likelihood during operation is given by an uncertainty wrapper (cf. DIN SPEC 92005 and [64]). In many cases this uncertainty is quite high so that remediating action would be triggered too often. An approach to deal with this issue is to aggregate uncertainties from different time steps [65]. Additionally, simple architectural patterns to handle uncertainty by increased safety margins are proposed in [66]. A complementary approach which we are investigating is based on statistical similarity metrics applied to training data on the one hand and to field/test data on the other [67, 68]. The general rationale here being that if the system encounters input data which is very dissimilar to the training data, there is a high level of uncertainty regarding the correctness of the output of the ML component.

Overall, more research is still required to facilitate safety assurance of systems with ML components. There are numerous approaches that can provide indications about potential flaws of a learned model (e.g. approaches from the field of explainable AI), but there is a lack of means to attain strong evidence that a ML component will always function as intended. Thus, it is necessary to work around this issue and assign safety responsibility to other redundant channels as illustrated in Fig. 2.

4 Summary and Conclusion

In this article we described key trends of future CPSoS, explained corresponding safety assurance challenges and summarized some of the existing work to tackle these challenges. Then we went on to introduce dynamic risk management, our vision for making future systems safety-aware. We introduced a conceptual architecture for DRM and presented some of our work regarding the key building blocks of that architecture.

DRM is not to be understood as a concrete one-size-fits- all solution, but rather as a vision and a framework for structuring various approaches into a comprehensive perspective. Consequently, DRM could be instantiated for many different application domains and applications scenarios, maybe using some of the approaches mentioned here, or others that are deemed suitable.

DRM touches all the challenges identified in Sect. 2. Overall, DRM is aimed at safety assurance and performance optimization in automated and autonomous systems. In this context, DRA is specifically dealing with the physical context complexity, DCA with safety modularization (which is also a means for dealing with system complexity) and interconnectedness, and perception has to deal with the uncertainties associated with ML components.

Each of the presented dimensions of challenges still requires additional research, whereas the assurance of ML components might be the one with the highest need. But also on the integration level, where DRM as a whole is located, we are still just at the beginning. Safety awareness and resilience are key characteristics regarding the competitiveness of future CPSoS products, because these not only need to be safe, but they also need to be available and performant. Indeed, there clearly is the potential for future systems to be better and safer than today's systems by a huge margin. Given the significant transformations the industries in the embedded domain are undergoing due to the described trends, those who have the best answers to these questions will be significantly strengthening their position in the global competition.

Looking even further into future as illustrated by Fig. 1 and, for instance, described in the safeTRANS roadmap "Safety, Security, and Certifiability of Future Man-Machine Systems." [10], in the long run systems are set to assume ever higher degrees of autonomy, interconnection and utilization of AI – requiring further research with respect to safety assurance and certifiability, significantly extending what has been presented here.

References

1. Thompson, H., Paulen, R., Reniers, M., Sonntag, C., Engell, S.: D2.4 Analysis of the State-of-the-Art and Future Challenges in Cyber-physical Systems of Systems; CPSoS - Cyber-physical Systems of Systems 2016. https://www.cpsos.eu/wpcontent/uploads/2015/02/D2-4-State-of-the-art-and-future-challengesin-cyber-physical-systems-of-2.pdf
2. The ECS SRIA. https://ecssria.eu/. Accessed 16 July 2023
3. Die Bundesregierung; Nationale Strategie für Künstliche Intelligenz. https://www.ki-strate gie-deutschland.de/files/downloads/201201_Fortschreibung_KI-Strategie.pdf, Accessed 16 July 2023
4. A European approach to artificial intelligence. https://digital-strategy.ec.europa.eu/en/pol icies/european-approach-artificial-intelligence. Accessed 16 July 2023
5. NSTC. National artificial intelligence research and development strategic plan 2023 update; National Science & Technology Council. https://www.whitehouse.gov/wp-content/uploads/ 2023/05/National-Artificial-Intelligence-Research-and-Development-Strategic-Plan-2023-Update.pdf, Accessed 16 July 2023
6. State Council. Next Generation Artificial Intelligence Development Plan. https://digich ina.stanford.edu/work/full-translation-chinas-new-generation-artificial-intelligence-develo pment-plan-2017/. Accessed 16 July 2023

7. Armengaud, E., et al.: DDI: a novel technology and innovation model for dependable, collaborative and autonomous systems. In 2021 Design, Automation & Test in Europe Conference & Exhibition (DATE), pp. 1626–1631. IEEE (2021).

8. Loebbecke, C., Amold, P.: Reflections on societal and business model transformation arising from digitization and big data analytics: a research agenda. J. Strateg. Inf. Syst. **24**(3), 149–157 (2015)

9. Seiberth, G.: Data-driven business models in connected cars, mobility services and beyond, BVDW research NO 01/18 (2018). https://www.bvdw.org/fileadmin/user_upload/20180509_bvdw_accenture_studie_datadrivenbusinessmodels.pdf

10. SafeTRANS, e.V.: Safety, Security, and Certifiability of Future Man-Machine Systems (2021). https://www.safetrans-de.org/de/Uploads/AK_2018_RLE_CPS/SafeTRANS_RM_SSC_FMMS_Roadmap_V2.pdf?m=1611136486 Accessed 16 July 2023

11. Reich, J., et al.: Engineering of runtime safety monitors for cyber-physical systems with digital dependability identities. In Computer Safety, Reliability, and Security: 39th International Conference, SAFECOMP 2020, Lisbon, Portugal, September 16–18, 2020, Proceedings 39, pp. 3–17. Springer International Publishing (2020)

12. DKE/AK 801.0.8, Anwendungsregel VDE AR 2842–61 Entwicklung und Vertrauenswürdigkeit von autonom/kognitiven Systemen, https://www.vde-verlag.de/normen/0800731/vde-ar-e-2842-61-2-anwendungsregel-2021-06.html, Accessed 16 July 2023

13. Adler, R., Reich, J., Hawkins, R.: Structuring research related to dynamic risk management for autonomous systems. In: Proceedings of 42nd International Conference on Computer Safety, Reliability and Security (SAFECOMP). Toulouse, France (2023).

14. Kuehn, M., Vogelpohl, T., Vollrath, M.: Takeover times in highly automated driving (Level 3) (2017). In: Proceedings of 25th International Technical Conference on the Enhanced Safety of Vehicles (ESV): Innovations in Vehicle Safety: Opportunities and Challenges. Detroit, US. https://trid.trb.org/view/1485287 Accessed 16 July 2023

15. Kagermann, H., et al.: Das Fachforum Autonome Systeme im Hightech-Forum der Bundesregierung –Chancen und Risiken für Wirtschaft. Wissenschaft und Gesellschaft. Final report, Berlin (2017)

16. Feth, P., Schneider, D., Adler, R.: A Conceptual Safety Supervisor Definition and Evaluation Framework for Autonomous Systems. 135–148 (2017).

17. Shalev-Shwartz, S., Shammah, S., Shashua, A.: On a formal model of safe and scalable self-driving cars. arXiv preprint arXiv:1708.06374 (2017)

18. Reich, J., Wellstein, M., Sorokos, I., Oboril, F., Scholl, K.U.: Towards a software component to perform situation-aware dynamic risk assessment for autonomous vehicles. In: European Dependable Computing Conference - EDCC 2021 Workshops (2021).

19. Saidi, S., Ziegenbein, D., Deshmukh, J.V., Ernst, R.: Autonomous systems design: charting a new discipline. IEEE Design & Test **39**(1), 8–23 (2021)

20. Huang, H.M., Pavek, K., Novak, B., Albus, J., Messin, E.: A framework for autonomy levels for unmanned systems (ALFUS). In: Proceedings of the AUVSI's unmanned systems North America, pp. 849–863 (2005).

21. Rasmus, A.: A model-based approach for exploring the space of adaptation behaviors of safety-related embedded systems. Dissertation. Fraunhofer Verlag (2013)

22. Adler, R., Förster, M., Junger, J., Trapp, M.: Runtime adaptation in safety-critical automotive systems. In Proc 25th IASTED International Multi-Conference Software Engineering, pp. 308–315 (2005)

23. Trapp, M., Schneider, D., Weiss, G.: Towards safety-awareness and dynamic safety management. In 2018 14th European Dependable Computing Conference (EDCC), pp. 107–111. IEEE (2018).

24. Damm, W., Hungar, H., Josko, B., Peikenkamp, T., Stierand, I.: Using contract-based component specifications for virtual integration testing and architecture design. In 2011 Design, Automation & Test in Europe, pp. 1–6. IEEE (2011).

25. Kaiser, B., Weber, R., Oertel, M., Böde, E., Nejad, B.M., Zander, J.: Contract-based design of embedded systems integrating nominal behavior and safety. Complex Syst. Inf. Model. Quart. **4**, 66–91 (2015)

26. Battram, P., Kaiser, B., Weber, R.: A modular safety assurance method considering multi-aspect contracts during cyber physical system design. In REFSQ Workshops, pp. 185–197 (2015)

27. Sljivo, I., Gallina, B., Carlson, J., Hansson, H.: Strong and weak contract formalism for third-party component reuse. In 2013 IEEE International Symposium on Software Reliability Engineering Workshops (ISSREW) (pp. 359–364). IEEE (2013).

28. Carlson, J., et al.: Delivery d132.2, generic component meta-model' of the safecer project. SafeCer project, Deliverable D **D132**(2), 1–100 (2014)

29. de la Vara, J.L., et al.: The AMASS approach for assurance and certification of critical systems. In Embedded World 2019 (2019).

30. Söderberg, A., Johansson, R.: Safety contract based design of software components. In 2013 IEEE International Symposium on Software Reliability Engineering Workshops (ISSREW), pp. 365–370. IEEE (2013).

31. Bate, R. Hawkins, McDermid, J.: A contract-based approach to designing safe systems. In: 8th Australian Workshop on Safety Critical Systems And Software (SCS) – vol. 33, pp. 25–36 (2003)

32. De La Vara, J.L., Parra, E., Ruiz, A., Gallina, B.: The AMASS tool platform: an innovative solution for assurance and certication of cyber-physical systems. In: REFSQ Workshops (2020)

33. Rushby, J.: Just-in-time certification. In 12th IEEE International Conference on Engineering Complex Computer Systems (ICECCS 2007), pp. 15–24. IEEE (2007).

34. Schneider, D., Trapp, M.: Conditional safety certification of open adaptive systems. ACM Trans. Autonom. Adaptive Syst. (TAAS) **8**(2), 1–20 (2013)

35. Cheng, B.H., et al.: Using models at runtime to address assurance for self-adaptive systems. Models@ run. Time: Foundations, Applications, And Roadmaps, 101–136 (2014).

36. Weyns, D., et al.: Perpetual assurances for self-adaptive systems. In: Software Engineering for Self-Adaptive Systems III. Assurances: International Seminar, Dagstuhl Castle, Germany, December 15–19, 2013, Revised Selected and Invited Papers, pp. 31–63. Springer International Publishing (2017).

37. Calinescu, R., Grunske, L., Kwiatkowska, M., Mirandola, R., Tamburrelli, G.: Dynamic QoS management and optimization in service-based systems. IEEE Trans. Software Eng. **37**(3), 387–409 (2010)

38. Schneider, D., Trapp, M.: B-space: dynamic management and assurance of open systems of systems. J. Internet Serv. Appl. **9**(1), 1–16 (2018)

39. Amorim, T., Ruiz, A., Dropmann, C., Schneider, D.: Multidirectional modular conditional safety certificates. In: Computer Safety, Reliability, and Security: SAFECOMP 2015 Workshops, ASSURE, DECSoS. ISSE, ReSA4CI, and SASSUR, Delft, The Netherlands, September 22, 2015, Proceedings 34, pp. 357–368. Springer International Publishing (2015).

40. Agirre, I., et al.: Safe and secure software updates on high-performance mixed-criticality systems: The UP2DATE approach. Microprocess. Microsyst. **87**, 104351 (2021)

41. https://ec.europa.eu/commission/presscorner/api/files/document/print/en/ip_22_7741/IP_22_7741_EN.pdf. Accessed 16 July 2023

42. https://ec.europa.eu/docsroom/documents/52376/attachments/1/translations/en/renditions/native. Accessed 16 July 2023

43. Kephart, J.O., Chess, D.M.: The vision of autonomic computing. Computer **36**(1), 41–50 (2003)
44. Brun, Y., et al.: Engineering self-adaptive systems through feedback loops. Software Eng. Self-Adaptive Syst. 48–70 (2009).
45. Naumann, M., et al.: On responsibility sensitive safety in car-following situations-A PARAMETER analysis on German highways. In: 2021 IEEE Intelligent Vehicles Symposium (IV), pp. 83–90. IEEE (2021).
46. Schneider, D., Trapp, M., Papadopoulos, Y., Armengaud, E., Zeller, M., Höfig, K.: WAP: digital dependability identities. In 2015 IEEE 26th International Symposium on Software Reliability Engineering (ISSRE), pp. 324–329. IEEE (2015).
47. https://www.aef-online.org/about-us/activities/tractor-implement-management-tim.html. Accessed 16 July 2023
48. Schmidt, A., Reich, J., Sorokos, I.: Live in conserts: model-driven runtime safety assurance on microcontrollers, edge, and cloud practical experience report. In 2021 17th European Dependable Computing Conference (EDCC), pp. 61–66. IEEE (2021).
49. Eggert, J.: Risk estimation for driving support and behavior planning in intelligent vehicles. at-Automatisierungstechnik, **66**(2), 119–131 (2018).
50. Machin, M., Guiochet, J., Waeselynck, H., Blanquart, J.P., Roy, M., Masson, L.: SMOF: a safety monitoring framework for autonomous systems. IEEE Trans. Syst. Man Cybern. Syst. **48**(5), 702–715 (2016)
51. Pek, C., Manzinger, S., Koschi, M., Althoff, M.: Using online verification to prevent autonomous vehicles from causing accidents. Nat. Mach. Intell. **2**(9), 518–528 (2020)
52. Khastgir, S., Sivencrona, H., Dhadyalla, G., Billing, P., Birrell, S., Jennings, P.: Introducing ASIL inspired dynamic tactical safety decision framework for automated vehicles. In: 2017 IEEE 20th International Conference on Intelligent Transportation Systems (ITSC), pp. 1–6. IEEE (2017)
53. Johansson, R., Nilsson, J.: The need for an environment perception block to address all asil levels simultaneously. In: 2016 IEEE Intelligent Vehicles Symposium (IV), pp. 1–4. IEEE (2016).
54. Hartsell, C., Ramakrishna, S., Dubey, A., Stojcsics, D., Mahadevan, N., Karsai, G.: Resonate: a runtime risk assessment framework for autonomous systems. In: 2021 International Symposium on Software Engineering for Adaptive and Self-Managing Systems (SEAMS), pp. 118–129. IEEE (2021).
55. Feth, P.: Dynamic Behaviour Risk Assessment for Autonomous Systems. Disserta-tion. TU Kaiserslautern, Germany (2020)
56. Westhofen, L., et al.: Criticality metrics for automated driving: a review and suitability analysis of the state of the art. Arch. Comput. Methods Eng. **30**(1), 1–35 (2023)
57. Lefèvre, S., Vasquez, D., Laugier, C.: A survey on motion prediction and risk assessment for intelligent vehicles. Robomech J. **1**(1), 1–14 (2014)
58. Reich, J., Trapp, M.: SINADRA: towards a framework for assurable situation-aware dynamic risk assessment of autonomous vehicles. In: 2020 16th European Dependable Computing Conference (EDCC), pp. 47–50. IEEE (2020).
59. Reich, J., Wellstein, M., Sorokos, I., Oboril, F., Scholl, K.U.: Towards a software component to perform situation-aware dynamic risk assessment for autonomous vehicles. In: Dependable Computing-EDCC 2021 Workshops: DREAMS, DSOGRI, SERENE 2021, Munich, Germany, September 13, 2021, Proceedings 17, pp. 3–11. Springer International Publishing (2021).
60. Wellstein, M.: Development of a Bayesian Network for Situation-Aware Lane Change Prediction based on the highD Dataset. Master Thesis, Technical University Kaiserslautern (2021)

61. Reich, J., et al.: Engineering dynamic risk and capability models to improve cooperation efficiency between human workers and autonomous mobile robots in shared spaces. In: International Symposium on Model-Based Safety and Assessment, pp. 237–251. Cham: Springer International Publishing (2022).

62. Reich, J., et al.: Concept and metamodel to support cross-domain safety analysis for ODD expansion of autonomous systems. In: Proceedings of 42nd International Conference on Computer Safety, Reliability and Security (SAFECOMP). Toulouse, France (2023).

63. Eykholt, K., et al.: Robust physical-world attacks on deep learning visual classification. In: Proceedings of the IEEE Conference on Computer Vision and Pattern Recognition, pp. 1625–1634 (2018).

64. Kläs, M., Sembach, L.: Uncertainty wrappers for data-driven models: increase the transparency of AI/ML-based models through enrichment with dependable situation-aware uncertainty estimates. In: Computer Safety, Reliability, and Security: SAFECOMP 2019 Workshops, ASSURE, DECSoS, SASSUR, STRIVE, and WAISE, Turku, Finland, September 10, 2019, Proceedings 38, pp. 358–364. Springer International Publishing (2019).

65. Groß, J., Kläs, M., Jöckel, L., Gerber, P.: Timeseries-aware Uncertainty Wrappers for Uncertainty Quantification of Information-Fusion-Enhanced AI Models based on Machine Learning. arXiv preprint arXiv:2305.14872 (2023).

66. Groß, J., Adler, R., Kläs, M., Reich, J., Jöckel, L., Gansch, R.: Architectural patterns for handling runtime uncertainty of data-driven models in safety-critical perception. In: International Conference on Computer Safety, Reliability, and Security, pp. 284–297. Cham: Springer International Publishing (2022).

67. Aslansefat, K., Sorokos, I., Whiting, D., Kolagari, R.T., Papadopoulos, Y.: SafeML: safety monitoring of machine learning classifiers through statistical difference measures. In: International Symposium on Model-Based Safety and Assessment (pp. 197–211). Springer, Cham (2020).

68. Akram, M.N., Ambekar, A., Sorokos, I., Aslansefat, K., Schneider, D.: StaDRe and StaDRo: reliability and robustness estimation of ML-based forecasting using statistical distance measures. In: International Conference on Computer Safety, Reliability, and Security (pp. 289–301). Cham: Springer International Publishing (2022).

69. Adler, R., Klaes, M.: Assurance cases as foundation stone for auditing AI-enabled and autonomous systems: workshop results and political recommendations for action from the examAI project. HCI International (2022). Lecture Notes in Computer Science, vol 13520. Springer, Cham (2022). https://doi.org/10.1007/978-3-031-18158-0_21

Toward Methodical Discovery
and Handling of Hidden Assumptions
in Complex Systems and Models

David Harel[1] , Uwe Aßmann[2] , Fabiana Fournier[3] , Lior Limonad[3] ,
Assaf Marron[1(✉)] , and Smadar Szekely[1]

[1] Department of Computer Science and Applied Mathematics,
Weizmann Institute of Science, Rehovot, Israel
`{david.harel,assaf.marron,smadar.szekely}@weizmann.ac.il`
[2] Technische Universität Dresden, Dresden, Germany
`uwe.assmann@tu-dresden.de`
[3] IBM Research, Haifa, Israel
`{fabiana,liorli}@il.ibm.com`

Abstract. Methodologies for development of complex systems and
models include external reviews by domain and technology experts.
Among others, such reviews can uncover undocumented built-in assumptions that may be critical for correct and safe operation or constrain
applicability. Since such assumptions may still escape human-centered
processes like reviews, agile development, and risk analyses, here, we
contribute toward making this process more methodical and automat-
able. We first present a blueprint for a taxonomy and formalization of
the problem. We then show that a variety of digital artifacts of the system or model can be automatically checked against extensive reference
knowledge. Since mimicking the breadth and depth of knowledge and
skills of experts may appear unattainable, we illustrate the basic feasibility of automation with rudimentary experiments using OpenAI's
ChatGPT. We believe that systematic handling of this aspect of system engineering can contribute significantly to the quality and safety
of complex systems and models, and to the efficiency of development
projects. We dedicate this work to Werner Damm, whose contributions
to modeling and model-based development, in industry and academia,
with a special focus on safety, helped establish a solid foundation to our
discipline and to the work of many scientists and professionals, including,
naturally, the approaches and techniques described here.

Keywords: System engineering · safety · requirements analysis ·
domain expertise · code review · risk analysis · languages · ontologies ·
explainability

1 Introduction

Many famous failures of complex systems may be related directly or indirectly
to unstated assumptions. For example, the famous Year-2000 (a.k.a Y2K) "bug"

M. Fränzle et al. (Eds.): Werner Damm Festschrift, LNCS 15471, pp. 147–162, 2026.
https://doi.org/10.1007/978-3-031-97537-0_9

is related to assumptions like "All calendar years mentioned in this system begin with '19' (or can be represented with two digits)." In 1999, the Mars Climate Orbiter spacecraft crashed because of the assumption that certain interacting software components used the same measurement units when in fact they were not. Some published reports and analyses of the tragic Tesla car crash in 2016 relate to human user assumptions about the automated driving capabilities of the vehicle, and/or to technical assumptions about which sensor input to rely on in which context [3]. In 1991, the Patriot missile defense system missed an incoming missile with fatal consequences, due to cumulative time drift resulting from data truncation. This bug may be retrospectively associated with an assumption that the system will be reset frequently, and/or that the speed of incoming missiles is below a certain threshold [19]. Environmental and biological models, e.g., with regard to global warming or Covid-19, may also contain explicit and implicit assumptions that affect their predictions. [7, 15, 21]. In the words of [24, p.11-1, col.2] *"it is possible that incorrect models were 'tuned' to observed behavior.".* The problem is of course general: despite thorough development and testing, many operational systems experience failures and near failures that are rooted in hidden assumptions.

Present system engineering methodologies and tools [5, 8, 16, 20] of course attempt to address some of these issues in elements of:

- requirements engineering and requirement traceability,
- risk analysis,
- testing and validation, including unit, feature and integration testing,
- test-driven development methods,
- agile development with repeated involvement of users and customers,
- Reviews of documents and code by stakeholders[1] with diverse backgrounds,

etc. In this paper, we are less interested in hidden assumptions that were missed because developers did not follow such practices or applied other risky development behavior (as in [9, page 7]), and focus on those that despite systematic use of existing methods remained in the "blind spot" of developers and other stakeholders. We also note that there may be considerations that stakeholders did consider explicitly, but incorrectly dismissed them as inapplicable without documentation.

Hidden assumptions are not always a negative factor. They often contribute to simplicity, efficiency, or cost reduction in design and development. For example, a critical business application that under certain conditions issues an error message and stops processing, may depend on a common undocumented assumption that it is used within a business process in which a human will handle these messages. For another example, consider useful low-cost electronic devices that are obviously not designed to operate in harsh conditions or in bright sunlight, but for which this fact is not documented. Thus, since deliberating an elicited

---

[1] We use the term stakeholders broadly, and include end users, customers/clients, developers/engineers, architects, product managers, project managers, etc.

hidden assumption may itself induce cost and complications, one should consider the benefit of bringing it to the forefront. For example, recognizing that a business application is dependent on a human operator may lead to actions like documenting the process and training the operators, or enhancing the system to handle such errors automatically; alerting users to limitations of low cost gadgets can promote the acquisition of alternative solutions; etc.

More generally, handling discovered hidden assumption includes documenting the missing details and alerting affected parties, or modifying the system and/or its testing such that the assumption is no longer valid or is replaced by other documented assumptions. Furthermore, each discovery of a hidden assumption should cause developers to evaluate whether this discovery should have been occurred earlier, leading to potential improvements in the development process.

While in the case of complex safety-critical systems the importance of the issues discussed here is manifest, the proposed methods and tools can apply in diverse contexts, including commercial and scientific, enterprise and personal, operational control and information processing, etc.

In Sect. 2 we contribute several perspectives and categories for studying hidden assumptions, toward constructing a systematic taxonomy. Then, in Sect. 3, we use such categories to outline methodical and automatic processes for discovery of hidden assumptions. In Sect. 4, we support and illustrate the argument that automation of this process, which when done by humans requires extensive knowledge and advanced analytical skills, is at all possible. We do this via rudimentary experiments using OpenAI's ChatGPT to compare documents, analyze code, and infer ontologies in the context of general and domain-specific knowledge.

The approaches discussed here tie directly into ideas and concepts introduced and/or developed by Werner Damm and colleagues including scenario-based programming (SBP) [10] and contract-based development of cyber-physical systems [4,25] and a variety of related topics in developing and verifying complex systems [6,11,16]. A common theme of this view of software and system engineering is the combination of separation of concerns and comprehensive coverage. Every requirement, expectation, constraint, etc., whether executable or not, deserves a concise specification, enabling full awareness of stakeholders and prioritization of systematic handling in implementing it or in using components that implement it.

In the spirit of the wise computing vision [14], researching and automating this aspect of requirements analysis and quality assurance can enhance the quality and safety of complex systems and the efficiency of development projects.

2   Perspectives on the Types of Hidden Assumptions

As illustrated by Lewis Carroll (Charles Lutwidge Dodgson) in "The Hunting of the Snark", when looking for something, it may be desirable to have a description of the object of the search, elusive as it may be. Most generally, we consider a hidden assumption to be a description of a system property or behavior that by its

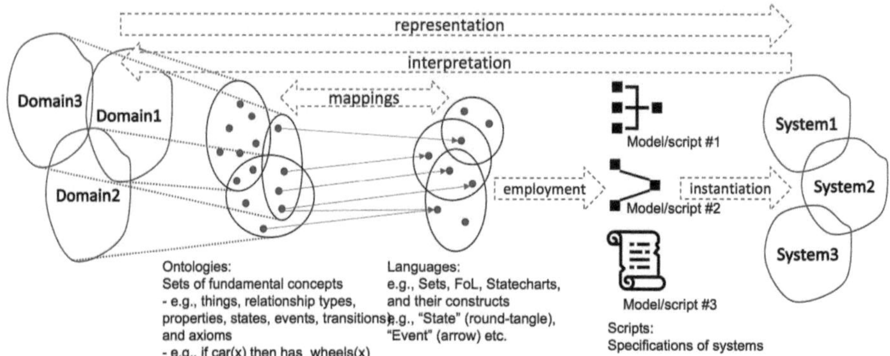

Fig. 1. From real-world problem domains to systems: Left to right (representation): application domains, formalization of concepts within these domains, representation of the formalization as entities in specification languages, programs/scripts/specifications that encode systems structure and behavior, systems. The inverse direction reflects the interpretation of systems and their specifications as real-world effects. Such artifacts can be scanned for indications of hidden assumptions at various levels of abstraction.

nature would fit in a requirements document or a similar object, and affects system development, testing, or use, but is not followed up on completely, because affected stakeholders are not aware of it. Toward a more specific definition, below we offer several perspectives, or views, for describing and categorizing such hidden assumptions. Figure 1 provides a high-level ontological context, highlighting the fact that a working system, concrete and real as it may be, depends on multiple stages of abstracted specifications of the relevant real-world entities. Furthermore, these specifications are digital and examining them for signs of hidden assumptions can be automated. For more details about domain, model and metamodel ontologies see [1,2].

2.1   A Domain-Dependent Perspective

Many hidden assumptions are based on the nature of the application. For example, in describing road intersections for an autonomous vehicle (AV), excluding a configuration that exists somewhere in the world may reflect a hidden assumption that it does not exist in the planned use cases. Domain-specific assumptions may be induced by the chosen abstraction levels, implying that different entities within a category should be treated uniformly, regardless of their differences. For example, specifying for an AV the concept of road lanes without reference to lane widths may imply the hidden assumption that the same behaviors hold for all lane widths. Such abstractions can range from highest-level use cases of the system, through function modules, to fine-grain physics of the nuts and bolts, literally, as well as electronics, in the system and its environment.

Domain-dependent assumptions often constitute ontological axioms. For example, typical rules of behavior of an AV in a city, expect that all road users move only in two dimensions, and do not fly. Such ontological axioms may also emerge from the language used. E.g., interior design for a home may depict a ceiling fan in the middle of a room; the existence of a ceiling is implied.

2.2 A Stakeholder-Centric Perspective of the Concept of Being Hidden

We offer the following broad classification of discovered hidden assumptions, based on who might not be aware of them:

- Not documented, but well known: The requirement, constraint, etc., is not documented, but, if asked, stakeholders of all roles would describe correctly how the system relates to the issue at hand;
- Assumed, and specified explicitly, but unknown to some stakeholders: The requirement is specified in code or documents that are not accessible to certain stakeholders (like customers) who should be aware of it.
- Undocumented and totally hidden from many relevant stakeholders: despite their diligent joint work, developers, system engineers, customers, and users have not specified and are not aware of a certain assumption that was built into the system.

Awareness of such categories can suggest discovery techniques (See Sect. 3), like focusing on esoteric area in the problem domain or in the development process.

2.3 Specification-Structure Perspective

In this perspective, we examine hidden assumptions that emanate from inherent traits and effects of the overall structure of the used formalism. Such formalisms include classical state machines, statecharts, Petri nets, class diagrams, sequence diagrams and live sequence charts (LSC), temporal logic, models in languages like Z notation and VDM, models of databases, MATLAB Simulink models, activity and use case diagrams, component diagrams, high-level architecture diagrams, etc. Sect. 2.4 discusses a programming language perspective.

Such formal specifications often represent intentions compactly using named entities. For example, in modeling an AV, one would define system states and conditions, like "engine on" vs "engine off" or the color of a traffic light; actions, like "stop", "start", or "slow down"; events, like "Police person signaled ego AV to pull over", etc. Below are examples of features of formal models that may induce hidden assumptions, especially when combined with such abstract representations.

1. *Abstraction, encapsulation and equivalence:* As discussed extensively in the context of contract-based design methodology [4], in many specifications, the

information provided about an entity often includes its interface and excludes important conditions and shared variables it relies on, effects it generates, parallel execution dependencies, division of responsibilities, etc. Similarly, in object-oriented design, all object instances belonging to the same class are assumed to be equivalent with regard to the class's attributes and methods; one may not readily know whether differences between subclasses of instances, and between individual instances, were or were not accounted for. If the relevant information is not proactively sought out and stated explicitly, it remains as hidden assumptions.

2. *Orthogonality and Set Disjointness:* Consider a program with parallel components or a statechart that divides a certain system state into several orthogonal regions. This may imply the assumption that all reachable combinations of substates are safe and are properly handled, and/or that unsafe combinations mysteriously become unreachable. Similarly, a modeled set may be divided into several disjoint subsets to communicate a certain abstraction, which in turn may hide important information. For example, a traffic simulation that initially distinguishes road users and stationary obstacles, clearly has to deal with objects that belong to both sets.

3. *Atomicity of transition:* Specifying a process as a state machine suggests that specified transitions are atomic, i.e., uninterruptible. Since in reality, some transitions do go through interim states that may require developers' attention, the specified atomicity of the transitions carries assumptions about this handling, like that it is not required, or is handled elsewhere.

4. *Implication by containment:* Formal models have clear relations of set containment. For example, in a statechart, an event exiting a containing state applies to all contained substates, as in the case of a transition leading to a safe state upon some critical event. This "For all" quantifier implies the assumption that all nested states can be interrupted and exited upon the occurrence of this event.

5. *Completeness:* Examining a state machine or an architecture/design diagram, a domain expert might be drawn into checking the correctness of existing specifications, and not notice that certain higher level aspects are altogether missing. Unless certain entities are mentioned but are not fully specified, there is little, if anything, in a system artifact to remind an expert reviewer to check against the their personal knowledge whether certain components, behaviors, external events, or inter-entity relations are entirely absent. While checking if a given complex system or a model is complete or correct with reference to some accepted scope and abstraction level, is a question in its own right [13], any gaps between a system's code and documents and an established domain knowledge, reflect hidden assumptions about the scope of the system.

2.4   A Language Perspective

Hidden assumptions can arise from characteristics of the language and specification method used like syntax, semantics, and visual layout, including aspects like order and containment relations [22]. Furthermore, user interpretations of specifications are often influenced by the users' mental models and expectations [23]. Below we offer attributes of language dependent hidden assumptions, also viewed as *specification deficits*. This list is based on an ontological paradigm adapted from [28, 29], which was created originally for assessing modeling grammars.

1. *Inexpressible:* A requirement that cannot be specified explicitly and directly due to language constraints. For example, in a rule-based language that is limited to statements like "Always, when condition C holds, do action A", one cannot readily state "Always, after event E occurs, do Action A".

2. *Unclear:* A part of a specification that may be misunderstood. For example, when an inexpressible requirement is implemented indirectly, or when using similar names for different entities.

3. *Unintended:* A specified requirement that, due to language semantics, carries an unintended meaning. For example: Consider the rule "Always, when condition C holds, do action A". If action A also makes condition C false, there may be a hidden assumption that A occurs exactly once whenever the system recognizes that C is true. But, depending on timing aspects of the system, the action may also occur several times.

4. *Opaquely encapsulated:* The discussion in Sect. 2.3 about abstraction and encapsulation of application components applies also to built-in language functions.

5. *Unspecified:* Information that is missing altogether. For example, the specification "Always, after event E, do action A" and "Always, after event E, do action B", does not detail whether A should be executed before, after, or concurrently with B.

6. *Implicit:* The requirement is not stated at all, but is implied by the language or the execution foundation. For example, continuing the previous one, the language semantics may dictate that rules are evaluated and executed to completion, one at a time, in the order in which they appear in the source files. This is also the case with parallelism and order of execution in synchronous languages [26], and conversely, nondeterminism in these issues as implied in other languages; still, one cannot be sure that the relevant stakeholders are fully aware of those definitions when writing or interpreting a model or a program, as part of their own actions.

7. *Perceived:* An explicit and clear specification that a human reader may misinterpret. E.g., the python statement A = B = C, assigning the value of C to B and then to A, may be misinterpreted as A = B == C, making A True if B is equal to C.

8. *Fragmented:* A requirement is stated only as a collection of sub-requirements. For example, an AV is designed per the undocumented requirement that it works only in certain geographies, and is programmed throughout only with the implied traffic laws, like driving on the right. Another example is when a requirement is implemented in two independent rules where one constitutes an overriding priority or exception to the other.

2.5   Program Code Perspective

In addition to the specification structure and language issues discussed in Sects. 2.3 and 2.4, almost every line of program code and every entity in a formal model (which is a form of program code) carries with it hidden assumptions. For example:

1. *Silent conditions, conjunctions and disjunctions:* Assume that the system checks condition C in certain contexts. For other contexts, a reviewer cannot readily tell whether the developer expects C to be always true, always false, or irrelevant, or if perhaps, not checking C is an omission.
2. *Fixed parameters and initial values:* All constants in a system represent requirements and assumptions. However, one should distinguish between the more conspicuous assumption, about the choice of value, and the more implicit one, that this value is fixed and applies equally in all cases where it is used.

2.6   Additional Perspectives

Additional categorization perspectives for hidden assumptions may be guided by the diverse development activities and related interests of stakeholders like those described in [17]: system logic and function view, software development process view, integration process view, physical component view, and scenarios view.

Time is of course another relevant dimension; requirements that were so obvious that they were left undocumented in early project stages can become obscure as time passes, people change roles, needs change, etc.

3   An Outline of Methodical Discovery of Hidden Assumptions

The categorizations in Sect. 2 suggests initial directions for creating methodologies and tools for discovery of hidden assumptions, as follows.

3.1   Semantic Analysis and Comparisons

We wish to mimic the ability of external experts to use their knowledge, experience, and analytical skills to examine programs, documents and demonstrations of a system and identify important omissions and hidden assumptions. Hence, we propose to create a mapping between generally available domain and technology knowledge on the one hand and the target system on the other. Gaps in either direction and any kind of difference in contents, perspective or level of detail will be very instructive regarding assumptions, omissions, and areas that require further attention.

For the system being studied one would collect: requirements documents; architectures; designs; models; software code; test cases; environment simulation software code; user guides and operation manuals (including "fine print" with disclaimers and warnings); technical specifications; sales brochures; questions-and-answers repositories; execution logs and even videos of the system execution in real-world tests; trouble and error reports by QA engineers and early users, including their resolutions or reasons for closure; development meeting summaries; long term development plans and deferred feature requests; etc.

Sources for general domain and technology knowledge would include text books, tutorials, and articles about the domain; examples of small systems, with problems and solutions; system artifacts, as listed in the previous paragraph, for related systems like prior versions of the target system, past solutions, competitive solutions, documentation of technical comparisons of such systems; regulations and industry standards; accounts of famous bugs and failures; ontology databases/schemas [30]; etc.

This specific mapping can be readily extended beyond the available materials. For example, if a tool examining a quantitative system aspect, say, a distance between two objects, raises a valid question about the fidelity of the associated measurement or precision of related program variables, human engineers, as well as tools, may be alerted to the need to review such aspects in other parts of the application.

3.2 Structure-Driven Question Generation

One approach for structure-driven discovery of hidden assumptions and unhandled contingencies is as follows. For each type of broad assumptions in the structure-driven perspective, select a set of relevant system artifacts, and state a question challenging the assumption. The questions may then be answered by a human, and in some cases, by tools based on domain and technology knowledge. For example:

1. For abstraction, encapsulation and equivalence assumptions: for any encapsulated function, check its interface definition and then ask about issues that are not directly manifested there. Candidate topics for software include performance, exception handling, resource consumption, and side effects. For hardware the missing aspects may emerge from part features and specifications in the domain. In an object-oriented design, for each class, find pairs of subclasses or object instances, ask about their differences, and then ask whether these difference should imply refinement of superclass methods, user interaction instructions, etc.

2. For Orthogonality and set disjointness assumptions: In a specification like that of a statecharts, for each state that is divided into parallel regions, run model checking to check if all composite states are reachable. If not, this can be instructive about the problem domain and application or about the design and the specification. In particular, if the unreachable composite state is an undesired one, the requirement that it is unreachable may reflect a hidden assumption, or its coding may be highly fragmented. For set definitions, select a member of some set and and ask whether it could also belong to another set. Continuing the example of road users vs. stationary obstacles, once asked

one realizes that what appears visually as a house may be a road user, when it is carried on a truck. Similarly, in a state machine, one assumes that all states are disjoint. An examination of pairs of named states with domain expertise in mind, may elicit situations where the system may satisfy the assumed conditions for two distinct states at the same time.

3. For state containment: for each sub-state and each event exiting the containing state, check if exiting at that substate requires special handling that was not specified.

4. For atomicity of transition: in a state-based formalism, for each transition, generate questions regarding its atomicity; domain knowledge may elicit interim states that need to be specified. Also, if there is an atomic transition between two system states that differ by more than one state variable, consider whether one needs to handle the case where only some of the variables have changed.

5. For Completeness: in a state-based formalism, generate questions that check whether the listed states and orthogonal regions indeed offer complete coverage. For each state, check if the incoming and outgoing transitions associated with this state are indeed the only ways to enter and exit the real-world condition represented in this formal state. For formalisms that include entities and relationships, systematically check pairs of entities, and ask whether they have relationships that have not been specified.

Such structure-based question generation may be intractable for large systems, due to large number of all pairs, or all possible sets, of such entities. Tools will have to employ additional techniques such as random sampling, heuristic choices of suspicious or promising subjects, automated learning from accumulated experience in such automated reviews, and incorporating interactive human assistance in the review process.

3.3   Explanation-Driven Search

Much insight can be gained from human explanations and commentaries that complement established system artifacts. Automated tools can look at questions and challenges like: trouble tickets that were closed without handling because "everything worked as expected"; challenges by stakeholders and external reviewers that were resolved by arduous explanations of what the system can and cannot do, or where and how it should and should not be used; analysis reports following major failures; etc. These explanations represent knowledge about the system that is not readily available to key stakeholders, and that perhaps should be better communicated and then re-evaluated. One should note though that such human-provided explanations are not necessarily correct or complete, and involve an intermediate step of human interpretation [23].

4   Experimenting with Semantic Analysis Using ChatGPT

Reliance on external knowledge for automated checking of complex systems may still be far away. It may also seem that automating key capabilities of knowledgeable, smart and dedicated external reviewers is a tantalizing goal. To illustrate that this goal is at all practical, below we list several small experiments that we have carried out with the help of OpenAI's ChatGPT, showing that building blocks for such methodological automation are already available. Whether or not these particular tools and methods will indeed constitute elements of the solutions is yet to be determined.

We conducted the experiments below with the general version of GPT 4.0 LLM, relying only on its built-in training corpus. Additional developments may use LLMs that are pre-trained and fine-tuned with materials that are more narrowly focused on certain application domains and software development methods, and/or take advantage of plugins to enable ad-hoc access to materials and create custom functionality.

From User-Defined Terms to Domain Ontologies. In order to map what an application does into available domain knowledge, one of the tasks is to associate system entities with well recognized domain entities. When interpreting the text of requirements document this mapping may be more direct [27], but what does one do when such documentation is not available or not complete? Can the source code of the application be interpreted and mapped in this manner? We presented ChatGPT with the following prompt[2] *"In software, program code, developers try to give meaningful names to variables, functions, etc. They often use "camel case" or acronyms, or other abbreviations, like dropping vowels, etc. For example, in a financial application, for a function that computes a final payment, an entity whose full name would be "final payment" can be named as "FinalPayment" or "finPayment" or "FinPmt" or "fPayment" etc. I wonder if you can guess the full name or meaning of fields named SlsTx, Bal, Refnd. tntvAmt.".* The LLM readily returned *Sales Tax, Balance, Refund,* and *Tentative Amount.* Note that hinting about the general problem domain (finance, in this case) is not only allowed, but perhaps even desired in such analyses by humans and machines. Clearly, with sufficient training data from source code, and some rules and examples about how programmers abbreviate and compound words to assign names to software entities, tools can infer the ontological associations of system entities.

From Computations and Programmed Scenarios to Domain Processes. To learn what a software application does and does not do, one needs to look at sequences of events and operations. ChatGPT has proven useful in interpreting computations that often call for human explanation. To our prompt *"Can you guess 'names' of variables, or 'their purpose' from the logic of the computation? Here is an example. Can you associate it with 'a story'? X1= 0.2; X5=0.06 X2=<Input>; X3=X2-X2*X1; X4=X3*(1+X5)"*, ChatGPT readily responded that this program snippet may be computing the final price of a product by

---

[2] We copy here our very first attempt; minor typos in the original (which ChatGPT readily overcame) were corrected here for readability of the article.

applying discounts and taxes to the base price. The same results were obtained also when such a computation was carried out in a single expression without interim variables.

Creating Ontological Checklists. While databases of curated ontologies exist, ChatGPT was able to readily retrieve compact lists of issues that must be handled in a given domain. Asking for the issues that must be handled in a rudimentary web store application it returned a long list of areas that must be handled, from creating invoices through managing stock to dealing with the fact that certain taxes depend on location. Such lists are an essential component of our intended analyses.

Comparing Lists. With components like those detailed above, one can create lists of issues and entities that are expected in the problem domain, and similar issues and entities as represented in the target system. Clearly, one does not need an LLM to compare well-structured lists, but it was nice to see that ChatGPT could compute set difference between two sets defined with text descriptions. In this way, one can discover problem domain issues that were not conspicuously handled in the application. The inverse can be used for enriching the ontology.

Using Domain Knowledge for Adding Function. Inspired by an ecological model described in [18], we presented to ChatGPT a simple scenario-based model of the interaction of bees and flowers. We explained to ChatGPT that this model is not part of a biological study but is part of a science education program that focuses on computational/algorithmic thinking skills, by describing natural phenomena as composition of compact, stand-alone rules/scenarios. The rules in the model revolved around monthly changes in patterns of average daylight hours and temperature, and the effect of these parameters on whether flowers bloom, and whether bees are active or hibernate. We asked ChatGPT to suggest additional topics that could enrich the model, while keeping with the goals: i.e., rather than adding all biological and ecological factors that could be relevant (and that ChatGPT was clearly aware of), suggest one or two other similar factors. ChatGPT suggested allowing for variations due to the changing weather that may change the temperature from the expected monthly averages. This is also an example of the potential duality between omissions and hidden assumptions: not accounting for changes in the weather may be a sheer omission or a conscious choice. This example also clarifies the importance of documenting assumptions as such: having a table for "the temperature" during each month does not necessarily mean that the model assumes that the temperature is fixed throughout each month—other computations may still modify the ambient temperature. Adding a sentence that states this assumption explicitly, clarifies the issue and supports a more thorough evaluation of the model.

Using Domain Knowledge for State Refinement. Along the lines of Sect. 3.2 we ran the following experiment.

Our Prompt: 'Vehicle State Refinement: In modeling complex systems, developers often establish orthogonal, alternative states. For example, in describing a vehicle's behavior or operation, one may recognize the Boolean state variables of S1 as "moving" vs. "standing", and S2 as "engine on" vs. "engine off". Then there are composite states with combinations of these state variables. Such composite states sometimes give the illusion of full coverage of possible states, but domain knowledge may require finer differentiation like S3 as "vehicle is parked" vs. "vehicle in use" or S4 as "vehicle in maintenance (as in being in the shop)" vs. "vehicle is operational". Can you suggest more state refinements like these for the case of vehicles? My goal is to show that "broad brush" discrete division of states is sometimes misleading, and may cause one to overlook important variables, and that domain expertise can help correct the situation'.

ChatGPT (pruned, with added numbering): *Absolutely, state refinements can lead to a more accurate understanding and modeling of complex systems. For a vehicle, there are indeed numerous additional states to consider that can provide a deeper understanding of the vehicle's condition and operation. Here are some examples: (1) Fuel Level: This state could be divided into categories such as "full", "partially full", "low", or "empty"... (2) Load: ... "empty", "partially loaded", or "fully loaded". This state has implications for fuel consumption, performance, and wear and tear on the vehicle. (3) Weather Conditions: ... "clear", "rain", "snow", "fog", etc. ... affect vehicle performance and safety ... (4) Road Conditions: ... "smooth", "bumpy", "icy", "wet", etc. ... (5) Vehicle Integrity: "Optimal", "minor damage", "major damage", "totalled"... (6) Occupancy: ..."unoccupied", "driver only", "passengers present", "at full capacity". Occupancy not only ...weight distribution,... driver behavior ... (7) Vehicle Mode: ... "electric mode", "fuel mode", or "mixed mode". Similarly, ... "manual driving", "assisted driving", or "fully autonomous" mode. (8) Maintenance Status: "Recently serviced", "due for service", "overdue for service". ... (9) Tire Condition: "New", "good", "worn", "balding", "flat". ... (10) Driver Condition:... "Alert", "fatigued", "impaired", "distracted". While not strictly a state of the vehicle, the driver's condition heavily influences the operation and safety of the vehicle. This list can be extended further depending on the level of granularity required and the specific use case. By identifying and considering these states, developers can model vehicles more accurately and comprehensively."*

Learning from Gaps in Tools. A tool designed to help uncover hidden assumptions, can also be helpful when it misses the mark. Failures can be studied in the context of a synthetic application constructed to challenge the tool, or by comparing the tool's output to comments by external reviewers, or as part of in a retrospective failure analysis. In such cases, having a concise description of a hidden assumption that tool has missed, can teach us about human thinking and development processes and guide incremental enhancements to the tool and the discovery processes. In another experiment we asked ChatGPT for the motion formulas of a pendulum. We then asked for assumptions in this model, and ChatGPT listed thirteen assumptions, including small angle, no friction, rigid rod, point mass, and more. However, ChatGPT did not list the assumptions that the pivot point is fixed in space, and that the pendulum motion is confined to two dimensions—a plane. A generalized enhancement that we derive from this experiment, which can be applied to other systems, is always adding questions "Which points in this configuration are fixed in space?", and "In how many dimensions is the problem considered?". With regard to the pendulum, in a separate experiment, ChatGPT answered these questions correctly, i.e., that the pivot is fixed and the motion is in two dimensions; furthermore, in response to another question, ChatGPT admitted that these two facts really belonged in the original list of assumptions. A similar discussion about insights to be gained from incremental development is made in [12] in the context of how humans develop increasingly sophisticated Turing Tests that aim to challenge the intelligence of machines that have passed lower level tests.

Human Interfaces for Discovery Tools. The above experiments suggest that in designing the interfaces for tools for discovery of hidden assumptions, developers can initially suffice with mimicking the interactions with external reviewers: feed the tools with the same development and marketing documents, code artifacts, simulation outputs and videos, etc., as would be shown to reviewers, and then ask the same kind of questions. To make-up for the expert's knowledge and experience which may not already be built into the tool (say, as databases or AI training), additional materials from other systems, reference documents, etc. may be fed for comparison and matching.

5   Conclusion

Reviews by external application specialists and technology professionals are considered of great potential value to a project. Here, and as part of the vision of wise computing [14], we presented initial steps for creating automated tools that help mimic some of the critical constructive reviews that outside experts can offer. Such tools can facilitate carrying out these reviews constantly, throughout development, and from the very beginning of the project, reducing costs and improving quality.

Future research directions include refining the definitions and types of hidden assumptions, building prototype tools, and enhancing the community-wide availability of relevant machine-accessible knowledge and expertise.

Acknowledgements. We thank the anonymous reviewers for their valuable comments and suggestions. This project has received funding from the European Union's Horizon research and innovation programme under grant agreements no 101094905 (AI4Gov). This research was funded in part by an NSFC-ISF grant issued jointly by the National Natural Science Foundation of China (NSFC) and the Israel Science Foundation (ISF grant 3698/21). Additional support was provided by a research grant from the Estate of Harry Levine, the Estate of Avraham Rothstein, Brenda Gruss, and Daniel Hirsch, the One8 Foundation, Rina Mayer, Maurice Levy, and the Estate of Bernice Bernath.

References

1. Aßmann, U., Zschaler, S., Wagner, G.: Ontologies, meta-models, and the model-driven paradigm. In: Calero, C., Ruiz, F., Piattini, M. (eds.) Ontologies for Software Engineering and Technology. Springer (2006). https://link.springer.com/chapter/10.1007/3-540-34518-3_9
2. Atkinson, C., Kühne, T.: Model-driven development: a metamodeling foundation. IEEE Softw. **20**(5), 36–41 (2003). http://csdl.computer.org/comp/mags/so/2003/05/s5036abs.htm
3. Banks, V.A., Plant, K.L., Stanton, N.A.: Driver error or designer error: using the perceptual cycle model to explore the circumstances surrounding the fatal tesla crash on 7th May 2016. Saf. Sci. **108**, 278–285 (2018)
4. Benveniste, A., et al.: Contracts for system design. Found. Trends® Electr. Des. Autom. **12**(2-3), 124–400 (2018)
5. Bialy, M., et al.: Software engineering for model-based development by domain experts. In: Handbook of System Safety and Security, pp. 39–64. Elsevier (2017)
6. Bienmüller, T., Damm, W., Wittke, H.: The statemate verification environment: making it real. In: Computer Aided Verification: 12th International Conference, CAV 2000, Chicago, IL, USA, July 15-19, 2000. Proceedings 12, pp. 561–567. Springer (2000)
7. Biggerstaff, M., Slayton, R.B., Johansson, M.A., Butler, J.C.: Improving pandemic response: employing mathematical modeling to confront coronavirus disease 2019. Clin. Infect. Dis. **74**(5), 913–917 (2022)
8. Buede, D.M., Miller, W.D.: The engineering design of systems: models and methods (2016)
9. Charette, R.N.: Why software fails [software failure]. IEEE Spectr. **42**(9), 42–49 (2005)
10. Damm, W., Harel, D.: LSCs: breathing life into message sequence charts. Formal Methods Syst. Des. **19**, 45–80 (2001)
11. Damm, W., Möhlmann, E., Peikenkamp, T., Rakow, A.: A formal semantics for traffic sequence charts. In: Principles of Modeling: Essays Dedicated to Edward A. Lee on the Occasion of His 60th Birthday, pp. 182–205 (2018)
12. Eisenstein, M.: A Test of Artificial Intelligence. Nature Outlook: Robotics and Artificial Intelligence. Accessed 10 2023
13. Harel, D.: A turing-like test for biological modeling. Nat. Biotechnol. **23**(4), 495–496 (2005)
14. Harel, D., Katz, G., Marelly, R., Marron, A.: Wise computing: toward endowing system development with proactive wisdom. Computer **51**(2), 14–26 (2018)
15. Kotwal, A., Yadav, A.K., Yadav, J., Kotwal, J., Khune, S.: Predictive models of COVID-19 in India: a rapid review. Med. J. Armed Forces India **76**(4), 377–386 (2020)
16. Kramer, B., Neurohr, C., Büker, M., Böde, E., Fränzle, M., Damm, W.: Identification and quantification of hazardous scenarios for automated driving. In: International Symposium on Model-based Safety and Assessment, pp. 163–178. Springer (2020)

17. Kruchten, P.: The 4+1 view model of architecture. IEEE Softw. 12(6), 42–50 (1995). https://doi.org/10.1109/52.469759
18. Marron, A., Cohen, I.R., Frankel, G., Harel, D., Szekely, S.: Challenges in modeling and unmodeling emergence, rule composition, and networked interactions in complex reactive systems. In: MODELSWARD, pp. 202–209 (2023). (Conference Proceedings: https://www.scitepress.org/ProceedingsDetails.aspx?ID=M/tLGjnMVq0=&t=1, https://doi.org/10.5220/0011728900003402)
19. Marshall, E.: Fatal error: how patriot overlooked a scud. Science **255**(5050), 1347–1347 (1992)
20. Masso, J., Pino, F.J., Pardo, C., García, F., Piattini, M.: Risk management in the software life cycle: a systematic literature review. Comput. Stand. Interfaces **71**, 103431 (2020)
21. Medhaug, I., Stolpe, M.B., Fischer, E.M., Knutti, R.: Reconciling controversies about the 'global warming hiatus'. Nature **545**(7652), 41–47 (2017)
22. Moody, D.: What makes a good diagram? Improving the cognitive effectiveness of diagrams in is development, vol. 2 (2007). https://doi.org/10.1007/978-0-387-70802-7_40
23. Norman, D.: The Design of Everyday Things (2016). https://doi.org/10.15358/9783800648108
24. Pelletier, J.D.: Coherence resonance and ice ages. J. Geophys. Res.: Atmos. **108**(D20) (2003)
25. Sangiovanni-Vincentelli, A., Damm, W., Passerone, R.: Taming Dr. Frankenstein: contract-based design for cyber-physical systems. Eur. J. Control **18**(3), 217–238 (2012)
26. Smyth, S., Schulz-Rosengarten, A., von Hanxleden, R.: Practical causality handling for synchronous languages. In: 2019 Design, Automation & Test in Europe Conference & Exhibition (DATE), pp. 1281–1284. IEEE (2019)
27. Tsarfaty, R., Pogrebezky, I., Weiss, G., Natan, Y., Szekely, S., Harel, D.: Semantic parsing using content and context: a case study from requirements elicitation. In: Proceedings of the 2014 Conference on Empirical Methods in Natural Language Processing (EMNLP), pp. 1296–1307 (2014)
28. Wand, Y., Weber, R.: On the ontological expressiveness of information systems analysis and design grammars. Inf. Syst. J. **3** (1993)
29. Wand, Y., Weber, R.: On the deep structure of information systems. Inf. Syst. J. **5** (1995)
30. Zipfl, M., Koch, N., Zöllner, J.M.: A comprehensive review on ontologies for scenario-based testing in the context of autonomous driving. arXiv preprint: arXiv:2304.10837 (2023)

Securing Future In-Vehicle Networks: Monitoring and Control for Ethernet Backbones

Timo Häckel[(✉)] [iD], Philipp Meyer [iD], Franz Korf [iD], and Thomas C. Schmidt [iD]

Department Computer Science, Hamburg University of Applied Sciences (HAW),
Hamburg, Germany
{timo.haeckel,philipp.meyer,franz.korf,t.schmidt}@haw-hamburg.de

Abstract. Vehicular systems are increasingly controlled by distributed software components that demand a reliable and secure communication infrastructure. Ethernet and Time-Sensitive Networking (TSN) offer a promising foundation for future high-bandwidth, real-time capable In-Vehicle Networks (IVNs). Current IVNs, however, are vulnerable to safety-compromising attacks, calling for a multifaceted security hardening. In this paper, we present a control and monitoring framework that guards safety and security throughout an attack-resistant real-time network architecture with effective misbehavior detection. Software-Defined Networking (SDN) control of TSN networks enables the enforcement of security policies and the dynamic adaptation to changing conditions. A combination of network monitoring techniques, traffic analysis, observable metrics, and detection algorithms tailored for automotive deployment enable the detection of anomalies and attacks. This network-centric approach enhances the security of existing automotive protocols and Electronic Control Units (ECUs) of limited computing power and remains extensible for future demands. Evaluation results from simulations, a hardware test bed, and a real prototypic vehicle demonstrate the feasibility, effectiveness, and remaining challenges of our approach.

Keywords: In-Vehicle Network · Time-Sensitive Networking · Software-Defined Networking · Network Security · Network Monitoring · Anomaly Detection

1 Introduction

Future software-defined vehicle systems [30] require the In-Vehicle Network (IVN) to support high-bandwidth cross-domain communication, *e.g.*, for Advanced Driver Assistance Systems (ADASs). To meet these demands, network and compute resources are centralized and shared among services. A real-time Ethernet backbone [50] with Time-Sensitive Networking (TSN) [22] ensures efficient and reliable communication.

This work was supported in part by the German Federal Ministry of Education and Research (BMBF) within the SecVI project.

M. Fränzle et al. (Eds.): Werner Damm Festschrift, LNCS 15471, pp. 163–182, 2026.
https://doi.org/10.1007/978-3-031-97537-0_10

Integrating vehicles into global communication (Vehicle-to-X (V2X)) increases the attack surface, which has been shown in real-world attacks [38], necessitating cybersecurity measures. Industry standards (*e.g.*, ISO/SAE 21434 [24]) and legislation (*e.g.*, the European Cyber Resilience Act [13]) demand security measures throughout the entire vehicle system and lifetime.

Securing vehicle access interfaces, performing plausibility checks, and encrypting critical data are important, but cannot rule out attacks reaching the IVN itself [44]. The IVN is the backbone of the vehicle system, as it connects all Electronic Control Units (ECUs) and enables communication. Manipulation of the IVN can impair the safety of critical driving functions and even harm passengers and other road users. Moreover, attacks can spread throughout the network, necessitating zero-trust in network nodes to prevent their propagation.

A multilayered defense strategy is crucial to secure the IVN against misbehavior [51]. A network-centric approach ensures security for resource-constrained ECUs and allows secure reuse of legacy devices. Precise control and isolation of network flows are required to minimize the attack surface and prevent attacks from spreading [51]. Effective network monitoring can detect misbehavior and enable a fast response. All security measures must meet the real-time demands of the IVN and remain compatible with existing protocols and devices.

In this work, we propose a monitoring and control architecture enhancing security in future IVN. Our approach encompasses traffic control and network monitoring, which we investigated in various previous work [14–18,35,36,47]. Implementing an attack-resistant real-time network architecture, we ensure both safety and security. To effectively detect misbehavior, we employ a combination of monitoring techniques, network traffic analysis, and observed metrics tailored for automotive deployment.

The remainder of this work is structured as follows: Sect. 2 reviews background and related work. In Sect. 3 we present our control and observe architecture for IVN security. Section 4 presents our evaluations in simulation, test beds, and a prototype vehicle, demonstrating the practical application of our approach. Finally, Sect. 5 concludes this work and outlines future work.

2   Background and Related Work

The race towards highly autonomous vehicles requires a new platform for their Electrical/Electronic (E/E) architecture, and the automotive industry is moving towards a software-defined vehicle architecture [30]. Multiple teams create components in a highly modular fashion with increasing complexity for integration and testing, and faster development cycles with frequent updates. The integration of such components requires contracts between components, formal timing specifications, and computation and interaction models [8]. This shift poses new challenges for the design of a secure IVN that can keep up with the increasing dynamics.

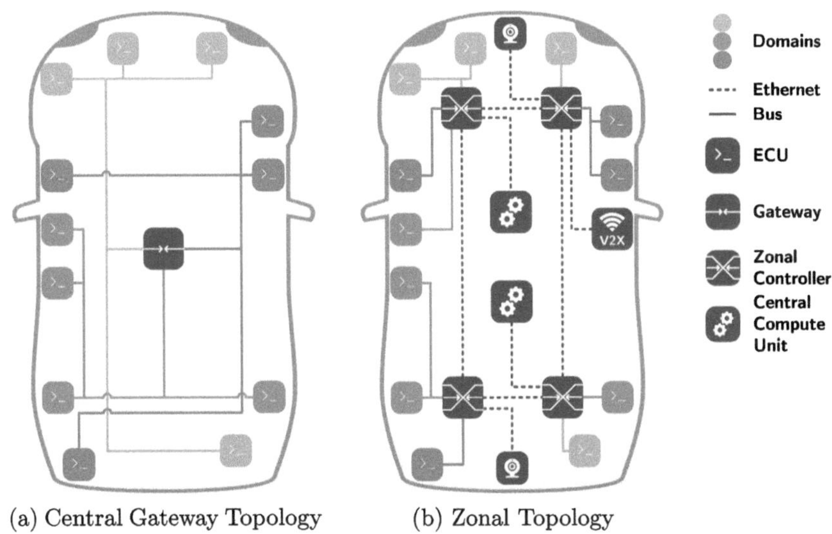

(a) Central Gateway Topology   (b) Zonal Topology

Fig. 1. acIVN topology evolution from separate domains connected via a central gateway (a), to a zonal topology (b) centralizing network and compute resources.

2.1 In-Vehicle Networks

Traditionally, vehicle control applications run on specialized ECUs that are connected via heterogeneous networks, such as Controller Area Network (CAN), Local Interconnect Network (LIN), Media Oriented System Transport (MOST), FlexRay, and Ethernet. The current automotive E/E architecture (*cf.*, Fig. 1a) is divided into separate domains that are interconnected via a central gateway [50]. Examples for these domains include the infotainment, drivetrain or passenger comfort. The increasing number of specialized ECUs required for new functions results in higher hardware overhead, leading to increased vehicle production costs. Additionally, the growing complexity and strict domain separation hinder flexibility.

Future applications require increased communication bandwidth and computational resources, *e.g.*, through high-resolution sensor fusion and ADAS for automated vehicles [10]. To cope with these demands central compute units interconnected via high-bandwidth Ethernet are proposed [6]. In a zonal topology (*cf.*, Fig. 1b), ECUs are grouped into zones according to their placement in the vehicle, *e.g.*, front, rear, and right, left side. Zonal controllers placed at the center of each zone can combine functions of gateways, switches, and applications.

On the software layer this shift to more centralized resources is accompanied by a shift to a dynamic Service-Oriented Architecture (SOA) [27]. Services utilizing virtualized resources in containerized applications are considered for improving the flexibility of the software architecture [27].

Figure 2 shows a future protocol stack for service-oriented communication via Ethernet [33]. *Scalable service-Oriented MiddlewarE over IP* (SOME/IP) [2] is the most widely deployed middleware for such an automotive SOA. Other candidates such as Data Distribution Service (DDS) [42] are also considered, and available in the AUTomotive Open System ARchitecture (AUTOSAR). Both utilize UDP/TCP/IP for adaptive communication. TSN [22] enhances Ethernet to meet the real-time requirements of the IVN. A header

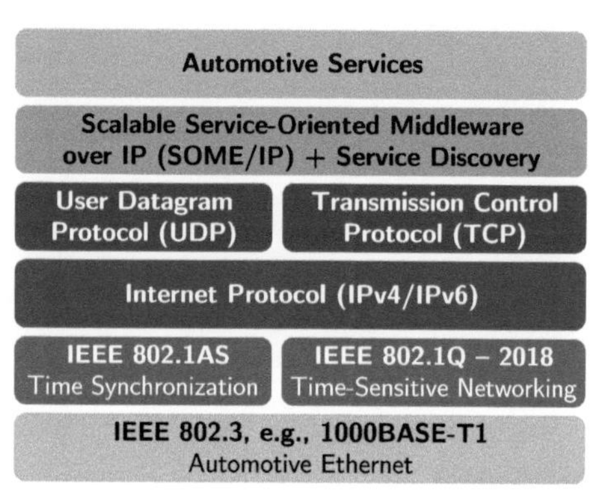

Fig. 2. The automotive protocol stack.

adds virtual LANs and a Priority Code Point (PCP) from zero to seven. TSN devices (*cf.*, Fig. 3) have gates at input and output ports. A closed input gate drops incoming frames and a closed output gate queues frames until it is open. Periodic Gate Control Lists (GCLs) determine opening and closing times of gates. Ingress meters and egress shapers complement the GCL functionality. It is important to note that while the ingress can operate on a per-stream basis, the egress operates on per-priority queuing using the frame PCP and shapes all streams of the same priority together. The switching fabric forwards frames that pass ingress control to the egress control of an outgoing port. Precise time synchronization via IEEE 802.1AS [23] is necessary to synchronize packet transmission and coordinate time between applications, ECUs, and network devices. We add fully programmable options for TSN flows (*cf.*, Sect. 3.1) and utilize the Per-Stream Filtering and Policing (PSFP) for network anomaly detection (*cf.*, Sect. 3.5).

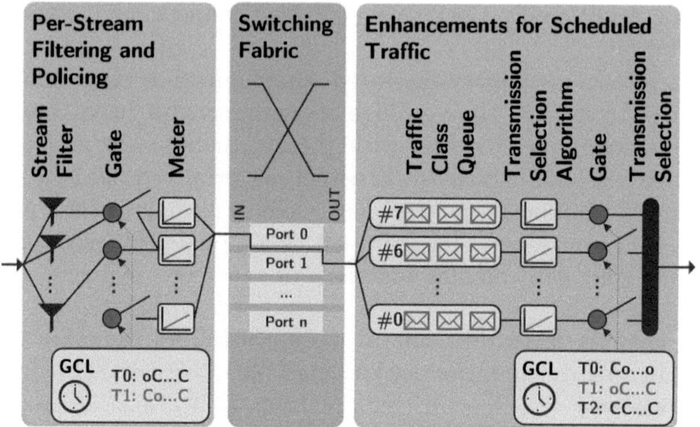

Fig. 3. TSN modules for ingress control and egress shaping.

2.2   In-Vehicle Network Security

Modern cars integrate with global communication, *e.g.*, for V2X communication, which in turn provides a multitude of new interfaces. Current vehicles are vulnerable to manipulation by third parties, which has been demonstrated in the field [38]. A fundamental analysis of the automotive attack surface showing how a variety of interfaces can be used to gain malicious access to in-car devices is provided by Checkoway *et al.* [7]. Manipulation of ECUs and the IVN can compromise the safety of the vehicle, putting passengers at risk.

Attacks on IVNs include Denial of Service (DoS), replay, spoofing, and falsified-information attacks [11,51]. They can be classified as *alter* attacks (modifying data), *listen* attacks (monitoring data), *disable* attacks (denying services), and *forge* attacks (inserting incorrect data) [39]. Defensive measures can be divided into attack prevention, detection, and mitigation [51]. The key security objectives for IVNs are *availability* (resources are accessible and deadlines are met), *integrity* (accuracy and completeness of data), and *authenticity* (verifiability of data sources and sinks) [11,39].

Integrity and authenticity can be ensured with authentication and encryption methods suitable for IVNs [21,40]. Encryption, however, cannot protect data flows from all network attacks, *e.g.*, DoS. Traditional in-vehicle ECUs have limited computing power which restricts resources for security features [40]. In addition, established and verified ECUs are used over many vehicle generations. Our network-centric approach establishes an attack-resistant network (*cf.*, Sect. 3.2) and enables intrusion detection (*cf.*, Sect. 3.4 and 3.5) that helps to secure ECUs with limited computing power and allows secure reuse of legacy devices.

2.3   Attack-Resistant Network Architectures

Current multiple-access bus systems such as the CAN bus are missing flow control capabilities and thus do not allow for safety and security measures on the network level. Gateways can only filter messages between different sections of the IVN [49]. Firewalls and access control mechanisms can prevent attacks with stateful inspection, rate limiting, and filtering [45]. Flexible security solutions, such as software-defined security elements, will be beneficial to cope with the growing dynamics of the IVN [11,48].

As software dynamics increase, they extend to the network level, where flows must be established during runtime. Software-Defined Networking (SDN) [34] is a promising solution to address the challenges posed by the growing dynamics of IVN [14,15,19]. In SDN, the control plane is separated from the data plane. While traditional networks focus on individual end devices, SDN takes a more holistic approach forwarding traffic based on flow rules that match packet headers from Layer 2 to Layer 4. The control plane is logically centralized in the SDN controller. This fundamental distinction allows SDN to offer greater flexibility, adaptability, robustness, and security compared to traditional networks [53].

The logically centralized control plane can be physically distributed for hot-standby and failover to avoid a single point of failure [20]. Additionally, security elements may be vulnerable, especially in software functions. A compromised SDN controller can be used to manipulate the network and bypass security measures. Protection mechanisms have been investigated in the past [20] and are out of scope of this work.

This work explores the potentials and challenges of a software-defined IVN in design, simulation (*cf.*, Sect. 4.1), and practical implementation in a prototype vehicle (*cf.*, Sect. 4.3). The discussed potentials include enforcing real-time guarantees even for dynamic flows (*cf.*, Sect. 3.1), protecting in-vehicle control flows from malicious actors (*cf.*, Sect. 3.2), incident response, and supporting specific automotive protocols within the network (*cf.*, Sect. 3.3).

2.4 Network Intrusion Detection

Network intrusion detection is a crucial component of any cybersecurity strategy [52]. It enables reactive measures to be taken to prevent extended damage. A multitude of approaches, datasets and challenges for network intrusion detection have been proposed [4,28]. In the context of in-vehicle Ethernet networks, the detection of malicious activities is particularly challenging due to domain specific requirements, protocols, and missing large-scale deployments. Thus, matching datasets for validation are rare and because there are just a handful of prototypes with future Ethernet-based IVNs out in the wild concrete attacks are rarely seen so far.

A comprehensive list of algorithm types can be used for network anomaly detection [46]. Many of them are based on Machine Learning (ML) techniques that require validation and verification similar to safety-critical applications for future cars [9]. The real-time pattern of TSN-based Ethernet communication can be beneficial for definition and learning of normal network behavior which can significantly improve performance compared to non-real-time Ethernet environments. It is promising that in-car traffic patterns effectively describe normal behavior and that network anomaly detection methods can be used efficiently to detect malicious activities and even zero-day attacks.

In this work, we explore two potential network anomaly detection architectures for in-vehicle Ethernet networks. Our dedicated network anomaly detection (*cf.*, Sect. 3.4) can leverage advanced ML methods, while network-integrated anomaly detection based on TSN ingress control (*cf.*, Sect. 3.5) exploits the already existing precise configuration of real-time patterns.

3   Control and Observe In-Vehicle Ethernet Backbones

Our control and observe architecture for in-vehicle Ethernet backbones is based on two key concepts: (*i*) Software-Defined Networking (SDN) enables dynamic reconfiguration for precisely separated traffic flows and builds an attack resistant network architecture. surface. (*ii*) Network Anomaly Detection Systems (NADSs) observe network flows, extract key traffic metrics, and detect misbehavior. Figure 4 illustrates the interaction between these components and the IVN.

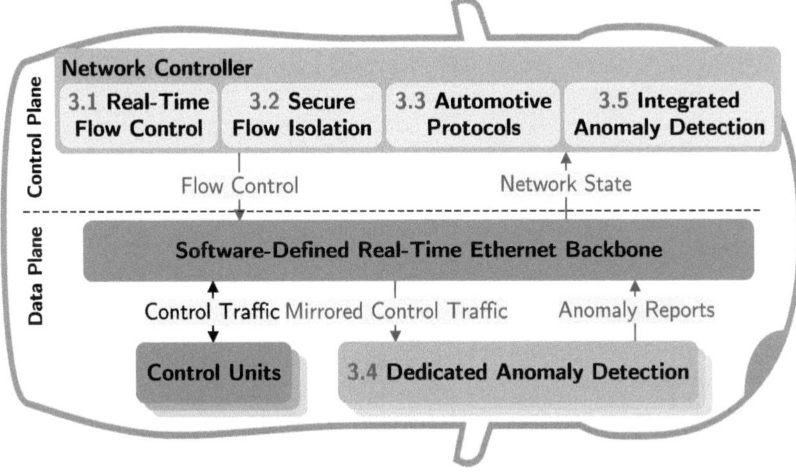

Fig. 4. Overview of the network observation and management architecture.

Deploying SDN in cars logically separates the *control plane* from the *data plane*, introducing a controller with global network knowledge. The control plane is responsible for situational awareness. Specialized control applications program the data plane based on the topology, traffic flows, Quality-of-Service (QoS) requirements and security policies of the IVN. This improves flexibility, with the ability to update and add services, and adapt to changes in communication. The data plane handles the actual data transmission and is subject to real-time requirements that demand deterministic and predictable behavior. It consists of a time-sensitive Ethernet backbone connecting vehicle control units via switches. We demonstrate the applicability of SDN in cars, including central control for real-time flows (*cf.*, Sect. 3.1), secure isolation of vehicle control flows (*cf.*, Sect. 3.2), and integration with existing automotive protocols (*cf.*, Sect. 3.3).

Anomaly detection systems are a subtype of Intrusion Detection Systems (IDSs) designed to identify deviations from normal behavior, which could indicate malicious activity [4]. NADSs monitor the data traffic to detect misbehavior. Defining or learning normal behavior presents a significant challenge. We present two approaches to fingerprint normal traffic and detect anomalies in IVNs: A dedicated NADS based on ML that allows evaluation of varying inputs, metrics, and algorithms (*cf.*, Sect. 3.4). An integrated NADS (*cf.*, Sect. 3.5) that exploits predefined TSN configurations as definition of normal behavior, TSN switch statistics as metrics, and a controller application as the detector.

The control and observe architecture can isolate traffic flows and their paths in the network, while at the same time detect anomalies in allowed traffic flows. Anomaly reports can be used along with network state information for dynamic reconfiguration to mitigate the impact of attacks and ensure that driver, passengers, and manufacturer can be informed [17].

3.1   Software-Defined Real-Time Flow Control

We proposed Time-Sensitive Software-Defined Networking (TSSDN) [14,15], which combines TSN (*cf.*, Fig. 3) and SDN for dynamic real-time communication. On the data plane, TSN endpoints are connected by switches that integrate SDN forwarding with TSN real-time control: (*i*)PSFP applies filters and time checks to incoming frames, (*ii*) passing frames undergo flow table lookup following SDN forwarding, and (*iii*)TSN egress control shapes outbound traffic.

Control plane functions are offloaded to a central controller that performs SDN address learning, routing, and TSN Stream Reservation (SR) and scheduling. Data plane devices forward packets that do not match any flow rule to the controller, where network applications decide whether to drop the packet, reply directly, or forward it. For forwarding, the application determines a route and installs flow rules on the data plane, enabling independent packet forwarding.

Reliable communication is crucial for safety-critical vehicle traffic. In the TSN configuration, priority queues can be partitioned into static and dynamic queues [32]. A static configuration with a separate flow table ensures correctness and serves as a fail-safe in case of controller failure or security incidents. The SDN controller handles additional dynamic traffic by identifying senders and receivers, verifying permissions, and creating precise flows. Therefore, the control plane requires adaptations for real-time traffic.

Asynchronous real-time flows in TSN use the Stream Reservation Protocol (SRP) to dynamically announce resource requirements across the network. Talkers announce streams with bandwidth demands, and interested listeners subscribe. Devices along the path reserve bandwidth and shape the stream accordingly in a transmission selection algorithm, *e.g.*, Credit Based Shaping (CBS).

We map the centralized SR model (802.1Qcc), which aligns well with SDN, to the OpenFlow protocol [14]. Talkers and listeners still use SRP for announcements, and network devices forward them to the SDN controller. A network application checks available bandwidth updates stream tables as listeners join or leave, programs the switches with flow entries and allocates bandwidth. This ensures correct identification, forwarding, and bandwidth control for the stream [14].

In simulation studies (*cf.*, Sect. 4.1), we found that the actual real-time data transfer performs equally well as with TSN [14]. The dynamic SR, which is not subject to real-time requirements in TSN, experiences a delay from traversing the controller. This delay is generally small compared to the latency introduced by cross-traffic and network schedules [15]. Since the SR is done in advance, real-time flows are already installed in the TSSDN switches, avoiding further controller inspection delays.

Synchronous real-time flows require coordination across multiple transmitters and links. A periodic GCL (*cf.*, Fig. 3) controls priority gates to schedule time slots for transmitters. Offline calculations determine fixed Time Division Multiple Access (TDMA) schedules that cannot accommodate changes in communication as services are activated, shutdown, or updated. In TSSDN, the controller can modify the GCL schedule and flow paths when synchronous traffic changes. Time slots can be added, removed, or shifted within the period, and complex operations such as rerouting can combine these actions.

At the egress, the GCL is scheduled per priority queue, not per flow. Adding flows to TDMA-scheduled priorities without updating the GCL can cause queue overflow and missed deadlines for critical traffic. This makes the dynamic nature of SDN flow control unsuitable. In contrast, the NETCONF protocol [12] is well-suited as it supports transaction-oriented configurations that ensure isolation and allow verifying modifications before applying changes.

In simulation studies (*cf.*, Sect. 4.1), we showed the need for multi-device transactions to avoid queue overflow and packet loss during schedule reconfiguration [15]. Moving a time slot in the schedule while a packet is in transit can cause the packet to miss its time slot on the next device, causing a delay for subsequent traffic of that priority. Therefore, updates must be executed simultaneously on all network devices or in a specific order based on the schedule modifications.

3.2   Secure Isolation of In-Vehicle Control Flows

Despite the shift from static configuration to dynamic SOA, control flows of automotive services remain well-defined. A control flow consists of related messages with the same unique identifier sent from one origin to one or more receivers via the network. Each control flow belongs to a specific domain (*e.g.*, drivetrain).

The automotive protocol stack (Fig. 2) incorporates a service-oriented middleware (*e.g.*, SOME/IP) that relies on transport, network, and data link layer protocols. Protocol headers include control flow information to indicate the transported information type to network and applications. SDN flows match header fields from layer 2 to 4. The choice of control flow embedding determines whether control flow information remains hidden or exposed to the network.

In the current state of the art, control flows are typically tunneled using an application layer protocol (*e.g.*, SOME/IP), which hides control flow identifiers from the network. This limits separation capabilities and prevents distinguishing between unique control flows in the network. However, control flow embedding can also be fully exposed, with the control flow context embedded in packet header fields used for network forwarding decisions. This enables perfect isolation with one network flow per vehicle control flow. Recently, Nayak *et al.* introduced a P4 [5] programmable data plane that supports SOME/IP [41], making the SOME/IP header information available for forwarding decisions.

In previous work [15,17,18,36], we assessed the impact of control flow embedding on traffic isolation in a realistic software-defined in-vehicle backbone (*cf.*, Sect. 4.3). When devices participate in a hidden application layer tunnel, *e.g.*, for an entire domain, they can receive and send all control flows of this domain. Exposed embedding enhances flow separation and isolation, restricting communication to known and permitted relations and reducing eavesdropping potential. Exposed embedding and SDN-based isolation require attackers to compromise the exact sender of each control flow on the Ethernet backbone to issue messages in that channel, which reduces the attack surfaces.

3.3 Network Support for Automotive Protocols

Support for automotive application layer protocols on the SDN control plane optimizes flow control. Intercepting network control protocols is a common practice to enhance networking objectives via SDN, such as improved Address Resolution Protocol (ARP) [1] or IP multicast routing [26]. Bertaux *et al.* [3] propose an initial design for an SDN application that dynamically allocates network resources for DDS services. Such an adaptation is needed for automotive protocols to extend the mentioned advantages of SDN to dynamically discovered services.

SOME/IP is the most widely accepted protocol for automotive SOA. It provides a complementary Service Discovery (SD) to resolve service locations during runtime. Further, it supports two communication schemes: (*i*) publish-subscribe for data required on every change or cyclically, and (*ii*) request-response if a service is needed once or sporadically.

Open challenges and opportunities remain for supporting the SD and communication on the network level. The SOME/IP discovery lacks trust mechanisms, access control, and authentication [25], which could be improved by network intelligence. Lower layer QoS, such as TSN shaping, is not supported by the dynamic SOA. A dedicated group management is needed for multicast groups, which is not supported by regular Ethernet switches. Multiple instances of the same service can exist, *e.g.*, for redundancy or distribution, requiring network-level support for resource reservation and fast handover.

We demonstrated that SDN can support the SOME/IP protocol using the controller as a rendezvous point for SD [16]. This enables in-network caching of service information and automatic path setup for communication, including multicast support. We evaluated the performance of our approach in simulation (*cf.*, Sect. 4.1), comparing its scalability to non-optimized SDN and standard Ethernet switching. The central controller processing all SOME/IP SD messages significantly reduces performance in both SDN approaches, but the delay only affects subscription setup, not the actual data transfer (*cf.*, TSN SR Sect. 3.1). Our approach improved SDN performance by up to 50% compared to the non-optimized SDN solution. This allows for potential extensions of a SOME/IP SD-aware SDN for optimizing SD, service mobility, and reconfiguration mechanisms to achieve improved robustness, QoS support, and security enhancements.

3.4 Dedicated Network Anomaly Detection with Mirrored Traffic

A dedicated NADS device can be easily added to existing networks. Integration solely requires that forwarding devices can mirror all incoming traffic to an interface with adequate bandwidth. The mirror interface links to a dedicated NADS device (*cf.*, Fig. 4) that captures and processes traffic at line rate.

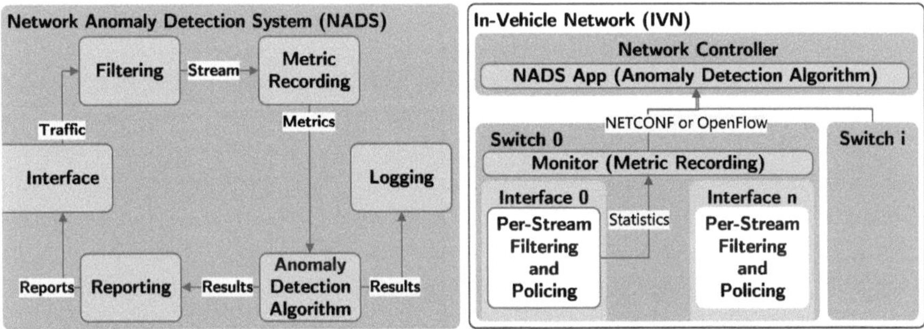

(a) Dedicated NADS for mirrored traffic.   (b) Integrated NADS based on PSFP.

Fig. 5. Different approaches for network anomaly detection in IVNs.

Figure 5a shows the architecture of a dedicated NADS instance. Multiple instances operate in parallel to observe multiple streams. Incoming traffic reaches a component that filters out non-observed traffic. Metrics (bandwidth, jitter, average frame size) are extracted from the stream over a certain observation interval and used as input for the anomaly detection component. The anomaly detection component uses ML models (*e.g.*, clustering, outlier detection, autoencoder) to train normal behavior and detect anomalies in the stream based on the chosen metrics. Results are logged and reported notifying the network. In our case, the reports are consumed by the SDN controller, which captures the network-wide threat situation and can initiate countermeasures to mitigate the anomaly, notify the driver, and forward reports to online infrastructure.

The dedicated NADS approach qualifies for straightforward deployment. A modular interface attaches to a simulation environment (*cf.*, Sect. 4.1), a test bed (*cf.*, Sect. 4.2), and a real-world IVN (*cf.*, Sect. 4.3). The different environments enable reproducible scenarios, automatic generation of labeled datasets, manageable real-world conditions, and real driving scenarios.

ML based NADSs with appropriate performance require a substantial number of datasets, which often must be labeled. Obtaining such datasets is challenging because there are no future IVN cars in the wild that experience real attacks. Datasets need to be build based on related data like *e.g.*, attacks on current IVNs or Ethernet networks in other domains. To overcome this, we use simulation (*cf.*, Sect. 4.1) to generate datasets with different IVN architectures and a variety of normal and attack scenarios.

False positives are a common issue in anomaly detection systems. They occur when the system identifies a normal behavior as anomalous. False positives can be particularly problematic in safety-critical in-car networks, as they lead to unnecessary alerts, avoidable countermeasures, and decreased trust in the system. Due to the short event cycles (*e.g.*, <1 s) in network communication, even small false positive rates (*e.g.*, 0,02%) can lead to multiple false positives during one trip with a vehicle (*e.g.*, >6 false positives in 8 h) [17].

3.5 Integrated Network Anomaly Detection with PSFP

Cyber-physical systems such as cars are subject to real-time requirements. This property carries over to the IVN and is implemented on the link layer through TSN protocols. Real-time Ethernet traffic shaping enforces patterns in communication behavior (*e.g.*, , timings, gaps, sequences) that provide the opportunity to fingerprint normal behavior [35].

Our integrated NADS approach utilizes the TSN standard Per-Stream Filtering and Policing (PSFP) to define and enforce stream behaviors of incoming traffic at forwarding devices (*cf.*, Fig. 3). PSFP consists of stream filters that match traffic to stream gates and flow meters. Stream gates can be open or closed, either with a static configuration or synchronized and defined in the associated Gate Control List (GCL). Meters utilize algorithms to enforce stream behavior, such as bandwidth, gap, and size. Individual frames that pass through these stages are forwarded, marked, or dropped dependent on gate states and meter results.

Figure 5b shows the NADS integration in a TSN network with a central controller. Switches monitor statistics of their active features, such as packet counters. In our case, we use PSFP statistics to monitor stream behavior, including the number of dropped frames. These statistics indicate violations of configured thresholds, *e.g.*, for timing, bandwidth, and frame size. When a frame violates configured thresholds in PSFP and is dropped the statistic for the number of dropped frames is increased. The SDN controller collects these statistics and employs an anomaly detection application to detect abnormal behavior in the specific stream. In a non-SDN network, central collection can still be implemented by traditional network management systems, *e.g.*, using SNMP.

The integrated approach is a lightweight solution for NADS as it utilizes existing network components. Nevertheless, it has certain system requirements: (*i*) the deployed switches must support PSFP, which filters incoming traffic and defines normal traffic per stream. (*ii*) The switches need to be manageable or SDN capable to record statistics, serving as the metric recording component of NADS. (*iii*) A central collector of statistics is required that acts as the detection algorithm of the NADS. In IVNs, all those requirements are in discussion for future deployment regardless of network anomaly detection [6,19]. Still, the availability of hardware and software components is currently limited. Therefore, we implemented the integrated approach in our simulation environment (*cf.*, Sect. 4.1).

PSFP is configured at design time for ensuring safety and security of critical traffic. It regulates normal traffic behavior, and a correctly configured PSFP will not drop frames during design compliant operation. By utilizing such distinct statistics for anomaly detection, the NADS operates without false positives [35], which is a significant advantage of the integrated approach. The absence of false positives is a solid foundation for automated countermeasures. Moreover, the performance of anomaly detection does not depend on training with large datasets but is solely influenced by the quality of the network design and sophistication of the PSFP configuration.

4 Evaluation in Simulation and Real-World Deployments

The evaluation of the proposed concepts necessitates a multifaceted approach. A simulation environment enables the exploration of new concepts, modifications to standards, and novel algorithms within a controlled setting. A test bed facilitates performance assessment in a controlled hardware environment and verifies their applicability in real-world deployments. A prototype based on a production vehicle serves to demonstrate the feasibility of the proposed concepts.

4.1 In-Car Network Simulation for Reproducible Evaluation

Simulation studies are essential for reproducible evaluations of concepts and algorithms in a controlled environment, particularly in the complex setting of the IVN. A key advantage is the ability to compare results by changing a single parameter at a time, such as topology, traffic patterns, shapers, network control algorithms, and anomaly detection algorithms.

Our simulation environment (*cf.*, Fig. 6) is based on OMNeT++, a discrete event simulator [43]. To simulate future IVNs, we use a combination of frameworks. The INET framework [43] is well-established in the networking community and provides networking modules for a full Ethernet-IP-UDP/TCP stack, including network devices, hosts, and

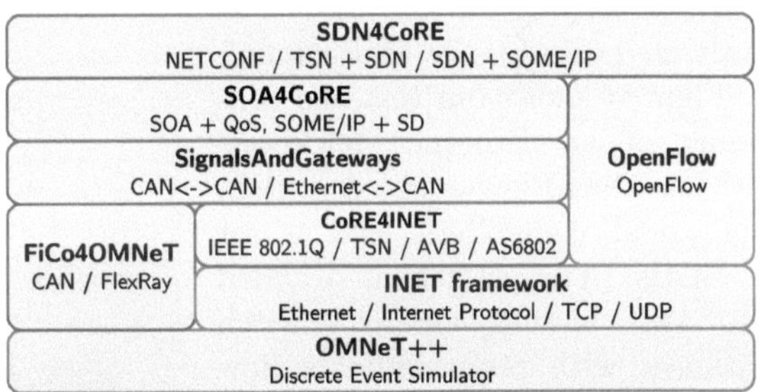

Fig. 6. In-car network simulation environment.

traffic generators. Our research group maintains a rich IVN simulation environment[1], including real-time Ethernet, bus systems such as CAN, and gateways to translation [37]. Recent extensions added Service-Oriented Architecture for Communication over Real-Time Ethernet (SOA4CoRE), and Software-Defined Networking for Communication over Real-Time Ethernet (SDN4CoRE) based on the OpenFlowOMNeTSuite [29].

A significant challenge arises from the gap between simulations and real-world implementations. The level of abstraction in the simulation models results in a trade-off between performance and realism. While our simulator accurately represents network communication and control, it does not simulate actual application behavior. To enhance realism, we incorporate traces collected in our prototype vehicle (*cf.*, Sect. 4.3) and model the communication relations of in-car ECUs. Noteworthily, not all simulated features are available in real-world implementations, and the performance of components such as the SDN controller can

---

vary significantly [47]. Further, we observed differences between standards and implementations, *e.g.*, possible combinations of shapers in TSN switches.

Simulations provide valuable insights and contribute to the advancement and validation of our research. We modeled our TSSDN architecture (*cf.*, Sect. 3.1), which is not yet available in hardware, and evaluated TSSDN performance for IVN [14,15]. We successfully implemented protocol extensions, *e.g.*, anomaly detection in the TSN ingress control [35] (*cf.*, Sect. 3.5) and a SOME/IP aware SDN control plane [16] (*cf.*, Sect. 3.3). To support anomaly detection, we collected network traces from our prototype vehicle [15] (*cf.*, Sect. 4.3), which were used for training, testing, and comparing ML algorithms.

4.2   Test Bed for Secure In-Vehicle Communication

Our test bed is a physical setup used to test and evaluate the performance and security of IVN designs. It enables the evaluation of security measures and validates real traffic characteristics for simulations (*cf.*, Sect. 4.3). Additionally, it serves as preparation for real car deployment (*cf.*, Sect. 4.3) and indicates hardware performance limitations.

Figure 7 shows our test bed consisting of an Ethernet backbone with an OpenFlow-enabled switch, four zonal controllers, and a set of ECUs. The switch is divided into two virtual Open vSwitch instances with programmable flow tables. Packet forwarding is controlled by a central ONOS[2] SDN controller equipped with custom applications following our concepts (*cf.*, Sect. 3.2). Each zonal controller is connected to a CAN bus that

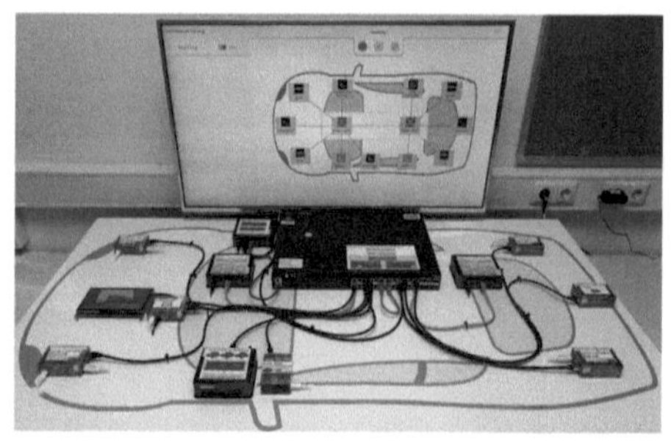

Fig. 7. IVN security test bed.

emulates messages of a production vehicle filtered for the corresponding zone. All CAN messages are tunneled through the backbone to the corresponding destination ECUs and intermediate zonal controllers. One ECU employs an IoT edge node for cloud connectivity [31], and additional ECUs generate high bandwidth traffic such as video and raw LIDAR streams. Incoming packets are mirrored in the switches to NADS instances for traffic monitoring in dedicated devices (*cf.*, Sect. 3.4). Each NADS employs ML to fingerprint communication behavior, and reports violations to the SDN controller.

Working with the test bed poses challenges in emulating real car traffic in various vehicle states. Realistic stimuli involve playing back captured data and creating additional stimuli such as LIDAR and camera streams. Finding suitable hardware, especially switches combining TSN and SDN, was difficult. Even

---

[2] Open Network Operating System (ONOS) by the Open Networking Foundation (ONF): https://opennetworking.org/onos/.

switches that implement either TSN or SDN had limitations in features and stability, restricting evaluations of concepts such as TSSDN and integrated NADS. Furthermore, real pre-compiled automotive ECUs with the AUTOSAR platform are impractical for prototyping. Consequently, we chose a reliable SDN-only switch and utilized established prototyping platforms such as the Raspberry Pi.

The test bed allows for experiments and evaluations, including long-term tests that emulate continuous operation without a physical car running 24/7 [17]. This enables comprehensive evaluation of SDN controllers [47], connectivity gateways, and anomaly detection strategies [17]. We performed penetration tests to assess vulnerabilities and potential attacks safely. Variants not supported by conventional automotive software and hardware can be evaluated, for example, security implications of different CAN-Ethernet embeddings [15,18] and gateway strategies that extend the SDN control plane to the gateways.

4.3 A Prototype with Security-Enhanced In-Vehicle Network

In collaboration with our industry partner, we integrated our concepts in a modified production vehicle (*cf.,* Fig. 8). The car includes a security-enhanced network in the trunk showcasing various aspects of the SecVI project[3].

Fig. 8. SecVI prototype with security-enhanced network installed in the trunk.

The installation follows a zone topology with four zonal controllers (left, right, front, rear) serving as gateways between the in-vehicle CAN buses and our Ethernet backbone. CAN messages are assigned to zones based on the position of the sending ECU as defined in the communication matrix of our vehicle. Additional communication along with powerful compute units for tasks such as infotainment and cloud connectivity align the setup with future cars. For monitoring and control of the Ethernet backbone, the prototype utilizes the same setup as the test bed (*cf.,* Sect. 4.2).

The integration of new concepts in the real car required significant effort, making prior evaluations through simulation and test bed essential. Adapting

---

[3] German BMBF project SecVI: [17,36].

an existing production vehicle for new concepts poses challenges, considering hardware limitations and uncertain future use cases such as ADAS, new sensor data streams, and cloud connectivity. Nonetheless, the prototype demonstrates concept applicability in real-world deployments while revealing challenges that need to be addressed before integration into production-grade vehicles.

Deploying our network control and monitoring architecture in the prototype vehicle revealed that our robust SDN architecture significantly reduces the attack surface of the IVN by limiting communication flows to the necessary ones while enhancing adaptability [15, 18]. Our ML-based NADS effectively detects attacks in real-world deployments with minimal false positives for control and video traffic [17, 36]. Additionally, our SDN successfully implements countermeasures through network reconfiguration and service mobility to combat attacks [17]. When applying direct countermeasures to anomaly reports, however, future in-car NADS should prioritize minimizing false positives over detection rates to avoid service disruptions in the vehicle.

5   Conclusion and Outlook

Future cars require a robust and secure communication infrastructure. We introduced our control and observe architecture for in-car Ethernet backbones. TSSDN enables real-time flow control, secure isolation of in-vehicle control flows, and network support for automotive protocols. A NADS provides a framework for intrusion detection through machine learning algorithms on dedicated devices, or integrated approaches using TSN ingress control (PSFP), for example.

To ensure accurate evaluations, we use a multifaceted approach: Simulations provide a controlled setting to evaluate new concepts, a test bed facilitates performance assessment in a hardware environment, and a prototype based on a production vehicle serves to demonstrate the feasibility in a real-world deployment. With this we demonstrated the effectiveness of our concepts and algorithms in multiple studies.

Several open issues need to be addressed in future work. Firstly, there is a need to optimize controllers for the vehicle use-case to ensure deterministic behavior in terms of latency and throughput [47]. Further investigation is required to develop protection mechanisms for the controller. Additionally, evaluating TSSDN in a hardware prototype is necessary. Real-time capable NADS algorithms should be developed to enhance intrusion detection, and filtering NADS false positives is important to reduce false alarms. A detailed comparison of different machine learning algorithms for various use cases and traffic types in the vehicle is needed. Furthermore, exploring incident response tailored to IVN is crucial for effectively addressing network security events. Finally, adaptive network configuration and observation are necessary for agile software development in future cars.

References

1. Alharbi, T., Portmann, M.: SProxy ARP - efficient ARP handling in SDN. In: 2016 26th International Telecommunication Networks and Applications Conference (ITNAC). IEEE (2016). https://doi.org/10.1109/atnac.2016.7878805
2. AUTomotive Open System ARchitecture (AUTOSAR) Consortium: SOME/IP Protocol Specification. Technical report. 696, AUTOSAR (2021). https://www.autosar.org/fileadmin/standards/R22-11/FO/AUTOSAR_PRS_SOMEIPProtocol.pdf
3. Bertaux, L., Hakiri, A., Medjiah, S., Berthou, P., Abdellatif, S.: A DDS/SDN based communication system for efficient support of dynamic distributed real-time applications. In: 2014 IEEE/ACM 18th International Symposium on Distributed Simulation and Real Time Applications, pp. 77–84. IEEE (2014). https://doi.org/10.1109/ds-rt.2014.18
4. Bhuyan, M.H., Bhattacharyya, D.K., Kalita, J.K.: Network anomaly detection: methods, systems and tools. IEEE Commun. Surv. Tutor. **16**(1), 303–336 (2014). https://doi.org/10.1109/SURV.2013.052213.00046
5. Bosshart, P., et al.: P4: programming protocol-independent packet processors. SIG-COMM Comput. Commun. Rev. **44**(3), 87–95 (2014). https://doi.org/10.1145/2656877.2656890
6. Brunner, S., Roder, J., Kucera, M., Waas, T.: Automotive E/E-architecture enhancements by usage of ethernet TSN. In: 2017 13th Workshop on Intelligent Solutions in Embedded Systems (WISES), pp. 9–13. IEEE (2017). https://doi.org/10.1109/WISES.2017.7986925
7. Checkoway, S., et al.: Comprehensive experimental analyses of automotive attack surfaces. In: Proceedings of the 20th USENIX Security Symposium, vol. 4, pp. 77–92. USENIX Association (2011). http://www.autosec.org/pubs/cars-usenixsec2011.pdf
8. Damm, W., et al.: Multi-layer time coherency in the development of ADAS/AD systems: design approach and tooling. In: Proceedings of the Workshop on Design Automation for CPS and IoT (DESTION 2019), pp. 20–30. ACM (2019). https://doi.org/10.1145/3313151.3313167
9. Damm, W., Fränzle, M., Gerwinn, S., Kröger, P.: Perspectives on the validation and verification of machine learning systems in the context of highly automated vehicles. In: 2018 AAAI Spring Symposia, pp. 512–515. AAAI Press (2018)
10. Damm, W., Fränzle, M., Lüdtke, A., Rieger, J.W., Trende, A., Unni, A.: Integrating neurophysiological sensors and driver models for safe and performant automated vehicle control in mixed traffic. In: Intelligent Vehicles Symposium (IV 2019), pp. 82–89. IEEE (2019). https://doi.org/10.1109/IVS.2019.8814188
11. Dibaei, M., et al.: Attacks and defences on intelligent connected vehicles: a survey. Digit. Commun. Netw. **6**(4), 399–421 (2020). https://doi.org/10.1016/j.dcan.2020.04.007
12. Enns, R., Bjorklund, M., Schoenwaelder, J., Bierman, A.: Network Configuration Protocol (NETCONF). RFC 6241, IETF (2011)
13. European Parliament, Council of the European Union: Regulation (EU) 2019/881 of the European Parliament and of the Council of 17 April 2019 on ENISA (the European Union Agency for Cybersecurity) and on information and communications technology cybersecurity certification and repealing Regulation (EU) No 526/2013 (Cybersecurity Act). Official Journal of the European Union (2019). https://eur-lex.europa.eu/eli/reg/2019/881/oj

14. Häckel, T., Meyer, P., Korf, F., Schmidt, T.C.: Software-defined networks supporting time-sensitive in-vehicular communication. In: Proceedings of the IEEE 89th Vehicular Technology Conference: VTC2019-Spring, pp. 1–5. IEEE (2019). https://doi.org/10.1109/VTCSpring.2019.8746473
15. Häckel, T., Meyer, P., Korf, F., Schmidt, T.C.: Secure time-sensitive software-defined networking in vehicles. IEEE Trans. Veh. Technol. **72**(1), 35–51 (2023). https://doi.org/10.1109/TVT.2022.3202368
16. Häckel, T., Meyer, P., Mueller, M., Schmitt-Solbrig, J., Korf, F., Schmidt, T.C.: Dynamic service-orientation for software-defined in-vehicle networks. In: Proceedings of the IEEE 97th Vehicular Technology Conference (VTC2023-Spring). IEEE (2023). https://doi.org/10.1109/VTC2023-Spring57618.2023.10199712
17. Häckel, T., et al.: A multilayered security infrastructure for connected vehicles – first lessons from the field. Presented at: 2022 IEEE Intelligent Vehicles Symposium Workshops (IV Workshops) (2022). arXiv preprint https://doi.org/10.48550/arXiv.2310.10336
18. Häckel, T., Schmidt, A., Meyer, P., Korf, F., Schmidt, T.C.: Strategies for integrating controls flows in software-defined in-vehicle networks and their impact on network security. In: 2020 IEEE Vehicular Networking Conference (VNC). IEEE (2020). https://doi.org/10.1109/VNC51378.2020.9318372
19. Haeberle, M., et al.: Softwarization of automotive E/E architectures: a software-defined networking approach. In: 2020 IEEE Vehicular Networking Conference (VNC), pp. 1–8. IEEE (2020). https://doi.org/10.1109/VNC51378.2020.9318389
20. Han, T., et al.: A comprehensive survey of security threats and their mitigation techniques for next-generation SDN controllers. Concurr. Comput. Pract. Experience **32**(16), 1–21 (2020). https://doi.org/10.1002/cpe.5300
21. Hu, Q., Luo, F.: Review of secure communication approaches for in-vehicle network. Int. J. Automot. Technol. **19**(5), 879–894 (2018). https://doi.org/10.1007/s12239-018-0085-1
22. IEEE 802.1 Working Group: IEEE Standard for Local and Metropolitan Area Network–Bridges and Bridged Networks. Standard Std 802.1Q-2018 (Revision of IEEE Std 802.1Q-2014), IEEE (2018). https://doi.org/10.1109/IEEESTD.2018.8403927
23. Institute of Electrical and Electronics Engineers: IEEE Standard for Local and Metropolitan Area Networks–Timing and Synchronization for Time-Sensitive Applications. Standard, IEEE (2020). https://doi.org/10.1109/IEEESTD.2020.9121845
24. International Organization for Standardization: Road vehicles – Cybersecurity engineering. Standard ISO/SAE DIS 21434, ISO, Geneva, CH (2020)
25. Iorio, M., Reineri, M., Risso, F., Sisto, R., Valenza, F.: Securing SOME/IP for in-vehicle service protection. IEEE Trans. Veh. Technol. **69**(11), 13450–13466 (2020). https://doi.org/10.1109/tvt.2020.3028880
26. Islam, S., Muslim, N., Atwood, J.W.: A survey on multicasting in software-defined networking. IEEE Commun. Surv. Tutor. **20**(1), 355–387 (2018). https://doi.org/10.1109/comst.2017.2776213
27. Kampmann, A., et al.: A dynamic service-oriented software architecture for highly automated vehicles. In: 2019 IEEE Intelligent Transportation Systems Conference (ITSC), pp. 2101–2108. IEEE (2019). https://doi.org/10.1109/ITSC.2019.8916841
28. Khraisat, A., Gondal, I., Vamplew, P., Kamruzzaman, J.: Survey of intrusion detection systems: techniques, datasets and challenges. Cybersecurity **2**(1), 1–22 (2019). https://doi.org/10.1186/s42400-019-0038-7

29. Klein, D., Jarschel, M.: An OpenFlow extension for the OMNeT++ INET framework. In: Proceedings of the 6th International ICST Conference on Simulation Tools and Techniques, SimuTools 2013, pp. 322–329. ICST (Institute for Computer Sciences, Social-Informatics and Telecommunications Engineering), Brussels, BEL (2013)
30. Laclau, P., Bonnet, S., Ducourthial, B., Li, X., Lin, T.: Predictive network configuration with hierarchical spectral clustering for software defined vehicles. In: Proceedings of the IEEE 97th Vehicular Technology Conference (VTC2023-Spring). IEEE (2023). https://doi.org/10.1109/VTC2023-Spring57618.2023.10199920
31. Langer, F., Schüppel, F., Stahlbock, L.: Establishing an automotive cyber defense center. In: 17th ESCAR Europe: Embedded Security in Cars (2019). https://doi.org/10.13154/294-6652
32. Leonardi, L., Bello, L.L., Patti, G.: Bandwidth partitioning for time-sensitive networking flows in automotive communications. IEEE Commun. Lett. **25**, 3258–3261 (2021). https://doi.org/10.1109/lcomm.2021.3103004
33. Matheus, K., Königseder, T.: Automotive Ethernet. Cambridge University Press, Cambridge (2015)
34. McKeown, N., et al.: OpenFlow: enabling innovation in campus networks. ACM SIGCOMM Comput. Commun. Rev. **38**(2), 69–74 (2008). https://doi.org/10.1145/1355734.1355746
35. Meyer, P., Häckel, T., Korf, F., Schmidt, T.C.: Network anomaly detection in cars based on time-sensitive ingress control. In: Proceedings of the IEEE 21th Vehicular Technology Conference: VTC2020-Fall. IEEE (2020). https://doi.org/10.1109/VTC2020-Fall49728.2020.9348746
36. Meyer, P., et al.: Demo: a security infrastructure for vehicular information using SDN, intrusion detection, and a defense center in the cloud. In: 2020 IEEE Vehicular Networking Conference (VNC). IEEE (2020). https://doi.org/10.1109/VNC51378.2020.9318351
37. Meyer, P., Korf, F., Steinbach, T., Schmidt, T.C.: Simulation of mixed critical in-vehicular networks. In: Virdis, A., Kirsche, M. (eds.) Recent Advances in Network Simulation. EICC, pp. 317–345. Springer, Cham (2019). https://doi.org/10.1007/978-3-030-12842-5_10
38. Miller, C., Valasek, C.: Remote exploitation of an unaltered passenger vehicle. Black Hat USA **2015**, 91 (2015). https://ericberthomier.fr/IMG/pdf/remote_car_hacking.pdf
39. Monteuuis, J.P., Boudguiga, A., Zhang, J., Labiod, H., Servel, A., Urien, P.: SARA: security automotive risk analysis method. In: Proceedings of the 4th ACM Workshop on Cyber-Physical System Security, CPSS 2018, pp. 3–14. ACM (2018). https://doi.org/10.1145/3198458.3198465
40. Mundhenk, P.: Security for Automotive Electrical/Electronic (E/E) Architectures. Cuvillier, Göttingen (2017). https://doi.org/10.32657/10220/45957
41. Nayak, N., et al.: Reimagining automotive service-oriented communication: a case study on programmable data planes. IEEE Veh. Technol. Mag. 69–79 (2023). https://doi.org/10.1109/mvt.2022.3225787
42. Object Management Group: Data Distribution Service. Standard DDS 1.4, OMG (2015). http://www.omg.org/spec/DDS/1.4
43. OpenSim Ltd.: OMNeT++ Discrete Event Simulator and the INET Framework. https://omnetpp.org/
44. Pekaric, I., Sauerwein, C., Haselwanter, S., Felderer, M.: A taxonomy of attack mechanisms in the automotive domain. Comput. Standards Interfaces **78**, 103539 (2021). https://doi.org/10.1016/j.csi.2021.103539

45. Pesé, M.D., Schmidt, K., Zweck, H.: Hardware/software co-design of an automotive embedded firewall. In: SAE Technical Paper. SAE International (2017). https://doi.org/10.4271/2017-01-1659
46. Rajbahadur, G.K., Malton, A.J., Walenstein, A., Hassan, A.E.: A survey of anomaly detection for connected vehicle cybersecurity and safety. In: 2018 IEEE Intelligent Vehicles Symposium (IV). IEEE (2018). https://doi.org/10.1109/ivs.2018.8500383
47. Rotermund, R., Häckel, T., Meyer, P., Korf, F., Schmidt, T.C.: Requirements analysis and performance evaluation of SDN controllers for automotive use cases. In: 2020 IEEE Vehicular Networking Conference (VNC). IEEE (2020). https://doi.org/10.1109/VNC51378.2020.9318378
48. Rumez, M., Grimm, D., Kriesten, R., Sax, E.: An overview of automotive service-oriented architectures and implications for security countermeasures. IEEE Access 8, 221852–221870 (2020). https://doi.org/10.1109/ACCESS.2020.3043070
49. Rumez, M., Duda, A., Grunder, P., Kriesten, R., Sax, E.: Integration of attribute-based access control into automotive architectures. In: 2019 IEEE Intelligent Vehicles Symposium (IV). IEEE (2019). https://doi.org/10.1109/ivs.2019.8814265
50. Steinbach, T.: Ethernet-basierte Fahrzeugnetzwerkarchitekturen für zukünftige Echtzeitsysteme im Automobil. Springer Vieweg, Wiesbaden (2018). https://doi.org/10.1007/978-3-658-23500-0
51. Thing, V.L.L., Wu, J.: Autonomous vehicle security: a taxonomy of attacks and defences. In: 2016 IEEE International Conference on Internet of Things (iThings) and IEEE Green Computing and Communications (GreenCom) and IEEE Cyber, Physical and Social Computing (CPSCom) and IEEE Smart Data (SmartData). IEEE (2016). https://doi.org/10.1109/ithings-greencom-cpscom-smartdata.2016.52
52. Waszecki, P., Mundhenk, P., Steinhorst, S., Lukasiewycz, M., Karri, R., Chakraborty, S.: Automotive electrical and electronic architecture security via distributed in-vehicle traffic monitoring. IEEE Trans. Comput. Aided Des. Integr. Circuits Syst. 36(11), 1790–1803 (2017). https://doi.org/10.1109/TCAD.2017.2666605
53. Yurekten, O., Demirci, M.: SDN-based cyber defense: a survey. Futur. Gener. Comput. Syst. 115, 126–149 (2021). https://doi.org/10.1016/j.future.2020.09.006

Finite-Memory Strategies for Petri Games

Paul Hannibal[1(✉)], Dennis Lisiecki[2], and Ernst-Rüdiger Olderog[1]

[1] Carl von Ossietzky University Oldenburg, Oldenburg, Germany
paul.jonathan.hannibal1@uni-oldenburg.de,
olderog@informatik.uni-oldenburg.de
[2] BTC Embedded Systems, Oldenburg, Germany
dennis.lisiecki@btc-embedded.com

Abstract. Petri games are a multi-player game model for distributed systems: the players are represented as tokens on a Petri net and grouped into environment players and system players. As long as the players move in independent parts of the net, they do not know of each other; when they synchronize at a joint transition, each player gets informed of the entire causal history of the other players. In the basic setting, the goal of the system players is to avoid a set of bad global states (markings) against any move of the environment. The question whether the system players have a winning strategy for achieving this goal is in general undecidable, but for a series of subclasses, it has been shown to be decidable, mostly in exponential time.

The causal history of places is represented by unfolding the underlying Petri net. A strategy is obtained by cutting decisions that are under control of the system out of the unfolding. In general, unfoldings and thus strategies are infinite. The idea for obtaining a finite winning strategy is to identify repetitions in the unfolding, where the strategy can copy previous decisions. In the context of checking the reachability of markings in Petri nets via unfoldings, *last known markings* are a suitable criterion for such repetitions. However, for Petri games this is in most cases too weak. In this paper, we propose a stronger notion of repetition, where last known markings are enhanced by equivalence classes of counters recording how many transitions one player has taken in the causal past of another player. We show that it is decidable whether a finite strategy obtained via this criterion of repetition is winning. The proof investigates a labeled transitions system corresponding to the given Petri game. The existence of a winning strategy is encoded as a SAT solving problem.

1   Introduction

The automatic synthesis of correct implementations of reactive systems from specifications is a dream that has inspired a huge amount of research and practical tool development. Of particular relevance is the synthesis of distributed reactive systems because their design is very error prone. Regrettably, these systems are "hard to synthesize" as Pnueli and Rosner proved in their setting: even

M. Fränzle et al. (Eds.): Werner Damm Festschrift, LNCS 15471, pp. 183–200, 2026.
https://doi.org/10.1007/978-3-031-97537-0_11

for seemingly simple distributed architectures the *realizability question* whether there exists an implementation satisfying a given temporal specification is undecidable [18]. The authors also show that for pipeline architectures, this question is decidable, albeit with nonelementary time complexity. Finkbeiner and Schewe sharpened these results by showing that the realizability question is undecidable if and only if the architecture has a so-called *information fork* [12,19].

Synthesis can be studied in terms of *games* because they model the interaction between a computer system and its environment. Specifications are interpreted as winning conditions, implementations as strategies. An implementation is correct if the strategy is *winning*, i.e., it ensures that the specification is met for all possible behaviors of the environment. Algorithms that determine the winner in the game between the system and its environment can be used to determine whether it is possible to implement a specification (*realizability* problem) and, if the answer is yes, to automatically construct a correct implementation (*synthesis* problem).

Searching for an alternative model of distributed reactive systems for which realizability can be solved with a better complexity, B. Finkbeiner and the third author introduced Petri games [11]. These are a multi-player game model for reactive distributed systems, where the players are represented as tokens on a Petri net and grouped into environment players and system players. Characteristic for Petri games is level of informedness of the players: as long as they move in independent parts of the net, they do not know of each other; when they synchronize at a joint transition, each player gets informed of the entire causal history of the other players. Formally, the causal history of places is represented by unfolding the underlying Petri net. A strategy is obtained by cutting decisions that are under control of the system out of the unfolding, with the aim of achieving the winning condition. In the simplest case, this is a local safety condition that forbids to visit certain 'bad' places of the Petri net.

Decidability results have been achieved for Petri games with restrictions on the number of players. In [11], it was shown that for Petri games with one environment player and a bounded number of system players, the question whether the system players have a deadlock avoiding winning strategy for a local safety winning condition is EXPTIME-complete. The proof proceeds by a reduction to a two-player Büchi game on a graph that in its nodes represents the causal information of the Petri game. A winning strategy for the Büchi game can be translated back into a winning strategy of the Petri game. In the meantime further results on (un-)decidability for Petri games have been obtained, e.g., [8,10,13,17], and tool support for the synthesis has been developed [9,14].

Asynchronous automata are another model for distributed reactive systems introduced by W. Zielonka [20]. Also here, various results on decidability and undecidability have been obtained, e.g., [15,16]. It turned out that Petri games can be translated into asynchronous automata such that their winning strategies are weakly bisimilar, albeit with an exponential blow-up of the places [2]. There is no tool support for asynchronous automata.

In this paper, we pursue an approach to finding winning strategies for Petri games that is applicable to an arbitrary number of system and environment players. We start by analyzing the unfolding of the Petri game directly. Note that for games with loops, this is infinite. The idea for obtaining a winning strategy with a finite memory is to identify repetitions in the unfolding, where the strategy can copy previous decisions. In the context of checking the reachability of markings in Petri nets via unfoldings, *last known markings* are a suitable criterion for such repetitions [5]. However, for Petri games this is in most cases too weak. In this paper, we propose a stronger notion of repetition, where last known markings are enhanced by equivalence classes of counters recording how many transitions one player has taken in the causal past of another player. We show that it is decidable whether a finite strategy obtained via this criterion of repetition is winning. The proof investigates a labeled transitions system corresponding to the given Petri game. The existence of a winning strategy is encoded as a SAT solving problem.

Our approach is related to *bounded synthesis* for Petri games [7,17], which is a semi-decision procedure to find small winning strategies for the system players. For a given bound b, this synthesis starts its search for a strategy from a less informative, 'bounded' unfolding that limits the number of copies of a place to bound b. If no strategy is found for a given bound, its value is increased.

Our paper is organized as follows. In Sect. 2 we recall the foundations of Petri games. In Sect. 3 we detail our approach to obtaining strategies with finite memory. In Sect. 4 we encode the approach as a SAT solving problem to obtain tool support. In Sect. 5 we conclude the paper.

Dedication. We dedicate our paper to Werner Damm. Ernst-Rüdiger met Werner for the first time in January 1979 during the Seminar "Formale Methoden und mathematische Hilfsmittel für die Softwarekonstruktion organized by Hans Langmaack, Erich J. Neuhold, and Manfred Paul at the Mathematical Research Institute in Oberwolfach, beautifully situated in the Black Forest, Germany[1]. Werner was then at the RWTH Aachen and Ernst-Rüdiger at the University of Kiel. This was the start of our life-long professional and personal friendship.

About ten years later, from October 1989 onwards, we were professorial colleagues at the University of Oldenburg, now both retired. In my view, the scientific highlight during our time in Oldenburg was the acquisition of the Collaborative Research Center AVACS (Automatic Verification and Analysis of Complex System) with Oldenburg, Freiburg, and Saarbrücken as partners. Werner envisaged the topic and guided all participating scientists through the maximal duration of 12 years of the Research Center from 2004 until 2015 [1]. It led to many wonderful cooperations with him, in particular within the project "Automatic Verification of Cooperating Traffic Agents" in the area of hybrid systems.

Werner's research interest span many topics, among them the Synthesis of Distributed Systems [3]. That is why we dedicate this paper to him. We thank Werner for shaping the scientific landscape of the Department for Computing Science in Oldenburg and wish him and his family all the best for the future.

---

[1] See https://www.mfo.de/occasion/7902/www_view.

2   Foundations

In this section, we define branching processes and unfoldings as in [4]. Also, we define Petri games and their winning strategies as in [11].

Some notation: the *power set* of a set A is denoted by $2^A = \{B \mid B \subseteq A\}$, the *set of nonempty finite subsets* of A by $2^A_{nf} = \{B \mid B \subseteq A$ and $B \neq \emptyset$ and finite$\}$, and the *set of finite subsets* of A by 2^A_f.

A *Petri net* or simply *net* is a structure $N = (\mathcal{P}, \mathcal{T}, pre, post, In)$, where $\mathcal{P}$ is the (possibly infinite) set of *places*, $\mathcal{T}$ is the (possibly infinite) set of *transitions*, *pre* and *post* are flow mappings, $In \subseteq \mathcal{P}$ is the *initial marking*, and the following properties hold: $\mathcal{P} \cap \mathcal{T} = \emptyset$, $pre : \mathcal{T} \rightarrow 2^{\mathcal{P}}_{nf}$, $post : \mathcal{T} \rightarrow 2^{\mathcal{P}}_f$. A Petri net is called finite if $\mathcal{P} \cup \mathcal{T}$ is a finite set. The flow mappings *pre* and *post* are extended to places as usual: $\forall p \in \mathcal{P} : pre(p) = \{t \in \mathcal{T} \mid p \in post(t)\}$ and $\forall p \in \mathcal{P} : post(p) = \{t \in \mathcal{T} \mid p \in pre(t)\}$. A *marking* M of a Petri net $\mathcal{N}$ is a multiset over $\mathcal{P}$. In particular, In is a marking. By convention, a net named N has the components $(\mathcal{P}, \mathcal{T}, pre, post, In)$, and analogously for net with decorated names like N_0, N_1, N_2.

A transition $t \in \mathcal{T}$ is *enabled* at marking M if $pre(t) \subseteq M$. If t is enabled, the transition t can be *fired*, such that the new marking is $M' = M - pre(t) + post(t)$ (standard multi-set operations). This is denoted by $M|t\rangle M'$. The marking M' is also denoted by $M|t\rangle$. This notation is extended to sequences of enabled transitions $M|t_1 \ldots t_n\rangle M'$ and $M|t_1 \ldots t_n\rangle$, respectively. A marking M is *reachable* if there exists a sequence of enabled transitions $(t_k)_{k=\{1,\ldots,n\}}$ and $In|t_1 \ldots t_n\rangle M$. This sequence can be empty. The set of all reachable markings of a net N is denoted as $\mathcal{R}(N)$. A Petri net N is called *safe*, if for all reachable markings $M(p) \leq 1$ holds for all $p \in \mathcal{P}$. Then, M is a subset of P.

A *node* x is a place or a transition $x \in \mathcal{P} \cup \mathcal{T}$. The binary *flow relation* $\mathcal{F}$ on nodes is defined as follows: $x \mathcal{F} y$ if $x \in pre(y)$. A node $x \in \mathcal{P} \cup \mathcal{T}$ is a *causal predecessor* of y, denoted as $x \leq y$, if $x \mathcal{F}^+ y$. We write $x < y$ if $x \leq y$ and $x \neq y$ holds. Furthermore, $x \leq x$ holds for all $x \in \mathcal{P} \cup \mathcal{T}$. Two nodes $x, y \in \mathcal{P} \cup \mathcal{T}$ are *causally related*, if $x \leq y$ or $y \leq x$ holds. We say x is a *causal successor* of y, if $y \leq x$ holds.

Two nodes $x_1, x_2 \in \mathcal{P} \cup \mathcal{T}$ are *in conflict*, denoted $x_1 \# x_2$, if there exist two transitions $t_1, t_2 \in \mathcal{T}$, $t_1 \neq t_2$ with $pre(t_1) \cap pre(t_2) \neq \emptyset$ and $t_i \leq x_i$, $i = 1, 2$. A node $x \in \mathcal{P} \cup \mathcal{T}$ is in self-conflict if $x \# x$. Informally speaking, two nodes are in conflict if two transitions exist that share some place in their presets and each node is a causal successor of one of those transitions. Two nodes $x, y \in \mathcal{P} \cup \mathcal{T}$ are *concurrent*, denoted $x \| y$, if they are neither causally related nor in conflict.

A Petri net N is *finitely preceded*, if for every node $x \in \mathcal{P} \cup \mathcal{T}$ the set $\{y \in \mathcal{P} \cup \mathcal{T} \mid y \leq x\}$ is finite. A Petri net N is *acyclic*, if the directed graph $(\mathcal{P} \cup \mathcal{T}, \mathcal{F})$ is acyclic. The following definitions lead to the definition of a branching process.

An *occurrence net* is a Petri net N with the following properties: N is acyclic, finitely preceded, $\forall p \in \mathcal{P} : |pre(p)| \leq 1$, no transition $t \in \mathcal{T}$ is in self-conflict, and $In = \{p \in \mathcal{P} \mid pre(p) = \emptyset\}$.

A homomorphism from one Petri net to another maps each node to a node such that the preset and postset relations are preserved including the initial marking. Formally, let N_1 and N_2 be two Petri nets. Then a *homomorphism* from N_1 to N_2 is a mapping $h : \mathcal{P}_1 \cup \mathcal{T}_1 \rightarrow \mathcal{P}_2 \cup \mathcal{T}_2$ with following properties: $h(\mathcal{P}_1) \subseteq \mathcal{P}_2$ and $h(\mathcal{T}_1) \subseteq \mathcal{T}_2$, for all transitions $t \in \mathcal{T}_1$, h restricted to $pre_1(t)$ is a bijection between $pre_1(t)$ and $pre_2(h(t))$, for all transitions $t \in \mathcal{T}_1$, h restricted to $post_1(t)$ is a bijection between $post_1(t)$ and $post_2(h(t))$, and the restriction of h to In_1 is a bijection between In_1 and In_2. An *isomorphism* is a bijective homomorphism.

A single computation of a Petri net is represented by a (possibly concurrent) run, which is obtained by resolving all nondeterministic choices while firing transitions. A branching process of a Petri net represents (possibly) multiple runs of the underlying Petri net, thereby keeping nondeterministic choices.

Branching Process. A *branching process* of a net N_0 is a pair $B = (N, \pi)$, where N is an occurrence net and π a homomorphism from N to N_0 such that:

(*) For all $t_1, t_2 \in \mathcal{T}$: if $pre(t_1) = pre(t_2)$ and $\pi(t_1) = \pi(t_2)$, then $t_1 = t_2$.

An example of a Petri net and a branching process is shown in Fig. 1. The property (*) of the definition of a branching process ensures that every run of the Petri net is represented at most once. Informally speaking, a run only consists of concurrent and causally related nodes and a node can be part of multiple runs. Nodes that are in conflict, cannot belong to the same run.

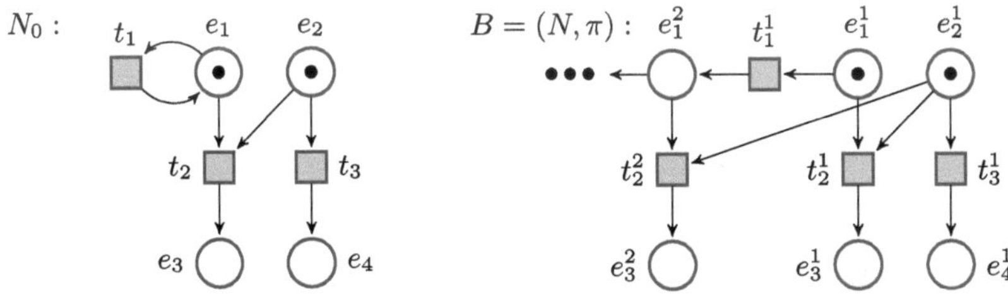

Fig. 1. A Petri net N_0 on the left and a branching process $B = (N, \pi)$ of N_0 on the right. Places are shown as circles, transitions as boxes and the preset and postset relations as arrows. The initial marking $\{e_1, e_2\}$ is represented by the black dots, the *tokens*. The homomorphism π from N to N_0 is given as $\pi(e_j^i) = e_j$ and $\pi(t_j^i) = t_j$. The transitions t_1^1 and t_2^1 are in conflict, i.e., $t_1^1 \# t_2^1$, and also $t_2^2 \# t_2^1$, $t_3^1 \# t_2^1$, $t_2^2 \# t_3^1$. Note that N exhibits infinitely many nondeterministic choices among the instances $t_2^1, t_2^2, \ldots$ of transition t_2.

Homomorphism on Branching Processes. Given two branching processes $B_1 = (N_1, \pi_1)$ and $B_2 = (N_2, \pi_2)$ of a Petri net N_0. A *homomorphism* from B_1 to B_2 is a homomorphism h from N_1 to N_2 such that $\pi_2 \circ h = \pi_1$. The branching processes B_1 and B_2 are *isomorphic* if there exists an isomorphism from B_1 to B_2 which is denoted as $B_1 \cong B_2$.

A natural partial order on branching processes is defined in the following.

188   P. Hannibal et al.

Subprocess Relation of Branching Processes. Let B_1 and B_2 be two branching processes of a Petri net N. Then B_1 approximates B_2, denoted by $B_1 \leq B_2$, if there exists an injective homomorphism from B_1 to B_2.

Now we define the unfolding of a Petri net as the maximal branching process that contains all (possibly infinite) runs of a Petri net.

Unfolding. The *unfolding* $unf(N_0)$ of a Petri net N_0 is the maximal branching process with respect to the subprocess relation $\leq$ of branching processes. This definition is unique up to isomorphism. We refer to the components of the unfolding as $\mathcal{T}_{unf(N_0)}$, $\mathcal{P}_{unf(N_0)}$, $pre_{unf(N_0)}$, $post_{unf(N_0)}$, and $In_{unf(N_0)}$.

In Fig. 1, the branching process B is the unfolding of the Petri net assuming that the dots to the left of the place e_1^2 indicate that the branching process continues infinitely in the same way.

We continue with the definition of a Petri game. We only consider Petri games on finite and safe Petri nets in this paper. We allow tokens to transit from a system place to an environment place and vice versa.

Definition 1 (Petri game). *A* Petri-game *on an underlying finite and safe Petri net N_0 is a tuple $G = (\mathcal{P}_0^S, \mathcal{P}_0^E, \mathcal{T}_0, pre_0, post_0, In_0, \mathcal{B})$, where the set $\mathcal{P}_0$ of places of N_0 are partitioned into disjoint sets of* system places, $\mathcal{P}_0^S$, *and* environment places, $\mathcal{P}_0^E$, *and where $\mathcal{B} \in 2^{\mathcal{P}_0}$ is the set of* bad markings.

Figure 2 shows an example of a Petri game in graphical representation.

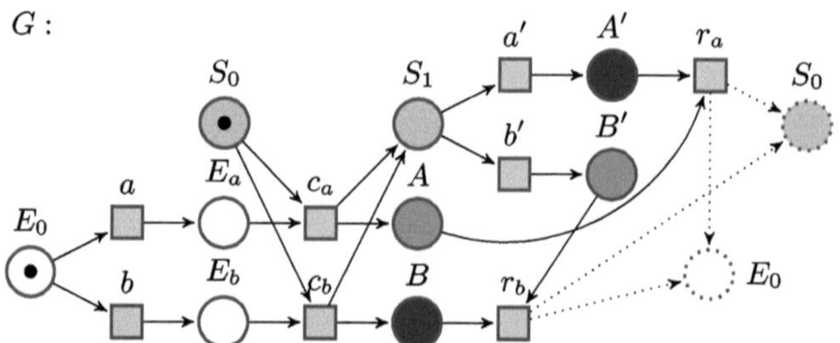

Fig. 2. A Petri game G. System places are shown in grey, environment places are kept white. The environment player chooses between the transitions a and b. The system player, after communicating with the environment player, must perform a corresponding choice between a' and b' to avoid the bad markings $\{A, B'\}$ (in red) and $\{B, A'\}$ (in purple). After the reset transitions r_a and r_b the players go back into the initial marking $\{S_0, E_0\}$. (The dotted places named E_0 and S_0 are identified with the initial places E_0 and S_0, to avoid drawing loops). (Color figure online)

Now we define a winning strategy of a Petri game as a branching process of the underlying Petri net of the game that satisfies four properties.

Definition 2 (Winning strategy). *A winning strategy σ_G of a Petri-game $G = (\mathcal{P}_0^S, \mathcal{P}_0^E, \mathcal{T}_0, pre_0, post_0, In_0, \mathcal{B})$ with underlying Petri-net N_0 is a branching process $\sigma_G = (N, \pi)$ of N_0 satisfying the following properties:*

1. ***Justified refusal**: Let $C \subseteq \mathcal{P}$ be a set of pairwise concurrent places and $t \in \mathcal{T}_0$ a transition with $\pi(C) = pre_0(t)$. If no $t' \in \mathcal{T}$ with $\pi(t') = t$ and $pre(t') = C$ exists, then there exists a place $p \in C$ with $\pi(p) \in \mathcal{P}_0^S$, such that $t \notin \pi(post(p))$.*
2. ***Safety**: For all reachable markings $M \in \mathcal{R}(N)$ it holds that $\pi(M) \notin \mathcal{B}$.*
3. ***Determinism**: For all $p \in \mathcal{P}$ with $\pi(p) \in \mathcal{P}_0^S$ and for all reachable markings M in N with $p \in M$ there exists at most one transition $t \in post(p)$, which is enabled in M.*
4. ***Deadlock avoiding**: For all reachable markings M in N there exists an enabled transition, if a transition is enabled in $\pi(M)$ in the underlying Petri-net N_0.*

We refer to a token on a system place as a system player, and a token on an environment place as an environment player. The *justified refusal* property ensures that a system player allows all instances of an outgoing transition or no instance at all. The *safety* property ensures that no bad markings are reachable. The *determinism* property ensures that for each system place at most one transition is enabled in every reachable marking. The *deadlock avoiding* property ensures that the system allows at least one transition in every reachable marking if an enabled transition exists in that marking.

The *unfolding* $unf(G)$ of a Petri game G is like the unfolding $unf(N_0) = (N, \pi)$ of the underlying Petri net N_0 of G, additionally keeping the distinction between system and environment: a place p in N is a system place if $\pi(p) \in \mathcal{P}_0^S$ and an environment place if $\pi(p) \in \mathcal{P}_0^E$. A winning strategy σ_G of G can be seen as a subprocess of $unf(G)$. In Fig. 3, an initial part of the unfolding of Petri game G in Fig. 2 is shown and inside as a subprocess, an initial part of a winning strategy is obtained by deleting the dashed arrows.

3   Finite Memory

In this section, we introduce the basic concepts of the finite memory used by the strategies we are looking for. Restricting the (possibly infinite) causal memory to a finite part allows us to construct a finite labelled transition system (LTS), in which we can effectively search for winning strategies via a SAT-encoding later in Sect. 4.

Since all players have their own causal memory representing their local view of the game, all players have their own finite memory. We restrict a player's finite memory to the marking reached by firing all transitions in her causal past. More precisely, we define the *last known cut* of a place x in a branching process of a Petri net as the marking that we reach from the initial marking by firing all transitions in the past of x. The finite memory is then the marking in the underlying Petri net which is defined as the *last known marking*.

$unf(G)$ and σ_G :

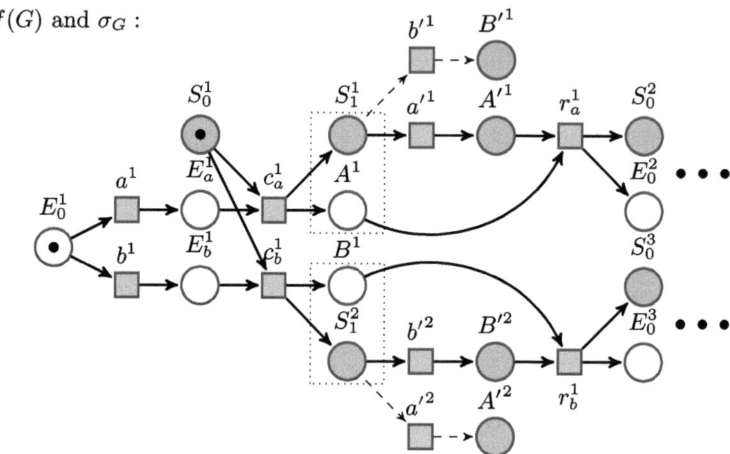

Fig. 3. An initial part of the unfolding $unf(G)$ of Petri game G in Fig. 2. Deleting the dashed arrows from the unfolding yields an initial part of a winning strategy σ_G. The homomorphism π from $unf(G)$ to G is given by dropping the superscripts of the names of places and transitions, e.g., $\pi(S_1^1) = \pi(S_1^2) = S_1$. The set of places in the upper dotted box $\{S_1^1, A^1\}$ is the last known cut of the place S_1^1 (and also A^1), which is defined later in Sect. 3 together with last known markings. The last known marking of the place S_1^1 is $\{S_1, A\}$. Analogous applies to the dotted lower box.

Definition 3 (Last known cut). *Consider a branching process $B = (N, \pi)$ and a place $x \in \mathcal{P}$. The* last known cut $LKC(x)$ *of x is defined as $LKC(x) = \{p \in \mathcal{P} \mid p \not< x \wedge \forall t \in pre(p) : t < x\}$. The* last known marking *of x is defined as $LKM(x) = \pi(LKC(x))$.*

This definition is equal to the definition of the cut of a local configuration in [6].

Note that $|pre(p)| \leq 1$ because N is an occurrence net and that $x \in LKC(x)$. We generally refer to markings in a branching process as *cuts*, which are maximal sets of places that are pairwise concurrent.

In Fig. 3, an example of a last known cut and its last known marking is shown. The last known markings $LKM(S_1^1) = \{S_1, A\}$ and $LKM(S_1^2) = \{S_1, B\}$ differ. The state of a player's finite memory is her last known marking. The idea of a strategy with finite memory is that a player's decisions are made based on the state of her finite memory. Since the states differ, the player is allowed to make different decisions, e.g. in place S_1^1 such a strategy may take transition a' and in S_1^2 transition b'. This also means that every time the last known marking of a place S_1^i is $\{S_1, A\}$ the previous decision to take transition a' must be repeated.

For formal simplicity reasons, we require the Petri net of the Petri game to keep a fixed number of tokens and each token moves only within its own partition.

Concurrency Preserving. A Petri net N is *concurrency preserving* if and only if $\forall t \in \mathcal{T}_N : |pre(t)| = |post(t)|$ holds. Furthermore, N is called *partitioning preserving* if N is concurrency preserving and there exists a partition $P = \bigcup_{i=1,\dots,n} P_i$ of $\mathcal{P}_N$ such that $\forall t \in \mathcal{T}_N : \forall i \in \{1,\dots,n\} : |pre(t) \cap P_i| = |post(t) \cap P_i|$. Such a partition P is called a *player partition* if and only if $|P_i \cap In_N| = 1$ for all $i = 1,\dots,n$. Then, N is *player-partition preserving*.

We denote the set of indices $\{1,\dots,n\}$ of a partition P as I_P. If P is player-partition preserving the indices refer to the players. For a set of places A, if it holds that $|A \cap P_i| = 1$ we write $p_i(A)$ for the single place $p_i \in A \cap P_i$. We extend a partition P of a Petri net N to the the places of the unfolding $\mathcal{P}_{unf(N)}$ as follows: $\forall p \in \mathcal{P}_{unf(N)} : p \in P_i \Leftrightarrow \pi(p) \in P_i$. We say a player i participates in a transition t if and only if $pre(t) \cap P_i \neq \emptyset$.

The last known markings are not a sufficient criterion to guarantee that the causal behaviour repeats. In particular, it is possible that in two cuts, where all players have equal last known markings, the last known markings differ after firing transitions t and t' with $\pi(t) = \pi(t')$, respectively, e.g., after firing the same transition in the underlying Petri net. In Fig. 4 is an example of this behaviour. After firing f^1 in the branching process B_1 and f' in the branching process B_2 the last known markings differ. Therefore, counters are introduced that capture

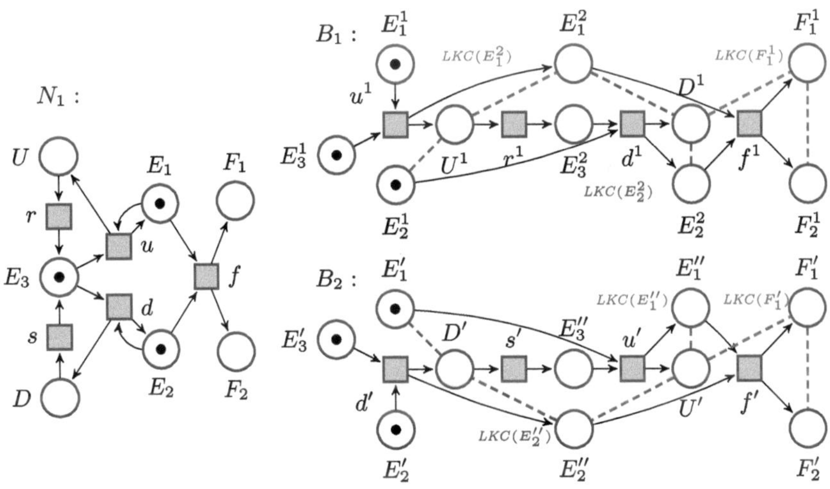

Fig. 4. A Petri net N_1 on the left and two branching processes B_1 and B_2 of N_1 on the right. The homomorphisms π_1 and π_2 from B_1 and B_2 to N_1 are given by dropping the superscripts, e.g., $\pi_1(E_1^1) = \pi_1(E_1^2) = E_1$. The partition of N_1 is given by $P_1 = \{E_1, F_1\}$, $P_2 = \{E_2, F_2\}$ and $P_3 = \{E_3, U, D\}$. In B_1 and B_2, the following last known markings are the same: $LKM(E_1^2) = LKM(E_1'') = \{U, E_1, E_2\}$ and $LKM(E_2^2) = LKM(E_2'') = \{D, E_1, E_2\}$. However, the last known markings $LKM(F_1^1) = \{D, F_1, F_2\}$ and $LKM(F_1') = \{U, F_1, F_2\}$ differ after firing f^1 and f' respectively. This shows that the last known markings are not a sufficient criterion for repetition.

the different possibilities of the causal successor relation between the places in the last known markings.

Definition 4 (Counter equivalence classes). *We define an equivalence relation $\sim \;\subseteq\; \mathbb{N}^n \times \mathbb{N}^n$ as follows. $(C_1, \ldots, C_n) \sim (C'_1, \ldots, C'_n) \Leftrightarrow \forall i, j \in \{1, \ldots, n\}:$ $C_i = C_j \Leftrightarrow C'_i = C'_j \wedge C_i < C_j \Leftrightarrow C'_i < C'_j$. The set of equivalence classes is denoted as EQ. The equivalence class of a tuple of numbers $(C_1, \ldots, C_n)$ is denoted as $[C_1, \ldots, C_n]$.*

For example, $[3, 2, 2, 4] = [5, 1, 1, 8]$. Now, we can define the LTS of a Petri game that restricts the memory to the players' last known markings. A state in the LTS can be associated with a cut in the unfolding, additionally storing the last known markings of each player and a counter equivalence class for each player.

Every player j has a counter for every player i showing how many transitions player i has taken in the causal memory of player j. We denote this counter as C^i_j. The counter equivalence class $\equiv_i \,\in EQ$ for a player i consists of one counter of each player counting the transitions of player i, i.e., it is of the form $\equiv_i = [C^i_1, C^i_2, \ldots, C^i_n]$. For example in Fig. 4, player 1 has seen 1 transition of player 3 on place E^2_1, thus $C^3_1 = 1$, and player 2 on place E^2_2 has seen 3 transitions of player 3, thus $C^3_2 = 3$. Therefore, player 3's place in the last known marking of place F^1_1 is place D of the last known marking of player 2 since she has seen more transitions of player 3 than player 1 has seen.

The constructed LTS is an extended reachability graph. However, the reached marking is not stored directly; it can be reconstructed from the last known markings since a player's own place in her last known marking is always part of the reached marking. The equivalence classes of counters are necessary overhead to determine the last known markings correctly. We choose an LTS as a representation since the states of an LTS are suitable for storing this finite set of information. Opposed to the unfolding, the players in the LTS have complete information (about the finite parts that are stored). Later, the SAT encoding ensures that a player's decisions are based solely on her own last known marking.

Definition 5 (Petri game to LTS). *Let $G = (\mathcal{P}^S, \mathcal{P}^E, \mathcal{T}, pre, post, In, \mathcal{B})$ be a player-partition preserving Petri game with partition $P = \bigcup_{i=1,\ldots,n} P_i$ of the Petri net N. We define the LTS $T_G = (Q, q_I, \Sigma, \delta)$ with last-known-marking-memory as follows:*

- *$Q = \mathcal{R}(N)^n \times EQ^n$ is the set of states. We refer to the marking of a state $q \in Q$ as $M(q) = \{p_i(LKM_i) \mid i \in I_P\}$.*
- *$q_I = (In, \ldots, In, [0, \ldots, 0], \ldots, [0, \ldots, 0]) \in Q$ is the initial state.*
- *$\Sigma = \mathcal{T}$ is the set of inputs.*
- *$\delta : Q \times \Sigma \rightharpoonup Q$ is a partial transition function, defined in (q, t) if and only if the transition t is enabled in $M(q)$. Then it is defined as:*

$$\delta((LKM_1, \ldots, LKM_n, \equiv_1, \ldots, \equiv_n), t) = (LKM'_1, \ldots, LKM'_n, \equiv'_1, \ldots, \equiv'_n),$$

where

$$LKM'_i = \{p'_1, \ldots, p'_n\}, p'_j = \begin{cases} p_j(LKM_i) & P_i \cap pre(t) = \emptyset \\ p_j(post(t)) & P_i \cap pre(t) \neq \emptyset \wedge P_j \cap pre(t) \neq \emptyset \\ p_j(LKM_k) & P_i \cap pre(t) \neq \emptyset \wedge P_j \cap pre(t) = \emptyset \\ \qquad \wedge C^j_k = max_{l:P_l \cap pre(t) \neq \emptyset}\{C^j_l\} \end{cases}$$

and $\equiv'_i = [C^{i}_1{}', \ldots, C^{i}_n{}']$, *with*

$$C^{i}_j{}' = \begin{cases} C^i_j & P_j \cap pre(t) = \emptyset \\ max\{C^i_j \mid P_j \cap pre(t) \neq \emptyset\} & P_j \cap pre(t) \neq \emptyset \wedge P_i \cap pre(t) = \emptyset \\ C^i_i + 1 & P_j \cap pre(t) \neq \emptyset \wedge P_i \cap pre(t) \neq \emptyset \end{cases}$$

We also refer to δ as a subset $\delta \subseteq Q \times T \times Q$.

The marking $M(q)$ consists of the players' own places in their last known marking. This equals the marking reached by firing the transitions leading to q.

LKM'_i is the updated last known marking of player i after transition t. We explain the three different cases of how to update a last known marking after firing a transition t.

In the first case, player i does not participate in t. Therefore the place of player j (and all other places in her last known marking) remains the same.

In the second case, both players i and j participate in t, so the place of player j in the last known marking of player i is set to the place of player j in the postset of t.

In the third case, player i participates in t and player j does not. Here, we look at the counters of all participating players and the player with the maximum counter has the most recent place (i.e. maximal with respect to the successor relation) of player j in its last known marking. Since player j does not participate, the place of player j is updated to the most recent place of the participating players. Note that the counters C^j_l are the counters before the update.

We refer to the last known marking LKM_i of a state $q \in Q$ as $q.LKM_i$.

C^i_j is the counter of how many transitions player i has taken in the causal memory of player j. We explain the three cases of how the counter $C^{i}_j{}'$ is updated after a transition t.

In the first case, if player j does not participate in a transition her counter remains the same.

In the second case, player j participates in t and player i does not participate in t. Then, the counter of how many transitions player i has taken in the causal memory of player j is updated to the maximum of the counters of the participating players.

In the third case, both players i and j participate in t. The counter C^i_i is always a maximal counter in an equivalence class of counters because the player i always has full knowledge about her own causal memory. So, the counter $C^{i}_j{}'$ is set to $C^i_i + 1$.

We show that the last known markings in the LTS coincide with the last known markings in the unfolding.

Theorem 1 (LTS and Petri game coherence). *Let $t_1, \ldots, t_n \in \mathcal{T}_{unf}$ be a firing sequence in $unf(G)$, $In[t_1 \ldots t_{n-1}\rangle M_{n-1}[t_n\rangle M_n$ the reached markings in $unf(G)$ and $g_0 \xrightarrow{\pi(t_1)} \cdots \xrightarrow{\pi(t_{n-1})} q_{n-1} \xrightarrow{\pi(t_n)} q_n$ the transition sequence of the LTS. Then the following holds:*

1. $\forall i \in P_I : LKM(p_i(M_n)) = q_n.LKM_i$ *and*
2. $\forall i \in P_I : C_j^i < C_k^i \Leftrightarrow p_i(LKC(p_j(M_n))) < p_i(LKC(p_k(M_n))) \wedge C_j^i = C_k^i \Leftrightarrow p_i(LKC(p_j(M_n))) = p_i(LKC(p_k(M_n)))$

Explanation *1. says that the last known marking of each player in a state of the LTS is the same as the last known marking of this player in the unfolding of the Petri game G after firing $\pi(t_1), \ldots, \pi(t_n)$ and $t_1, \ldots, t_n$, respectively. 2. says that the counter relation $\equiv_i$ in a state in the LTS coincides with the causal successor relation of the places of player i in the last known cuts of each player.*

Proof. We prove the correctness of the construction by induction over the sequences of transitions.

Starting with the empty transition sequence, the initial state of the LTS is $q_I = (In, \ldots, In, [0, \ldots, 0], \ldots, [0, \ldots, 0])$ and 1. and 2. hold.

Induction step: Let $t_1, \ldots, t_n \in \mathcal{T}_{unf}$ be a sequence of transitions in $unf(G)$ and $In[t_1 \ldots t_{n-1}\rangle M_{n-1}[t_n\rangle M_n$ the reached markings in $unf(G)$ and $g_0 \xrightarrow{\pi(t_1)} \cdots \xrightarrow{\pi(t_{n-1})} q_{n-1} \xrightarrow{\pi(t_n)} q_n$ the transition sequence of the LTS.

According to the induction assumption 1. and 2. hold for the pair M_{n-1} and q_{n-1}.

To show 1.: We distinguish the three cases of how the place $p_j(q_{n-1}.LKM_i)$ is updated by firing t_n. We refer to the original place as $p_j = p_j(q_{n-1}.LKM_i)$ and the updated place as $p_j' = p_j(q_n.LKM_i)$.

First case: player i does not participate in t_n such that $LKM(p_i(M_n)) = LKM(p_i(M_{n-1}))$. Therefore, it is correct that $p_j'(q_n.LKM_i) = p_j(q_{n-1}.LKM_i)$ remains unchanged.

Second case: both players i and j participate. As per definition of the last known cut it holds that $post(t_n) \subseteq LKC(p_i(M_n))$ (in the unfolding). Therefore, $p_j(LKM(p_i(M_n)))$ is the place of player j in the postset of $\pi(t_n)$, which is equal to the definition $p_j' = p_j(post(\pi(t_n)))$.

Third case: player i participates in t_n and player j does not. The counter $C_k^j = max_{l:P_l \cap pre(\pi(t_n)) \neq \emptyset}\{C_l^j\}$ is defined as the maximum of the counters of all players participating in t_n. This player is player k. According to 2. of the induction assumption the order of the counters refers to the causal successor relation. Thus, $p_j' = p_j(q_{n-1}.LKM_k)$ is equal to $p_j(LKM(p_i(M_n)))$.

To show 2.: We distinguish the cases of how two counters C_j^i and C_k^i are updated to $C_j^{i\prime}$ and $C_k^{i\prime}$, respectively.

First case: player j and player k do not participate in t_n. Then, both counters remain unchanged and 2. holds by the induction assumption.

Second case: w.l.o.g. player j participates in t_n and player k does not. The counter $C_k^{i\prime} = C_k^i$ remains unchanged. There are two subcases to distinguish:

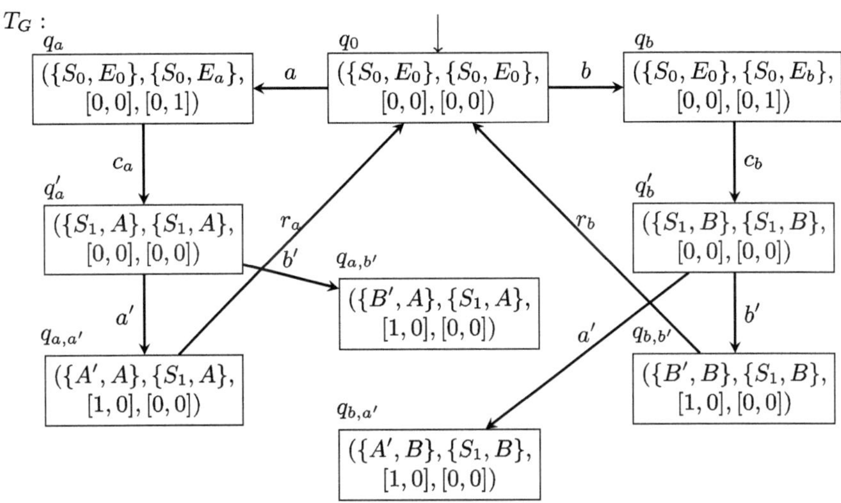

Fig. 5. LTS T_G of the Petri game G of Fig. 2. The partition P consists of $P_1 = \mathcal{P}^S$ and $P_2 = \mathcal{P}^E$. We have assigned additional labels to the states (e.g. q_a) to refer to them in the SAT encoding. For example, in q_a the last known marking of the system player is $\{S_0, E_0\}$ and the last known marking of the environment player is $\{S_0, E_a\}$. The counters from left to right: number of transitions taken by the system player in the causal memory of the system player, number of transitions taken by the system player in the causal memory of the environment player, number of transitions taken by the environment player in the causal memory of the system player and the number of transitions taken by the environment player in the causal memory of the environment player, which is 1 since a is a purely environmental transition. Note that $[0,0] = [1,1] = [2,2]$ in q_a' after firing c_a.

In the first subcase, player i does not participate in t_n. Here, $C_j^{i'} = max\{C_j^i \mid P_j \cap pre(t) \neq \emptyset\}$ is the maximum of the counters of the players participating in t_n. Let l be a player with a maximum counter in $\{C_j^i \mid P_j \cap pre(t) \neq \emptyset\}$. By induction assumption the relation of the counters used in $max\{C_j^i \mid P_j \cap pre(t) \neq \emptyset\}$ and $C_k^{i'} = C_k^i$ coincides with the causal successor relation. By definition of the last known cut $p_i(LKC(p_j(M_n))) = p_i(LKC(p_l(M_{n-1})))$ holds and 2. follows.

In the second subcase, player i participates in t_n. From the induction assumption follows that C_i^i is maximal, since a player always knows the exact number of transitions that she has participated in. Therefore, it holds that $C_j^{i'} = C_i^i + 1 > C_k^i = C_k^{i'}$. Since player k does not participate in t_n, it holds that $p_i(LKC(p_j(M_n))) > p_i(LKC(p_k(M_n)))$ and 2. follows here.

Third case: Player j and player k participate in t_n. Again, we distinguish between the two subcases whether player i participates in t_n or not. In the former subcase, $C_j^{i'} = max\{C_j^i \mid P_j \cap pre(t) \neq \emptyset\} = C_k^{i'}$. As both player j and player k participate in t_n it holds that $p_i(LKC(p_j(M_n))) = p_i(LKC(p_k(M_n)))$ and 2. holds. In the second subcase we get $C_j^{i'} = C_i^i + 1 = C_k^{i'}$. Here also

$p_i(LKC(p_j(M_n))) = p_i(LKC(p_k(M_n)))$ holds as all three players participate in t_n.   □

In Fig. 5, the LTS of the Petri game G of Fig. 2 is shown. A bad marking is reached in the states $q_{a,b'}$ ($\{B', A\}$) and $q_{b,a'}$ ($\{A', B\}$).

4   Petri Game as SAT

In this section, a strategy of an LTS is defined and if it is a winning strategy. Afterwards, the existence of a winning strategy is encoded as a SAT solving problem.

Strategy in LTS. Let T_G be the LTS of a Petri game G of the net N. An *LTS-strategy* σ in T_G is a family of mappings: $\sigma = \{\sigma_i \mid i \in I_P\}$, where $\sigma_i :$ $\mathcal{R}(N) \mapsto 2^{\mathcal{T}}$ and $\sigma_i(M) = \mathcal{T}$ if $p_i(M) \in \mathcal{P}^E$.

Player i's strategy σ_i of an LTS-strategy σ determines for every possible last known marking a set of transitions that are allowed to be fired. If and only if all players in a preset of a transition allow that transition, it is added to the unfolding of the LTS-strategy. Purely environmental transitions (all players in the preset are environment players) cannot be restricted.

Unfolding of LTS-Strategy. Let σ be a an LTS-strategy of T_G. The *unfolding* $unf(\sigma)$ is defined as the branching process B, such that

$$\forall t \in \mathcal{T}_{unf(G)} : t \in \mathcal{T}_{unf(\sigma)} \Leftrightarrow pre(t) \subseteq \mathcal{P}_{unf(\sigma)} \wedge \forall p_i \in pre(t) : t \in \sigma_i(LKM(p_i)).$$

Winning LTS-Strategy. An LTS-strategy σ of a Petri game G is a winning strategy if and only if $unf(\sigma)$ is a winning strategy in G.

Now, we encode the existence of a winning LTS-strategy as a SAT solving problem. Each boolean assignment of the encoding can be interpreted as an LTS-strategy.

Definition 6 (Encoding existence of winning LTS-strategy as SAT). *Let T_G be the LTS of a Petri game G of the net N. Before defining the encoding ϕ_{T_G} of the existence of a winning LTS-strategy we define several auxiliary variables.*

The set of boolean variables is $\mathcal{V}(\phi_{T_G}) = \{M_i(t) \mid M \in \mathcal{R}(N), i \in I_P, t \in \mathcal{T}\}$.

$M_i(t)$ encodes that transition t is allowed in all places of player i with last known marking M if and only if this variable is true. We require $M_i(t) = true$ if $p_i(M) \in \mathcal{P}^E$.

For all outgoing transitions t of a state $q = (LKM_1, \ldots, LKM_n, \equiv_1, \ldots, \equiv_n)$ we define $t_q = \bigwedge_{P_i \cap pre(t) \neq \emptyset} LKM_i(t)$.

We introduce auxiliary variables for the three winning properties that must hold in a state q: B_q for safety, D_q for determinism, and A_q for deadlock avoidance.

$$B_q = \begin{cases} true & M(q) \notin \mathcal{B} \\ false & else \end{cases}$$

$$D_q = \bigwedge_{(q,t,q') \in \delta \wedge p \in pre(t) \cap \mathcal{P}^S} \left(t_q \Rightarrow \bigwedge_{(q,t',q'') \in \delta \wedge p \in pre(t) \wedge t' \neq t} \neg t'_q \right)$$

$$A_q = \begin{cases} true & \neg \exists (q,t,q') \in \delta \\ \bigvee_{(q,t,q') \in \delta} t_q & else \end{cases}$$

Note that we always assume an empty conjunction to be true.

Now, we define an auxiliary variable F_q for each state q. F_q is the conjunction $F_q = B_q \wedge D_q \wedge A_q$.

The SAT-formula ϕ_{T_G} is now defined as follows: starting from the initial state q_0 the formula F_{q_0} must hold, and for every outgoing transition t that is allowed by the strategy, the formula of the subsequent state must also hold.

$$\phi_{T_G} = F_{q_0} \bigwedge_{(q,t,q') \in \delta} (F_q \wedge t_q \Rightarrow F_{q'})$$

The *justified refusal* property is always satisfied by the unfolding of an LTS-strategy since a decision of a strategy is interpreted as a set of transitions that is always allowed (in those places with the same finite memory state).

Strategy of a Boolean Assignment. We define the LTS-strategy of a boolean assignment $\mathcal{A} : \mathcal{V}(\phi_{T_G}) \mapsto \{true, false\}$ of the variables in ϕ_{T_G} as follows:

$$\forall i \in I_P : \forall M \in \mathcal{R}(N) : \sigma_i(M) = \{t \in \mathcal{T} \mid \mathcal{A}(M_i(t)) = true\}$$

Theorem 2 (LTS to SAT). *There is a SAT-formula ϕ_G such that ϕ_G is satisfiable if and only if there is a winning LTS-strategy σ in T_G.*

Proof. Let T_G be the LTS of the Petri game G and ϕ_{T_G} the SAT encoding. If there is no boolean assignment $\mathcal{A}$ of the variables satisfying ϕ_{T_G} there exists no winning LTS-strategy in G. For $\mathcal{A} \not\models \phi_{T_G}$ there exists a sequence $t^0_{q_0}, t^1_{q_1}, \ldots, t^n_{q_n}$ such that $\mathcal{A} \models t^1_{q_0} \wedge \ldots \wedge t^n_{q_n}$ and $\mathcal{A} \not\models F_{q_n}$. Let $t_0, \ldots, t_n \in \mathcal{T}_{unf(\sigma_\mathcal{A})}$ be the transition sequence in $unf(\sigma_\mathcal{A})$ such that $\pi(t_i) = t^i$ for all $i \in \{1, \ldots, n\}$. From Theorem 1 follows that $M_n = In[t_0 \ldots t_n\rangle = M(q_n)$. Since F_q encodes the properties of a winning strategy, this implicates that one of the winning properties must be violated in M_n. Thus, there is no winning LTS-strategy in G.

Let $\mathcal{A}$ be a boolean assignment of ϕ_{T_G} such that $\mathcal{A} \models \phi_{T_G}$ and $t_0, \ldots, t_n$ a firing sequence in $unf(\sigma_\mathcal{A})$. We show that the winning properties hold in the marking $M_n = In[t_0 \ldots t_n\rangle$. Let $q_0 \xrightarrow{\pi(t_0)} \ldots \xrightarrow{\pi(t_n)} q_n$ be the transition sequence in the LTS. From the encoding together with Theorem 1 follows that $\mathcal{A} \models F_{q_n} = B_{q_n} \wedge D_{q_n} \wedge A_{q_n}$. This means that the three properties safety, determinism, and deadlock avoidance hold in M_n. Note that the justified refusal property holds as Theorem 1 states that the last known markings are correct, such that a strategy always allows all instances of a transition or no instance at all. $\qquad\square$

Corollary 1. *The existence of an LTS-strategy of a Petri game G with underlying net N can be determined in NEXP in the number of players n and in NP in the size $(|\mathcal{P} \cup \mathcal{T}|)$ of the net N for a fixed number of players.*

Proof. The LTS T_G has exponentially many states: there are at most $|\mathcal{P}|^n$ different reachable markings in a safe net and at most 2^{3n^2} equivalence classes of counters. The exponent is $3n^2$ since we have at most n^2 pairs of counters C_i, C_j

and their relation is $C_i < C_j$, $C_i = C_j$ or $C_i > C_j$. Thus, the LTS has at most $|Q| \leq |\mathcal{P}|^{n^2} \cdot 2^{3n^3}$ states.

Since the length of the SAT encoding is in $\mathcal{O}(|Q| \cdot |\mathcal{T}|)$ ($\Sigma = \mathcal{T}$) and the satisfiability problem of SAT is in NP the resulting complexity is NEXP in the number of players. Fixing the number of players n results in complexity NP in the size of the net. $\qquad\square$

Example. The encoding for the LTS T_G in Fig. 5 is carried out as follows: $\phi_{T_G} = F_{q0} \wedge (F_{q0} \wedge a_{q0} \Rightarrow F_{q_a}) \wedge \ldots$ with $F_{q0} = B_{q0} \wedge D_{q0} \wedge A_{q0}$, $a_{q0} = In_2(a) = true$, $B_{q0} = true$, $D_{q0} = true$, $A_{q0} = a_{q0} \vee b_{q0}$, and $F_{q_a} = B_{q_a} \wedge D_{q_a} \wedge A_{q_a}$ with $B_{q_a} = true$, $D_{q_a} = c_{a_{q_a}} \Rightarrow true$, $A_{q_a} = c_{a_{q_a}}$, and so on.

A boolean assignment $\mathcal{A}$ satisfies ϕ_{T_G} if $\mathcal{A} \models In_1(c_a) \wedge In_1(c_b) \wedge \{S_2, A\}_1(a') \wedge \neg\{S_2, A\}_1(b') \wedge \{S_2, B\}_1(b') \wedge \neg\{S_2, A\}_1(a') \wedge \{A', A\}_1(r_a) \wedge \{B', B\}_1(r_b)$. Then, $\mathcal{A}$ yields a winning LTS-strategy, and thus a winning strategy. This winning strategy is the same as in Fig. 3.

5   Conclusion

Distribute synthesis is an active research area. We investigate it in the formal model of Petri games. These are interesting because they exploit the concepts of Petri nets to formalise a notion of partial informedness of the players that rests on the causal memory.

Multi-player Petri games have been solved by translating them into two-player Büchi games on finite graphs [11]. The states in the graph represent the markings of the Petri game, enriched by additional information that is able to mimic their causal memory. This approach yields algorithms for deciding the realizability problem and solving the synthesis problem of Petri game under restricted distributions of players with exponential time complexity. Despite of this complexity, various benchmarks can be solved using this approach [9].

For Petri games not satisfying the restrictions on the distribution of players, the semi-decision approach of bounded synthesis can be applied [7,17]. This approach analyses the Petri game directly: for each given bound b, a finite and partially unfolded version of the Petri game is checked for a winning strategy.

Our approach also analyses finite unfoldings of the given Petri game. However, these are obtained differently, by defining a criterion when the players have the same memory of their past. The idea is that in states with the same memory a strategy can take the same decision. A first step is to take the *last known marking* as such a memory. In [10], it was shown that this criterion is sufficient to solve Petri games, where multiple environment players interacting with a *single* system player. However, this is not sufficient in general. In our paper, we extend last known markings by equivalence classes of *counters* recording how many transitions one player has taken in the causal past of another player. This works for more Petri games without restriction on the distribution of players.

As future work, we envisage a stepwise refinement of this memory by using nested last known markings to gradually cover more and more Petri games.

References

1. Becker, B., Podelski, A., Damm, W., Fränzle, M., Olderog, E.R., Wilhelm, R.: SFB/TR 14 AVACS – automatic verification and analysis of complex systems (Der Sonderforschungsbereich/Transregio 14 AVACS – Automatische Verifikation und Analyse komplexer Systeme). Inf. Technol. **49**(2), 118–126 (2007). https://doi.org/10.1524/itit.2007.49.2.118
2. Beutner, R., Finkbeiner, B., Hecking-Harbusch, J.: Translating asynchronous games for distributed synthesis. In: Fokkink, W.J., van Glabbeek, R. (eds.) 30th International Conference on Concurrency Theory, CONCUR 2019. LIPIcs, vol. 140, pp. 26:1–26:16. Schloss Dagstuhl – Leibniz-Zentrum für Informatik (2019). https://doi.org/10.4230/LIPIcs.CONCUR.2019.26
3. Damm, W., Finkbeiner, B.: Automatic compositional synthesis of distributed systems. In: Jones, C., Pihlajasaari, P., Sun, J. (eds.) FM 2014. LNCS, vol. 8442, pp. 179–193. Springer, Cham (2014). https://doi.org/10.1007/978-3-319-06410-9_13
4. Engelfriet, J.: Branching processes of Petri nets. Acta Inf. **28**(6), 575–591 (1991). https://doi.org/10.1007/BF01463946
5. Esparza, J.: Model checking using net unfoldings. Sci. Comput. Program. **23**(2), 151–195 (1994). https://doi.org/10.1016/0167-6423(94)00019-0
6. Esparza, J., Römer, S., Vogler, W.: An improvement of Mcmillan's unfolding algorithm. Formal Methods Syst. Des. **20**(3), 285–310 (2002). https://doi.org/10.1023/A:1014746130920
7. Finkbeiner, B.: Bounded synthesis for Petri games. In: Meyer, R., Platzer, A., Wehrheim, H. (eds.) Correct System Design. LNCS, vol. 9360, pp. 223–237. Springer, Cham (2015). https://doi.org/10.1007/978-3-319-23506-6_15
8. Finkbeiner, B., Gieseking, M., Hecking-Harbusch, J., Olderog, E.R.: Global winning conditions in synthesis of distributed systems with causal memory. In: Manea, F., Simpson, A. (eds.) 30th EACSL Annual Conference on Computer Science Logic, CSL 2022. LIPIcs, vol. 216, pp. 20:1–20:19. Schloss Dagstuhl – Leibniz-Zentrum für Informatik (2022). https://doi.org/10.4230/LIPIcs.CSL.2022.20
9. Finkbeiner, B., Gieseking, M., Olderog, E.R.: Adam: causality-based synthesis of distributed systems. In: Kroening, D., Păsăreanu, C.S. (eds.) Computer Aided Verification, pp. 433–439. Springer, Cham (2015)
10. Finkbeiner, B., Gölz, P.: Synthesis in distributed environments. In: Lokam, S., Ramanujam, R. (eds.) 37th IARCS Annual Conference on Foundations of Software Technology and Theoretical Computer Science (FSTTCS 2017). LIPIcs, vol. 93, pp. 28:1–28:14. Schloss Dagstuhl – Leibniz-Zentrum für Informatik (2017). https://doi.org/10.4230/LIPIcs.FSTTCS.2017.28
11. Finkbeiner, B., Olderog, E.R.: Petri games: synthesis of distributed systems with causal memory. Inf. Comput. **253**, 181–203 (2017). https://doi.org/10.1016/j.ic.2016.07.006
12. Finkbeiner, B., Schewe, S.: Uniform distributed synthesis. In: 20th IEEE Symposium on Logic in Computer Science (LICS 2005), Proceedings, pp. 321–330. IEEE Computer Society (2005). https://doi.org/10.1109/LICS.2005.53
13. Gieseking, M.: Correctness of data flows in asynchronous distributed systems: model checking and synthesis. Ph.D. thesis, University of Oldenburg, Germany (2022). http://oops.uni-oldenburg.de/5688
14. Gieseking, M., Hecking-Harbusch, J., Yanich, A.: A web interface for petri nets with transits and Petri games. In: TACAS 2021. LNCS, vol. 12652, pp. 381–388. Springer, Cham (2021). https://doi.org/10.1007/978-3-030-72013-1_22

15. Gimbert, H.: On the control of asynchronous automata. In: Lokam, S.V., Ramanujam, R. (eds.) 37th IARCS Annual Conference on Foundations of Software Technology and Theoretical Computer Science (FSTTCS 2017). LIPIcs, vol. 93, pp. 30:1–30:15. Schloss Dagstuhl – Leibniz-Zentrum für Informatik (2017). https://doi.org/10.4230/LIPIcs.FSTTCS.2017.30
16. Gimbert, H.: Distributed asynchronous games with causal memory are undecidable. Log. Methods Comput. Sci. **18**(3) (2022). https://doi.org/10.46298/lmcs-18(3:30)2022
17. Hecking-Harbusch, J.: Synthesis of asynchronous distributed systems from global specifications. Ph.D. thesis, Saarland University, Saarbrücken, Germany (2021). https://publikationen.sulb.uni-saarland.de/handle/20.500.11880/32108
18. Pnueli, A., Rosner, R.: Distributed reactive systems are hard to synthesize. In: 31st Annual Symposium on Foundations of Computer Science, vol. 2, pp. 746–757. IEEE Computer Society (1990). https://doi.org/10.1109/FSCS.1990.89597
19. Schewe, S.: Synthesis of distributed systems. Ph.D. thesis, Saarland University, Saarbrücken, Germany (2008). http://react.cs.uni-sb.de/publications/S08c.html
20. Zielonka, W.: Notes in finite asynchronous automata. RAIRO - Theor. Inform. Appl. - Informatique Théorique et Applications **21**(2), 99–135 (1987)

On the Verification of Parametric Systems

Dennis Peuter, Philipp Marohn, and Viorica Sofronie-Stokkermans[(✉)]

University of Koblenz, Koblenz, Germany
{dpeuter,pmarohn,sofronie}@uni-koblenz.de

Abstract. We present an approach to the verification of systems for whose description some elements – constants or functions – are underspecified and can be regarded as parameters, and, in particular, describe a method for automatically generating constraints on such parameters under which certain safety conditions are guaranteed to hold. We present an implementation and illustrate its use on several examples.

We dedicate this paper to Werner Damm, whose contributions to the area of modeling and verification of complex systems were a source of inspiration for the work described here.

1   Introduction

Many reasoning problems in mathematics or program verification can be reduced to checking satisfiability of ground formulae w.r.t. a theory. More interesting however is to consider problems – in mathematics or verification – in which the properties of certain function symbols are underspecified (these symbols are considered to be parametric) and (weakest) additional conditions need to be derived under which given properties hold. In this paper we study this type of problems from the perspective of deductive verification: We consider parametric reactive and linear hybrid systems – modeled by transition constraints or generalizations thereof. A classical problem in verification is to check whether a safety property – expressed by a suitable formula – is an invariant, or holds for paths of bounded length, *for given instances of the parameters*, or *under given constraints on parameters*. If the formula expressing the safety property is not an inductive invariant, we analyze two possibilities:

- We present a method that allows us to *derive constraints on the parameters* which guarantee that the property is an inductive invariant of the system.
- We show how this method can be adapted, in certain cases, for *strengthening* the formula in order to obtain an inductive invariant.

We present a method for property-directed symbol elimination in local theory extensions proposed in [47,48] and illustrate its applicability to solving the tasks above using an ongoing implementation in the system SEH-PILoT [35].

Related Work. Among the approaches to the verification of parametric reactive infinite state systems and timed automata we mention [8,20,23]; for parametric hybrid automata [1,9,18,39,51]. However, most papers only consider situations

M. Fränzle et al. (Eds.): Werner Damm Festschrift, LNCS 15471, pp. 201–221, 2026.
https://doi.org/10.1007/978-3-031-97537-0_12

in which the parameters are constants. Approaches to invariant strengthening or invariant synthesis were proposed in e.g. [5,6,14,17,21,29,30,36].

In this paper we present a survey of our work in the verification of parametric systems using hierarchical reasoning and hierarchical symbol elimination, work which was strongly influenced by the inspiring collaboration of the last author with Werner Damm in the AVACS project and by his papers (e.g. [10,40], cf. also [3,31] to mention only a few). We first used hierarchical reasoning and quantifier elimination for obtaining constraints on the parameters (constants or functions) in [28] where we analyzed possibilities for the verification of controllers for systems of trains. We then further developed these ideas in [45,46] and [49]; in [12,13] we analyzed possible applications to the verification of hybrid systems and, in [38], we showed that similar ideas can also be used e.g. for invariant strengthening (the method for invariant strengthening we proposed extends the results in [6,14] to more general theories and is orthogonal to the approach in [30,36]). After giving an overview of those results, we describe a new, ongoing, implementation in the system SEH-PILoT, and illustrate on some examples how SEH-PILoT can be used for the generation of constraints under which a given formula is guaranteed to be an inductive invariant and for invariant strengthening. Another new application is the use of SEH-PILoT for deriving constraints on parameters under which linear hybrid systems are chatter-free. Further details and more examples can be found in the extended version of this paper [37].

Structure of the Paper. In Sect. 2 we give the main definitions needed in the paper and present existing results on local theory extensions, a method for symbol elimination which turns out to be useful in the verification of parametric systems, and an implementation. In Sect. 3 we introduce a class of parametric systems described by transition constraint systems, we identify situations in which decision procedures exist for invariant checking and bounded model checking of such systems, as well as methods for obtaining constraints on the parameters which guarantee that certain properties are invariants, and methods for invariant strengthening in transition constraint systems. In Sect. 4 we study similar problems for some classes of parametric hybrid automata. In Sect. 5 we present the conclusions and mention some plans for future work.

2   Local Theory Extensions

In this section we introduce a class of logical theories used for modeling reactive, real time and hybrid systems for which we can obtain decidability results.

We consider signatures of the form $\Pi = (\Sigma, \mathsf{Pred})$ or many-sorted signatures of the form $\Pi = (S, \Sigma, \mathsf{Pred})$, where S is a set of sorts, Σ is a family of function symbols and Pred a family of predicate symbols. If Π is a signature and C is a set of new constants, we will denote by Π^C the expansion of Π with constants in C, i.e. the signature $\Pi^C = (\Sigma \cup C, \mathsf{Pred})$.

We assume known standard definitions from first-order logic. In this paper we refer to (finite) conjunctions of clauses also as "sets of clauses", and to (finite) conjunctions of formulae as "sets of formulae". Thus, if N_1 and N_2 are finite sets of formulae then $N_1 \cup N_2$ will stand for the conjunction of all formulae in $N_1 \cup N_2$. All free variables of a clause (resp. of a set of clauses) are considered to be universally quantified. We denote "verum" with $\top$ and "falsum" with $\bot$.

Theories can be defined by specifying a set of axioms, or by specifying a set of structures (the models of the theory). In this paper, (logical) theories are simply sets of sentences.

If F, G are formulae and $\mathcal{T}$ is a theory we write: $F \models G$ to express the fact that every model of F is a model of G; $F \models_{\mathcal{T}} G$ – also written as $\mathcal{T} \cup F \models G$ and sometimes $\mathcal{T} \wedge F \models G$ – to express the fact that every model of F which is also a model of $\mathcal{T}$ is a model of G. $F \models \bot$ means that F is unsatisfiable; $F \models_{\mathcal{T}} \bot$ means that there is no model of $\mathcal{T}$ in which F is true. If there is a model of $\mathcal{T}$ which is also a model of F we say that F is satisfiable w.r.t. $\mathcal{T}$. If $F \models_{\mathcal{T}} G$ and $G \models_{\mathcal{T}} F$ we say that F and G are equivalent w.r.t. $\mathcal{T}$.

A theory $\mathcal{T}$ over a signature Π allows quantifier elimination if for every formula ϕ over Π there exists a quantifier-free formula ϕ^* over Π which is equivalent to ϕ w.r.t. $\mathcal{T}$.

Example 1. *Presburger arithmetic with congruence modulo n, rational linear arithmetic $LI(\mathbb{Q})$ and real linear arithmetic $LI(\mathbb{R})$, the theories of real closed fields (real numbers) and of algebraically closed fields, the theory of finite fields, the theory of absolutely free algebras, and the theory of acyclic lists in the signature $\{\mathrm{car}, \mathrm{cdr}, \mathrm{cons}\}$ ([7, 19, 22, 32, 50]) allow quantifier elimination.*

Theory Extensions. Let $\Pi_0 = (\Sigma_0, \mathsf{Pred})$ be a signature, and $\mathcal{T}_0$ be a "base" theory with signature Π_0. We consider extensions $\mathcal{T} := \mathcal{T}_0 \cup \mathcal{K}$ of $\mathcal{T}_0$ with new function symbols Σ (*extension functions*) whose properties are axiomatized using a set $\mathcal{K}$ of (universally closed) clauses in an extended signature $\Pi = (\Sigma_0 \cup \Sigma, \mathsf{Pred})$, such that each clause in $\mathcal{K}$ contains function symbols in Σ. Let $\Sigma_P \subseteq \Sigma$ be a set of parameters.

Let G be a set of ground Π^C-clauses. We want to check whether G is satisfiable w.r.t. $\mathcal{T}_0 \cup \mathcal{K}$ or not and – if it is satisfiable – to automatically generate a weakest universal $\Pi_0 \cup \Sigma_P$-formula Γ such that $\mathcal{T}_0 \cup \mathcal{K} \cup \Gamma \cup G$ is unsatisfiable.

In what follows we present situations in which hierarchical reasoning is complete and weakest constraints on parameters can be computed.

Local Theory Extensions. Let Ψ be a map which associates with every finite set T of ground terms a finite set $\Psi(T)$ of ground terms. A theory extension $\mathcal{T}_0 \subseteq \mathcal{T}_0 \cup \mathcal{K}$ is Ψ-local if it satisfies the condition:

(Loc_f^{Ψ}) For every finite set G of ground Π^C-clauses (for an additional set C of constants) it holds that $\mathcal{T}_0 \cup \mathcal{K} \cup G \models \bot$ if and only if $\mathcal{T}_0 \cup \mathcal{K}[\Psi_{\mathcal{K}}(G)] \cup G$ is unsatisfiable.

where, for every set G of ground Π^C-clauses, $\mathcal{K}[\Psi_{\mathcal{K}}(G)]$ is the set of instances of $\mathcal{K}$ in which the terms starting with a function symbol in Σ are in $\Psi_{\mathcal{K}}(G) = \Psi(\mathsf{est}(\mathcal{K}, G))$, where $\mathsf{est}(\mathcal{K}, G)$ – the set of ground extension subterms of $\mathcal{K}$ and G – is the set of ground terms starting with an *extension function* (i.e. a function in Σ) occurring in G or $\mathcal{K}$. If T is a set of ground terms, we use the notation $\Psi_{\mathcal{K}}(T) := \Psi(\mathsf{est}(\mathcal{K}, T))$.

In [24, 26, 42] we proved that if $\Psi_{\mathcal{K}}$ is a term closure operator, i.e. the following conditions hold for all sets of ground terms T, T':

(1) $\mathsf{est}(\mathcal{K}, T) \subseteq \Psi_{\mathcal{K}}(T)$,
(2) $T \subseteq T' \Rightarrow \Psi_{\mathcal{K}}(T) \subseteq \Psi_{\mathcal{K}}(T')$,
(3) $\Psi_{\mathcal{K}}(\Psi_{\mathcal{K}}(T)) \subseteq \Psi_{\mathcal{K}}(T)$,
(4) $\Psi_{\mathcal{K}}$ is invariant under constant renaming: for any map $h : C \to C$, $\bar{h}(\Psi_{\mathcal{K}}(T)) = \Psi_{\bar{h}(\mathcal{K})}(\bar{h}(T))$, where $\bar{h}$ is the extension of h to ground terms and formulae.

then Ψ-local extensions can be recognized by showing that certain partial models embed into total ones. Especially well-behaved are theory extensions with the property $(\mathsf{Comp}_f^{\Psi})^1$ which requires that every partial model of $\mathcal{T}$ whose reduct to Π_0 is total and the "set of defined terms" is finite and closed under Ψ, embeds into a total model of $\mathcal{T}$ *with the same support* (cf. e.g. [24]). If Ψ is the identity, we denote (Loc_f^{Ψ}) by (Loc_f) and (Comp_f^{Ψ}) by (Comp_f); an extension satisfying (Loc_f) is called *local*. The link between embeddability and locality allowed us to identify many classes of local theory extensions (cf. e.g. [24, 42, 44]):

Example 2 (Extensions with free/monotone functions [24, 42]). *The following types of extensions of a theory $\mathcal{T}_0$ are local:*

(1) Any extension of $\mathcal{T}_0$ with uninterpreted function symbols ((Comp_f) holds).
(2) Any extension of a theory $\mathcal{T}_0$ for which $\leq$ is a partial order with functions monotone w.r.t. $\leq$ (condition (Comp_f) holds if all models of $\mathcal{T}_0$ are lattices w.r.t. $\leq$).

Example 3 (Extensions with definitions [24, 27]). *Consider an extension of a theory $\mathcal{T}_0$ with a new function symbol f defined by axioms of the form:*

$$\mathsf{Def}_f := \{\forall \overline{x}(\phi_i(\overline{x}) \to F_i(f(\overline{x}), \overline{x})) \mid i = 1, \dots, m\}$$

(definition by "case distinction") where ϕ_i and F_i, $i = 1, \dots, m$, are formulae over the signature of $\mathcal{T}_0$ such that the following hold:

(a) $\phi_i(\overline{x}) \wedge \phi_j(\overline{x}) \models_{\mathcal{T}_0} \bot$ for $i \neq j$ and
(b) $\mathcal{T}_0 \models \forall \overline{x}(\phi_i(\overline{x}) \to \exists y(F_i(y, \overline{x})))$ for all $i \in \{1, \dots, m\}$.

Then the extension is local (and satisfies (Comp_f)). Examples:

(1) Any extension with a function f defined by axioms of the form:

$$\mathsf{D}_f := \{\forall \overline{x}(\phi_i(\overline{x}) \to f(\overline{x}) = t_i) \mid i = 1, \dots, n\}$$

where ϕ_i are formulae over the signature of $\mathcal{T}_0$ such that (a) holds.
(2) Any extension of $\mathcal{T}_0 \in \{\mathsf{LI}(\mathbb{Q}), \mathsf{LI}(\mathbb{R})\}$ with functions satisfying axioms:

$$\mathsf{Bound}_f := \{\forall \overline{x}(\phi_i(\overline{x}) \to s_i \leq f(\overline{x}) \leq t_i) \mid i = 1, \dots, n\}$$

where ϕ_i are formulae over the signature of $\mathcal{T}_0$, s_i, t_i are $\mathcal{T}_0$-terms, condition (a) holds and $\models_{\mathcal{T}_0} \forall \overline{x}(\phi_i(\overline{x}) \to s_i \leq t_i)$ [24].

---

[1] We use the index f in (Comp_f) in order to emphasize that the property refers to completability of partial functions with a finite domain of definition.

2.1   Hierarchical Reasoning in Local Theory Extensions

Consider a Ψ-local theory extension $\mathcal{T}_0 \subseteq \mathcal{T}_0 \cup \mathcal{K}$. Condition (Loc_f^Ψ) requires that for every finite set G of ground Π^C-clauses: $\mathcal{T}_0 \cup \mathcal{K} \cup G \models \perp$ if and only if $\mathcal{T}_0 \cup \mathcal{K}[\Psi_\mathcal{K}(G)] \cup G \models \perp$. In all clauses in $\mathcal{K}[\Psi_\mathcal{K}(G)] \cup G$ the function symbols in Σ only have ground terms as arguments, so $\mathcal{K}[\Psi_\mathcal{K}(G)] \cup G$ can be flattened and purified by introducing, in a bottom-up manner, new constants $c_t \in C$ for subterms $t = f(c_1, \ldots, c_n)$ where $f \in \Sigma$ and c_i are constants, together with definitions $c_t = f(c_1, \ldots, c_n)$, all included in a set Def. We thus obtain a set of clauses $\mathcal{K}_0 \cup G_0 \cup \mathsf{Def}$, where $\mathcal{K}_0$ and G_0 do not contain Σ-function symbols and Def contains clauses of the form $c = f(c_1, \ldots, c_n)$, where $f \in \Sigma$, $c, c_1, \ldots, c_n$ are constants.

Theorem 4 ([24,42]). *Let $\mathcal{K}$ be a set of clauses. Assume that $\mathcal{T}_0 \subseteq \mathcal{T}_1 = \mathcal{T}_0 \cup \mathcal{K}$ is a Ψ-local theory extension. For any finite set G of ground clauses, let $\mathcal{K}_0 \cup G_0 \cup \mathsf{Def}$ be obtained from $\mathcal{K}[\Psi_\mathcal{K}(G)] \cup G$ by flattening and purification, as explained above. Then the following are equivalent to $\mathcal{T}_1 \cup G \models \perp$:*

(1) $\mathcal{T}_0 \cup \mathcal{K}[\Psi_\mathcal{K}(G)] \cup G \models \perp$.

(2) $\mathcal{T}_0 \cup \mathcal{K}_0 \cup G_0 \cup \mathsf{Con}_0 \models \perp$, where $\mathsf{Con}_0 = \{ \bigwedge\limits_{i=1}^{n} c_i = d_i \rightarrow c = d \mid \begin{matrix} f(c_1, \ldots, c_n) = c \in \mathsf{Def} \\ f(d_1, \ldots, d_n) = d \in \mathsf{Def} \end{matrix} \}$.

We can also consider chains of theory extensions:

$$\mathcal{T}_0 \subseteq \mathcal{T}_1 = \mathcal{T}_0 \cup \mathcal{K}_1 \subseteq \mathcal{T}_2 = \mathcal{T}_0 \cup \mathcal{K}_1 \cup \mathcal{K}_2 \subseteq \cdots \subseteq \mathcal{T}_n = \mathcal{T}_0 \cup \mathcal{K}_1 \cup \ldots \cup \mathcal{K}_n$$

in which each theory is a local extension of the preceding one. For a chain of n local extensions a satisfiability check w.r.t. the last extension can be reduced (in n steps) to a satisfiability check w.r.t. $\mathcal{T}_0$. The only restriction we need to impose in order to ensure that such a reduction is possible is that at each step the clauses reduced so far need to be ground – this is the case if each variable in a clause appears at least once under an extension function. This instantiation procedure for chains of local theory extensions has been implemented in H-PILoT [25].[2]

2.2   Hierarchical Symbol Elimination

In [48] we proposed a method for property-directed symbol elimination described in Algorithm 1.

Theorem 5 ([47,48]). *Let $\mathcal{T}_0$ be a Π_0-theory allowing quantifier elimination, Σ_P be a set of parameters (function and constant symbols) and Σ a set of function symbols such that $\Sigma \cap (\Sigma_0 \cup \Sigma_P) = \emptyset$. Let $\mathcal{K}$ be a set of flat and linear[3]*

---

[2] H-PILoT allows the user to specify a chain of extensions: if a function symbol f occurs in $\mathcal{K}_n$ but not in $\bigcup_{i=1}^{n-1} \mathcal{K}_i$ it is declared as level n.

[3] A clause is flat if the arguments of extension symbols are variables. A clause is linear if a variable does not occur under different function symbols or twice in a term.

Algorithm 1. Symbol elimination in theory extensions [47,48]

Input: $\mathcal{T}_0 \subseteq \mathcal{T}_0 \cup \mathcal{K}$ theory extension with signature $\Pi = \Pi_0 \cup (\Sigma \cup \Sigma_P)$
　　　　where Σ_P is a set of parameters and $\mathcal{K}$ is a set of flat clauses.
　　　　G set of flat ground clauses; T set of flat ground Π^C-terms s.t. $\mathsf{est}(\mathcal{K}, G) \subseteq T$
Output: $\forall \overline{y}.\, \Gamma_T(\overline{y})$ (constraint on parameters; universal $\Pi_0 \cup \Sigma_P$-formula)

Step 1 Purify $\mathcal{K}[T] \cup G$ as described in Theorem 4 (with set of extension symbols Σ_1).
　　　 Let $\mathcal{K}_0 \cup G_0 \cup \mathsf{Con}_0$ be the set of Π_0^C-clauses obtained this way.

Step 2 Let $G_1 = \mathcal{K}_0 \cup G_0 \cup \mathsf{Con}_0$. Among the constants in G_1, we identify
　　　 (i)　the constants c_f, $f \in \Sigma_P$, where c_f is a constant parameter or c_f is introduced
　　　　　 by a definition $c_f = f(c_1, \ldots, c_k)$ in the hierarchical reasoning method,
　　　 (ii)　all constants $\overline{c}_p$ which are not parameters occurring as arguments of functions
　　　　　 in Σ_P in such definitions.
　　　 Replace all the other constants $\overline{c}$ with existentially quantified variables $\overline{x}$ (i.e.
　　　 replace $G_1(\overline{c}_p, \overline{c}_f, \overline{c})$ with $\exists \overline{x}.\, G_1(\overline{c}_p, \overline{c}_f, \overline{x})$).

Step 3 Construct a formula $\Gamma_1(\overline{c}_p, \overline{c}_f)$ equivalent to $\exists \overline{x}.\, G_1(\overline{c}_p, \overline{c}_f, \overline{x})$ w.r.t. $\mathcal{T}_0$ using a
　　　 method for quantifier elimination in $\mathcal{T}_0$ and let $\Gamma_2(\overline{c}_p, \overline{c}_f)$ be $\neg \Gamma_1(\overline{c}_p, \overline{c}_f)$.

Step 4 Replace (i) each constant c_f introduced by definition $c_f = f(c_1, \ldots, c_k)$ with the
　　　 term $f(c_1, \ldots, c_k)$ and (ii) $\overline{c}_p$ with universally quantified variables $\overline{y}$ in $\Gamma_2(\overline{c}_p, \overline{c}_f)$.
　　　 The formula obtained this way is $\forall \overline{y}.\, \Gamma_T(\overline{y})$.

---

clauses in the signature $\Pi_0 \cup \Sigma_P \cup \Sigma$ in which all variables occur also below functions in $\Sigma_1 = \Sigma_P \cup \Sigma$ and G a set of flat ground clauses (i.e. the arguments of the extension functions are constants). Assume $\mathcal{T}_0 \subseteq \mathcal{T}_0 \cup \mathcal{K}$ satisfies condition (Comp_f^{Ψ}) for a suitable closure operator Ψ. Let $T = \Psi_{\mathcal{K}}(G)$. Then Algorithm 1 yields a universal $\Pi_0 \cup \Sigma_P$-formula $\forall \overline{x}.\, \Gamma_T(\overline{x})$ s.t. $\mathcal{T}_0 \cup \forall \overline{x}.\, \Gamma_T(\overline{x}) \cup \mathcal{K} \cup G \models \bot$, and s.t. $\forall \overline{x}.\, \Gamma_T(\overline{x})$ is entailed by every universal formula Γ with $\mathcal{T}_0 \cup \Gamma \cup \mathcal{K} \cup G \models \bot$.

Algorithm 1 yields a formula $\forall \overline{x}.\, \Gamma_T(\overline{x})$ with $\mathcal{T}_0 \cup \forall \overline{x}.\, \Gamma_T(\overline{x}) \cup \mathcal{K} \cup G \models \bot$ also if the extension $\mathcal{T}_0 \subseteq \mathcal{T}_0 \cup \mathcal{K}$ is not Ψ-local or $T \neq \Psi_{\mathcal{K}}(G)$, but in this case there is no guarantee that $\forall \overline{x}.\, \Gamma_T(\overline{x})$ is the weakest universal formula with this property.

A similar result holds for chains of local theory extensions; for details cf. [48].

2.3   Implementation

Hierarchical Reasoning: H-PILoT. The method for hierarchical reasoning in (chains of) local extensions of a base theory described before was implemented in the system H-PILoT [25]. H-PILoT carries out a hierarchical reduction to the base theory. Standard SMT provers (e.g. CVC4 [2] or Z3 [4]) or specialized provers (e.g. Redlog [15]) are used for testing the satisfiability of the formulae obtained after the reduction. H-PILoT uses eager instantiation, so provers like CVC4 or Z3 might in general be faster in proving unsatisfiability. The advantage of using H-PILoT is that knowing the instances needed for a complete instantiation allows us to correctly detect satisfiability (and generate models) in situations in which standard SMT provers return "unknown", and also to use property-directed symbol elimination to obtain additional constraints on parameters which ensure unsatisfiability, as explained in what follows.

Symbol Elimination: SEH-PILoT (Symbol Elimination with H-PILoT). For obtaining *constraints on parameters* we use Algorithm 1 [48] which was implemented in SEH-PILoT (cf. also [35]) for the case in which the theory can be structured as a local theory extension or a chain of local theory extensions. For the hierarchical reduction, SEH-PILoT uses H-PILoT. The symbol elimination is handled by Redlog. The supported base theories are currently limited to the theory of real closed fields and the theory of Presburger arithmetic.

Input. SEH-PILoT is invoked with an input file that specifies the tasks and all options. A task is a description of a problem (constraint generation or invariant strengthening cf. Sect. 3.1). It contains a list of parameters (or, alternatively, of symbols to be eliminated), optionally a list of conditions on the parameters, and the formalization of the actual problem in the syntax of H-PILoT[4].

Execution. SEH-PILoT follows the steps of Algorithm 1. It uses H-PILoT for the hierarchical reduction and writes the result in a file which can be used as input for Redlog. Optionally formulae can be simplified using Redlog's interface to the external QEPCAD-based simplifier SLFQ or with a list of assumptions. The obtained formula then gets translated from the syntax of Redlog back to the syntax of H-PILoT. Depending on the chosen mode this is then either the final result of the task (i.e. a constraint) or the input for the next iteration (i.e. invariant strengthening cf. Sect. 3.1).

Output: The output is a file containing the results of each task. Depending on the chosen options the file contains in addition a list of the various steps that have taken place and their results as well as a small statistic showing the amount of time each step has required and the number of atoms before and after a simplification was applied.

3   Verification Problems for Parametric Systems

We identify situations in which decision procedures for the verification of parametric systems exist and in which methods for obtaining constraints on the parameters which guarantee that certain properties are invariant can be devised.

We specify a reactive system S as a tuple $(\Pi_S, \mathcal{T}_S, \mathsf{Tr}_S)$ where $\Pi_S = (\Sigma_S, \mathsf{Pred}_S)$ is a signature, $\mathcal{T}_S$ is a Π_S-theory (describing the data types used in the specification and their properties), and $\mathsf{Tr}_S = (V, \Sigma, \mathsf{Init}, \mathsf{Update})$ is a transition constraint system which specifies: the variables (V) and function symbols (Σ) whose values change over time, where $V \cup \Sigma \subseteq \Sigma_S$; a formula Init specifying the properties of initial states; a formula Update with variables in $V \cup V'$ and function symbols in $\Sigma \cup \Sigma'$ (where V' and Σ' are new copies of V resp. Σ, denoting the variables resp. functions after the transition) specifying the relationship

---

[4] A detailed description of the form of such input files can be found in [25].

between the values of variables x (functions f) before and their values x' (f') after a transition.

We consider *invariant checking* and *bounded model checking* problems, cf. [34]:

Invariant Checking. A formula Φ is an inductive invariant of a system S with theory $\mathcal{T}_S$ and transition constraint system $\mathsf{Tr}_S = (V, \Sigma, \mathsf{Init}, \mathsf{Update})$ if:

(1) $\mathcal{T}_S \wedge \mathsf{Init} \models \Phi$ and
(2) $\mathcal{T}_S \wedge \Phi \wedge \mathsf{Update} \models \Phi'$, where Φ' results from Φ by replacing each $x \in V$ by x' and each $f \in \Sigma$ by f'.

Bounded Model Checking. We check whether, for a fixed k, states not satisfying a formula Φ are reachable in at most k steps. Formally, we check whether:

$$\mathcal{T}_S \wedge \mathsf{Init}_0 \wedge \bigwedge_{i=0}^{j-1} \mathsf{Update}_i \wedge \neg\Phi_j \models \bot \quad \text{for all } 0 \leq j \leq k,$$

where Update_i is obtained from Update by replacing every $x \in V$ by x_i, every $f \in \Sigma$ by f_i, and each $x' \in V'$, $f' \in \Sigma'$ by x_{i+1}, f_{i+1}; Init_0 is Init with x_0 replacing $x \in V$ and f_0 replacing $f \in \Sigma$; Φ_i is obtained from Φ similarly.

3.1   Verification, Constraint Generation and Invariant Strengthening

We consider transition constraint systems $\mathsf{Tr}_S = (V, \Sigma, \mathsf{Init}, \mathsf{Update})$ in which $\Sigma_S = \Sigma_0 \cup V \cup \Sigma \cup \Sigma_P$, the formulae in Update contain variables in X and functions in Σ and possibly parameters in Σ_P. We assume that $\Sigma_P \cap \Sigma = \emptyset$.

We consider universal formulae Φ which are conjunctions of clauses of the form $\forall \overline{x}(C(\overline{x}, \overline{f}(\overline{x})))$, where C is a flat clause over Σ_S.[5] Such formulae describe "global" properties of the function symbols in Σ_S at a given moment in time, e.g. equality of two functions (possibly representing arrays), or monotonicity of a function. They can also describe properties of individual elements (ground formulae are considered to be in particular universal formulae). If the formula Φ is not an inductive invariant, our goals are to:

- generate constraints on Σ_P under which Φ becomes an inductive invariant;
- obtain a universally quantified inductive invariant I in a specified language (if such an inductive invariant exists) such that $I \models_{\mathcal{T}_S} \Phi$, or a proof that there is no universal inductive invariant (in that language) that entails Φ.

Verification and Constraint Generation. We make the following assumptions: Let $\mathsf{LocSafe}$ be a class of universal formulae over Σ_S.

(A1) There exists a chain of local theory extensions $\mathcal{T}_0 \subseteq \cdots \subseteq \mathcal{T}_S \cup \mathsf{Init}$ such that in each extension all variables occur below an extension function.

---

[5] We use the following abbreviations: $\overline{x}$ for $x_1, \ldots, x_n$; $\overline{f}(\overline{x})$ for $f_1(\overline{x}), \ldots, f_n(\overline{x})$.

(A2) For every $\Phi \in \mathsf{LocSafe}$ there exists a chain of local theory extensions $\mathcal{T}_0 \subseteq \cdots \subseteq \mathcal{T}_S \cup \Phi$ such that in each extension all variables occur below an extension function.

(A3) $\mathsf{Update} = \{\mathsf{Update}_f \mid f \in F\}$ consists of update axioms for functions in a set F, where, for every $f \in F$, Update_f has the form

$\mathsf{Def}_f := \{\forall \overline{x}(\phi_i^f(\overline{x}) \rightarrow C_i^f(\overline{x}, f'(\overline{x}))) \mid i \in I\},$

such that (i) $\phi_i(\overline{x}) \wedge \phi_j(\overline{x}) \models_{\mathcal{T}_S} \bot$ for $i \neq j$, (ii) $\mathcal{T}_S \models \bigvee_{i=1}^{n} \phi_i$, and (iii) C_i^f are conjunctions of literals and $\mathcal{T}_S \models \forall \overline{x}(\phi_i(\overline{x}) \rightarrow \exists y(C_i^f(\overline{x}, y)))$ for all $i \in I$. [6]

In what follows, for every formula ϕ containing symbols in $V \cup \Sigma$ we denote by ϕ' the formula obtained from ϕ by replacing every variable $x \in V$ and every function symbol $f \in \Sigma$ with the corresponding symbols $x', f' \in V' \cup \Sigma'$.

Theorem 6 ([24,45,49]). *The following hold under assumptions* $(\mathbf{A1}) - (\mathbf{A3})$:

(1) If ground satisfiability w.r.t. $\mathcal{T}_0$ is decidable, then for every $\Phi \in \mathsf{LocSafe}$ (i) the problem of checking whether Φ is an inductive invariant of the system S is decidable; (ii) bounded model checking Φ for a fixed bound k is decidable.

(2) If $\mathcal{T}_0$ allows quantifier elimination and the initial states or the updates contain parameters, the symbol elimination method in Algorithm 1 yields constraints on these parameters that guarantee that Φ is an inductive invariant. If in $(\mathbf{A1}), (\mathbf{A2})$ we additionally assume that all local extensions in $\mathsf{LocSafe}$ satisfy condition (Comp_f), then the constraint generated with Algorithm 1 is the weakest among all universal constraints on the parameters under which Φ is an inductive invariant.

Invariant Strengthening. We now consider the problem of inferring – in a goal-oriented way – universally quantified inductive invariants. The method we proposed in [38] is described in Algorithm 2.

In addition to assumptions $(\mathbf{A1}), (\mathbf{A2}), (\mathbf{A3})$ we now consider the following assumptions (where $\mathcal{T}_0$ is the base theory in assumptions $(\mathbf{A1})$–$(\mathbf{A3})$):

(A4) Ground satisfiability in $\mathcal{T}_0$ is decidable; $\mathcal{T}_0$ allows quantifier elimination.

(A5) All candidate invariants I computed in the while loop in Algorithm 2 are in $\mathsf{LocSafe}$, and all local extensions in $\mathsf{LocSafe}$ satisfy condition (Comp_f).

Under assumptions $(\mathbf{A1}) - (\mathbf{A5})$ the algorithm is partially correct:

Theorem 7 ([38]). *The following hold:*

(1) If Algorithm 2 terminates and returns a formula I, then I is an invariant of the system S containing only function symbols in Σ_P that entails Φ.

---

⁶ The update axioms describe the change of the functions in a set $F \subseteq \Sigma$, depending on a finite set $\{\phi_i \mid i \in I\}$ of mutually exclusive conditions over non-primed symbols. In particular we can consider updates of the form $\mathsf{D}_{f'}$ or $\mathsf{Bound}_{f'}$ as in Example 3.

Algorithm 2. Iteratively strengthening a formula to an inductive invariant [38]

Input: System $S = ((\Sigma_S, \mathsf{Pred}_S), \mathcal{T}_S, \mathsf{Tr}_S)$, where $\mathsf{Tr}_S = (V, \Sigma, \mathsf{Init}, \mathsf{Update})$;
$\qquad\qquad \Sigma_P \subseteq \Sigma_S$; $\Phi \in \mathsf{LocSafe}$ (over Σ_P).
Output: Inductive invariant I of S that entails Φ and contains only symbols in Σ_P
$\qquad\qquad$ (if such an invariant exists).

1: $I := \Phi$
2: **while** I is not an inductive invariant for S **do:**
$\quad$ **if** $\mathsf{Init} \not\models I$ **then return** "no universal inductive invariant for S over Σ_P entails Φ"
$\quad$ **if** I is not preserved under Update **then** Let Γ be obtained by eliminating
$\quad$ all primed variables and symbols not in Σ_P from $I \wedge \mathsf{Update} \wedge \neg I'$;
$\quad$ $I := I \wedge \Gamma$
3: **return** I is an inductive invariant

---

(2) Under assumptions **(A1)**–**(A5)**, *if there exists a universal inductive invariant J containing only function symbols in Σ_P that entails Φ, then J entails every candidate invariant I generated in the while loop of Algorithm 2.*

(3) Under assumptions **(A1)**–**(A5)**, *if Algorithm 2 terminates because the candidate invariant I generated so far is not entailed by* Init *then no universal inductive invariant entails Φ.*

Theorem 8 (Partial Correctness, [38]). *Under assumptions* **(A1)**–**(A5)**, *if Algorithm 2 terminates, then its output is correct.*

In [38] we identified situations in which assumption **(A5)** holds (i.e. does not have to be stated explicitly) and conditions under which the algorithm terminates.

3.2 Example

We show how SEH-PILoT can be used for two variants of an example from [38].

Example 1: Invariant checking and constraint generation. Consider the following program, using subprogram $\mathsf{add1}(a)$, which adds 1 to every element of array a. We check whether $\Phi := d_2 \geq d_1$ is an inductive invariant, and, if not, generate additional conditions s.t. Φ is an inductive invariant.

```
d1 = 1; d2 = 1; i:= 0;
while (nondet()) {
    a = add1(a);
    d1 = a[i]; d2 = a[i+1];
    i:= i + 1}
```

Φ holds in the initial states described by $\mathsf{Init} := d_1 = 1 \wedge d_2 = 1 \wedge i = 0$; it is an inductive invariant of the while loop iff the formula

$$\forall j(a'[j] = a[j] + 1) \wedge d_1' = a'[i] \wedge d_2' = a'[i+1] \wedge i' = i+1 \wedge \ d_1 \leq d_2 \ \wedge \ d_1' > d_2'$$

is unsatisfiable. As this formula is satisfiable, Φ is not an inductive invariant.

Constraint Generation. We use SEH-PILoT to generate constraints on the parameters $\Sigma_P = \{a, d_1, d_2\}$ under which Φ is an inductive invariant.

```
tasks:
    example constraint generation:
        mode: GENERATE_CONSTRAINTS
        options:
            parameter: [a, d1, d2]
            slfq_query: true
        specification_type: HPILOT
        specification_theory: REAL_CLOSED_FIELDS
        specification:
            file: |
                Base_functions := {(+,2), (-,2), (*,2)}
                Extension_functions := {(b, 1, 1), (a, 1, 2), (ap, 1, 3)}
                Relations := {(<=,2), (<,2), (>=,2), (>,2)}

                Clauses := % ---Phi(d1,d2)-
                    d1 <= d2;
                    % --Update--
                    (FORALL j). ap(j) = a(j) + _1;
                    d1p = ap(i); d2p = ap(i + _1);
                    ip = i + _1;
                Query :=    d1p - d2p > _0; %--not Psi(d1p,d2p)--
```

SEH-PILOT gives the following output:

```
Metadata:
    Date: '2023-07-24 15:04:48'
    Number of Tasks: 1
    Runtime Sum: 0.4339
example constraint generation:
    Runtime: 0.4339
    Result: (FORALL i). OR(a(i + _1) - a(i) >= _0, d1 - d2 > _0)
```

Since ap (i.e. a') is defined by a clause satisfying the requirements for an extension by definitions, it defines a local extension satisfying (Comp_f), therefore $\forall i(a[i] \leq a[i+1])$ is the weakest condition under which $\varPhi$ is an inductive invariant.

Example 2: Invariant strengthening. Consider the program below, using subprograms $\mathsf{copy}(a, b)$, which copies the array b into array a, and $\mathsf{add1}(a)$, which adds 1 to every element of array a. The task is to prove that if b is an array with its elements sorted in increasing order then the formula $\varPhi := d_2 \geq d_1$ is an invariant of the program. It can be checked that $\varPhi$ holds in the initial states $\mathsf{Init} := d_1 = 1 \wedge d_2 = 1 \wedge i = 0 \wedge \forall l(a(l) = b(l)) \wedge \forall l, j(l \leq j \rightarrow b(l) \leq b(j))$. We can prove that it is not an inductive invariant.

```
d1 = 1; d2 = 1; i:= 0;
copy(a, b);
while (nondet()) {
    a = add1(a);
    d1 = a[i]; d2 = a[i+1];
    i:= i + 1}
```

We strengthen $\varPhi$ using SEH-PILoT. As explained before, we obtain:

$$\Gamma = \forall i(a(i) \leq a(i+1) \vee d_1 > d_2)$$

Combining the constraint computed by SEH-PILoT with $\varPhi$ we obtain the following candidate invariant:

$$\varPhi_1 = d_1 \leq d_2 \wedge \forall i \, a(i+1) \geq a(i)$$

To check whether $\varPhi_1$ is an inductive invariant, we first check whether it holds in the initial states using H-PILoT (with external prover Z3), and prove that this is the case, then we prove that Φ_1 is also invariant under updates. We could therefore strengthen Φ to obtain an inductive invariant after one iteration.

4   Systems Modeled Using Hybrid Automata

Hybrid automata were introduced in [1] to describe systems with discrete control (represented by a finite set of control modes); in every control mode certain variables can evolve continuously in time according to precisely specified rules.

Definition 1 ([1]). *A hybrid automaton* $S = (X, Q, \mathsf{flow}, \mathsf{Inv}, \mathsf{Init}, E, \mathsf{guard},$ $\mathsf{jump})$ *is a tuple consisting of:*

(1) A finite set $X = \{x_1, \ldots, x_n\}$ of real valued variables (regarded as functions $x_i : \mathbb{R} \to \mathbb{R}$) and a finite set Q of control modes;
(2) A family $\{\mathsf{flow}_q \mid q \in Q\}$ of predicates over the variables in $X \cup \dot{X}$ ($\dot{X} = \{\dot{x}_1, \ldots, \dot{x}_n\}$, where $\dot{x}_i$ is the derivative of x_i) specifying the continuous dynamics in each control mode[7]; a family $\{\mathsf{Inv}_q \mid q \in Q\}$ of predicates over the variables in X defining the invariant conditions for each control mode; and a family $\{\mathsf{Init}_q \mid q \in Q\}$ of predicates over the variables in X, defining the initial states for each control mode.
(3) A finite multiset E with elements in $Q \times Q$ (the control switches), where every $(q, q') \in E$ is a directed edge between q (source mode) and q' (target mode); a family of guards $\{\mathsf{guard}_e \mid e \in E\}$ (predicates over X); and a family of jump conditions $\{\mathsf{jump}_e \mid e \in E\}$ (predicates over $X \cup X'$, where $X' = \{x'_1, \ldots, x'_n\}$ is a copy of X consisting of "primed" variables).

A *state* of S is a pair (q, a) consisting of a control mode $q \in Q$ and a vector $a = (a_1, \ldots, a_n)$ that represents a value $a_i \in \mathbb{R}$ for each variable $x_i \in X$. A state (q, a) is *admissible* if Inv_q is true when each x_i is replaced by a_i. There are two types of *state change*: (i) A *jump* is an instantaneous transition that changes the control location and the values of variables in X according to the jump conditions; (ii) In a *flow*, the state can change due to the evolution in a given control mode over an interval of time: the values of the variables in X change continuously according to the flow rules of the current control location; all intermediate states are admissible. A *run* of S is a finite sequence $s_0 s_1 \ldots s_k$ of admissible states such that (i) the first state s_0 is an initial state of S (the values of the variables satisfy Init_q for some $q \in Q$), (ii) each pair (s_j, s_{j+1}) is either a jump of S or the endpoints of a flow of S.

Notation. In what follows we use the following notation. If $x_1, \ldots, x_n \in X$ we denote the sequence $x_1, \ldots, x_n$ with $\bar{x}$, the sequence $\dot{x}_1, \ldots, \dot{x}_n$ with $\dot{\bar{x}}$, and the sequence of values $x_1(t), \ldots, x_n(t)$ of these variables at a time t with $\bar{x}(t)$.

We identify the following verification problems:

Invariant checking is the problem of checking whether a quantifier-free formula Φ in real arithmetic over the variables X is an inductive invariant in a hybrid automaton S, i.e.:

---

[7] We assume that the functions $x_i : \mathbb{R} \to \mathbb{R}$ are differentiable during flows.

(1) Φ holds in the initial states of mode q for all $q \in Q$;
(2) Φ is invariant under jumps and flows:
- For every flow in a mode q, the continuous variables satisfy Φ both during and at the end of the flow.
- For every jump, if the values of the continuous variables satisfy Φ before the jump, they satisfy Φ after the jump.

Bounded model checking is the problem of checking whether the truth of a formula Φ is preserved under runs of length bounded by k, i.e.:

(1) Φ holds in the initial states of mode q for every $q \in Q$;
(2) Φ is preserved under runs of length j for all $1 \leq j \leq k$.

A hybrid automaton S is a linear hybrid automaton (LHA) if it satisfies the following two requirements:

1. **Linearity** For every control mode $q \in Q$, the flow condition flow_q, the invariant condition Inv_q, and the initial condition Init_q are convex linear predicates. For every control switch $e = (q, q') \in E$, the jump condition jump_e and the guard guard_e are convex linear predicates. In addition, we assume that the flow conditions flow_q are conjunctions of *non-strict* inequalities.
2. **Flow independence** For every control mode $q \in Q$, the flow condition flow_q is a predicate over the variables in $\dot{X}$ only (and does not contain any variables from X). This requirement ensures that the possible flows are independent from the values of the variables, and depend only on the control mode.

4.1   Verification, Constraint Generation and Invariant Strengthening

We now study possibilities of verification, constraint generation and invariant strengthening for linear hybrid automata; we assume that the properties Φ and ϕ_{safe} to be checked are convex linear predicates over X.

Theorem 9 ([13]). *The following are equivalent for any LHA:*

(1) Φ is an inductive invariant of the hybrid automaton;
(2) For every $q \in Q$ and $e = (q, q') \in E$, the following formulae are unsatisfiable:

$I_q \qquad \mathsf{Init}_q \wedge \neg\Phi(\overline{x})$

$F_{\mathsf{flow}}(q) \qquad \Phi(\overline{x}(t_0)) \wedge \mathsf{Inv}_q(\overline{x}(t_0)) \wedge \underline{\mathsf{flow}}_q(t_0, t) \wedge \mathsf{Inv}_q(\overline{x}(t)) \wedge \neg\Phi(\overline{x}(t)) \wedge t \geq t_0$

$F_{\mathsf{jump}}(e) \qquad \Phi(\overline{x}(t)) \wedge \mathsf{Jump}_e(\overline{x}(t), \overline{x}'(0)) \wedge \mathsf{Inv}_{q'}(\overline{x}'(0)) \wedge \neg\Phi(\overline{x}'(0))$
where if $\mathsf{flow}_q = \bigwedge_{j=1}^{n_q}(\sum_{i=1}^{n} c_{ij}^q \dot{x}_i \leq_j c_j^q)$ *then:*
$\underline{\mathsf{flow}}_q(t, t') = \bigwedge_{j=1}^{n_q}(\sum_{i=1}^{n} c_{ij}^q (x_i' - x_i) \leq_j c_j^q(t' - t))$, *where* $x_i' = x_i(t'), x_i = x_i(t)$.

Theorem 9 shows that linear hybrid automata can be modeled as a type of constraint transition systems, in which the updates are due to flows and jumps, and can be described by the formulae $F_{\mathsf{flow}}(q)$ and $F_{\mathsf{jump}}(e)$ above – in $F_{\mathsf{flow}}(q)$ the value of the variable x_i before the update is $x_i(t_0)$ and the value of the variable x_i after the update is $x_i(t)$, for $t \geq t_0$. Therefore the results in Theorems 6 and 8 apply here in a simplified form, since only the variables X are updated; there are no updates of function symbols, so we can use Algorithm 1 and SEH-PILoT to check whether a formula is an inductive invariant and, if not, for computing constraints on the parameters under which this is guaranteed to be the case.

4.2 Chatter-Freedom and Time-Bounded Reachability

In [13] we also considered properties stating that for every run σ in the automaton S, if ϕ_{entry} holds at the beginning of the run, then ϕ_{safe} becomes true in run σ at latest at time t, i.e. properties of the form $\phi = \Box(\phi_{\mathsf{entry}} \to \Diamond_{\leq t} \phi_{\mathsf{safe}})$.

For *chatter-free hybrid automata*, i.e., automata in which mode entry conditions are chosen with sufficient safety margin (by specifying inner envelopes described by formulae $\mathsf{InEnv}_q, q \in Q$) such that a minimal dwelling time ε_t in each mode is guaranteed, checking such properties can be reduced to bounded model checking.

Definition 10 ([13]).   *A hybrid automaton S is* chatter-free *with minimal dwelling time ε_t (where $\varepsilon_t > 0$) if:*

(i) all transitions lead to an inner envelope, i.e. for all $q \in Q, (q, q') \in E$ the following formula is valid:

$$\forall \overline{x}(\mathsf{Inv}_q(\overline{x}) \wedge \mathsf{guard}_{(q,q')}(\overline{x}) \wedge \mathsf{jump}_{(q,q')}(\overline{x}, \overline{x}') \to \mathsf{InEnv}_{q'}(\overline{x}'));$$

(ii) for any flow starting in the inner envelope of a mode q, no guard of a mode switch (q, q') will become true in a time interval smaller than ε_t.

A hybrid automaton S is chatter-free *iff S is chatter-free with minimal dwelling time ε_t for some $\varepsilon_t > 0$.*

Conditions similar to chatter-freedom (e.g. finite or bounded variability in real-time logics) were also studied e.g. in [33,52].

Theorem 11 ([13]).   *Let S be an LHA.*
Condition (i) in Definition 10 holds iff the following formula is unsatisfiable:

$$\mathsf{Inv}_q(\overline{x}) \wedge \mathsf{guard}_{(q,q')}(\overline{x}) \wedge \mathsf{jump}_{(q,q')}(\overline{x}, \overline{x}') \wedge \neg \mathsf{InEnv}_{q'}(\overline{x}').$$

Condition (ii) in Definition 10 holds iff $\bigwedge_{(q,q') \in E} F_{q,q'}$ is unsatisfiable, where:
$$F_{q,q'} : \quad \mathsf{InEnv}(x_1(0), \ldots, x_n(0)) \wedge \mathsf{Inv}_q(\overline{x}(0)) \wedge \underline{\mathsf{flow}_q}(0, t) \wedge$$
$$\mathsf{guard}_{(q,q')}(x_1(t), \ldots, x_n(t)) \wedge t \leq \varepsilon_t.$$
For any LHA S, checking whether S is chatter-free with time-dwelling ε_t is decidable.

If some of the constants used in specifying these conditions are considered to be parametric, we can use Algorithm 1 to obtain the weakest condition on the parameters under which conditions (i) resp. (ii) hold.

4.3 Example

We consider the following (very simplified) chemical plant example (considered also in [13]), modeling the situation in which we control the reaction of two substances, and the separation of the substance produced by the reaction. Let x_1, x_2 and x_3 be variables which describe the evolution of the volume of substances 1 and 2, and the substance 3 generated from their reaction, respectively. The plant is described by a hybrid automaton with four modes:

Mode 1: Fill. In this mode the temperature is low, and hence the substances 1 and 2 do not react. The substances 1 and 2 (possibly mixed with a very small quantity of substance 3) are filled in the tank in equal quantities up to a given error margin. This is described by the following invariants and flow conditions:

$\mathsf{Inv}_1 :\ x_1 + x_2 + x_3 \leq L_f \ \wedge\ \bigwedge_{i=1}^{3} x_i \geq 0 \ \wedge\ -\varepsilon_a \leq x_1 - x_2 \leq \varepsilon_a \ \wedge\ x_3 \leq \mathsf{min}$

$\mathsf{flow}_1 : \mathsf{dmin} \leq \dot{x}_1 \leq \mathsf{dmax} \wedge \mathsf{dmin} \leq \dot{x}_2 \leq \mathsf{dmax} \wedge \dot{x}_3 = 0 \wedge -\delta_a \leq \dot{x}_1 - \dot{x}_2 \leq \delta_a$

If the proportion is not kept the system jumps into mode 4 (**Dump**); if the total quantity of substances exceeds level L_f the system jumps into mode 2 (**React**).

Mode 2: React. In this mode the temperature is high, and the substances 1 and 2 react. The reaction consumes equal quantities of substances 1 and 2 and produces substance 3.

$\mathsf{Inv}_2 :\ L_f \leq x_1 + x_2 + x_3 \leq L_{\mathsf{overflow}} \wedge \bigwedge_{i=1}^{3} x_i \geq 0 \wedge -\varepsilon_a \leq x_1 - x_2 \leq \varepsilon_a \wedge x_3 \leq \mathsf{max}$

$\mathsf{flow}_2 : \dot{x}_1 \leq -\mathsf{dmin} \wedge \dot{x}_2 \leq -\mathsf{dmin} \wedge \dot{x}_3 \geq \mathsf{dmin} \wedge \dot{x}_1 = \dot{x}_2 \wedge \dot{x}_3 + \dot{x}_1 + \dot{x}_2 = 0$

If the proportion between substances 1 and 2 is not kept the system jumps into mode 4 (**Dump**); if the total quantity of substances 1 and 2 is below some minimal level min the system jumps into mode 3 (**Filter**).

Mode 3: Filter. In this mode the temperature is low again and the substance 3 is filtered out.

$\mathsf{Inv}_3 :\ x_1 + x_2 + x_3 \leq L_{\mathsf{overflow}} \ \wedge\ \bigwedge_{i=1}^{3} x_i \geq 0 \wedge -\varepsilon_a \leq x_1 - x_2 \leq \varepsilon_a \ \wedge\ x_3 \geq \mathsf{min}$

$\mathsf{flow}_3 : \dot{x}_1 = 0 \wedge \dot{x}_2 = 0 \wedge \dot{x}_3 \leq -\mathsf{dmin}$

If the proportion between substances 1 and 2 is not kept the system jumps into mode 4 (**Dump**). Otherwise, if the concentration of substance 3 is below some minimal level min the system jumps into mode 1 (**Fill**).

Mode 4: Dump. In this mode the content of the tank is emptied. We assume that this happens instantaneously, i.e. $\mathsf{Inv}_4 : \bigwedge_{i=1}^{3} x_i = 0$ and $\mathsf{flow}_4 : \bigwedge_{i=1}^{3} \dot{x}_i = 0$.

Jumps. The automaton has the following jumps:

- $e_{12} = (1, 2)$ with $\mathsf{guard}_{e_{12}} = x_1 + x_2 + x_3 \geq L_f$; $\mathsf{jump}_{e_{12}} = \bigwedge_{i=1}^{3} x_i' = x_i$;
- $e_{23} = (2, 3)$ with $\mathsf{guard}_{e_{23}} = x_1 + x_2 \leq \mathsf{min}$; $\mathsf{jump}_{e_{23}} = \bigwedge_{i=1}^{3} x_i' = x_i$;
- $e_{31} = (3, 1)$ with $\mathsf{guard}_{e_{31}} = -\varepsilon_a \leq x_1 - x_2 \leq \varepsilon_a \wedge 0 \leq x_3 \leq \mathsf{min}$; $\mathsf{jump}_{e_{31}} = \bigwedge_{i=1}^{3} x_i' = x_i$;
- Two edges e_{14}^1, e_{14}^2 from 1 to 4, and two edges e_{24}^1, e_{24}^2 from 2 to 4, with:
  $\mathsf{guard}_{e_{j4}^1} = x_1 - x_2 \geq \varepsilon_a$, $\mathsf{guard}_{e_{j4}^2} = x_1 - x_2 \leq -\varepsilon_a$; $\mathsf{jump}_{e_{j4}^i} = \bigwedge_{i=1}^{3} x_i' = 0$;
- Two edges e_{34}^1, e_{34}^2 from 3 to 4, with $\mathsf{guard}_{e_{34}^1} = x_3 \leq \mathsf{min} \wedge x_1 - x_2 \geq \varepsilon_a$;
  $\mathsf{guard}_{e_{34}^2} = x_3 \leq \mathsf{min} \wedge x_1 - x_2 \leq -\varepsilon_a$, and $\mathsf{jump}_{e_{34}^i} = \bigwedge_{i=1}^{3} x_i' = 0$ for $j = 1, 2$.

Tests with SEH-PILoT. We illustrate how SEH-PILoT can be used for testing whether formulae are invariant resp. for constraint generation and invariant strengthening for variants of the linear hybrid system described before. We assume that $\mathsf{min}, \mathsf{max}, \mathsf{dmin}, \mathsf{dmax}, L_{\mathsf{safe}}, L_f, L_{\mathsf{overflow}}, \varepsilon_{\mathsf{safe}}, \varepsilon_a, \delta_a$ are parameters. In order to have only linear constraints, we here often assume that $\mathsf{dmin}, \mathsf{dmax}$ and δ_a are constant (for experiments we choose $\mathsf{dmin} = \delta_a = 1, \mathsf{dmax} = 2$).

1. Invariant checking and constraint generation. Consider the property $\Phi = (x_1 + x_2 + x_3 \leq L_{\mathsf{safe}})$. Without knowing the link between L_{safe}, L_f and L_{overflow} we cannot prove that Φ is an inductive invariant. Due to the form of jump conditions, it is easy to see that Φ is always invariant under the jumps $(1, 2), (2, 3)$ and $(3, 1)$ which do not change the values of the variables and it is invariant under the jumps to mode 4 (which reset all variables to 0) iff $L_{\mathsf{safe}} \geq 0$.

We used SEH-PILoT to generate the weakest constraint on the parameters under which Φ is invariant under flows. This constraint is obtained by eliminating $\{x_1, x_2, x_3, x_1', x_2', x_3', t\}$ from the formulae $\underline{\mathsf{flow}}_q(0, t)$, $q \in \{1, 2, 3, 4\}$. Details about the tests (including, for instance, the input file for SEH-PILoT for mode 1 for $\mathsf{dmin} = \delta_a = 1$) can be found in [37]. SEH-PILoT generates the following constraint for mode 1:

$$\Gamma = (\mathsf{min} < 0 \vee L_{\mathsf{safe}} < 0 \vee L_f - L_{\mathsf{safe}} \leq 0 \vee \varepsilon_a \leq 0)$$

If we know that $\mathsf{min} \geq 0, L_{\mathsf{safe}} \geq 0$ and $\varepsilon_a > 0$ we obtain the condition $L_f \leq L_{\mathsf{safe}}$.

2. Invariant strengthening. Consider now a variant of the hybrid system described before in which the condition $\bigwedge_{i=1}^{3} x_i \geq 0$ was left out from the mode invariants for modes 1 and 2. We here assume in addition – for simplifying the presentation – that $L_f = L_{\mathsf{overflow}}$. Consider the property $\Phi = x_1 + x_2 \leq L_f$.

We want to show that Φ holds on all runs in the automaton which start in mode 1 in the state $\mathsf{Init} := x_1 = 0 \wedge x_2 = 0 \wedge x_3 = 0$. Φ clearly holds in Init if $L_f > 0$. However, without the additional conditions $\bigwedge_{i=1}^{3} x_i \geq 0$ in the invariants of modes 1 and 2, we cannot prove that Φ is invariant under flows. We try to strengthen Φ in order to obtain an inductive invariant – e.g. by obtaining an additional condition on x_3, by eliminating $\{x_1, x_2, x_1', x_2', x_3', t\}$ from the formulae $\underline{\mathsf{flow}}_q(0, t)$, $q \in \{1, 2, 3, 4\}$. The input file for SEH-PILoT for mode 1 (again for $\mathsf{dmin} = \delta_a = 1$) can be found in [37]. SEH-PILoT generates the following constraint for mode 1:

$$\Gamma := x_3 \geq 0 \vee x_3 > \mathsf{min} \vee \varepsilon_a \leq 0.$$

Since we assume that $\varepsilon_a > 0$ and the invariant of mode 1 contains the condition $x_3 \leq \mathsf{min}$, it follows that we can consider the candidate invariant:

$$\Phi_1 := (x_1 + x_2 \leq L_f) \wedge x_3 \geq 0.$$

It can be checked that Φ_1 holds in the initial state described by the formula Init and it is preserved under jumps (under the assumption that $L_f \geq 0$) and flows.

3. Constraint generation for guaranteeing chatter-freedom. We now analyze the property of chatter-freedom. Let $\varepsilon > 0$ and consider the following parametrically described inner envelope for mode 1:

$$\mathsf{InEnv}_1 := x_1 + x_2 + x_3 \leq L_{\mathsf{safe}} \wedge \bigwedge_{i=1}^{3} x_i \geq 0 \wedge |x_1 - x_2| \leq \varepsilon_{\mathsf{safe}} \wedge x_3 \leq \mathsf{min}$$

where $0 < L_{\mathsf{safe}} \leq L_f$, $0 < \varepsilon_{\mathsf{safe}} \leq \varepsilon_a$, $\mathsf{dmin} < \mathsf{dmax}$, and $\mathsf{min}, \mathsf{dmax}, \mathsf{dmin} > 0$. We show how SEH-PILoT can be used to generate constraints on the parameters $L_f, L_{\mathsf{overflow}}, \varepsilon_a, \delta_a, \mathsf{min}, \mathsf{dmin}$ and $L_{\mathsf{safe}}, \varepsilon_{\mathsf{safe}}$ under which a minimal dwelling time ε in mode 1 is guaranteed. We analyze the jumps $e \in \{e_{12}, e_{14}^1, e_{14}^2\}$ from mode 1 to modes 2 and 4, and for each such e use SEH-PILoT to generate constraints Γ_e on the parameters under which the following formula is unsatisfiable:

$$\mathsf{InEnv}_1(\overline{x}) \wedge \mathsf{Inv}_1(\overline{x}) \wedge \underline{\mathsf{flow}}(\overline{x}, \overline{x}', t) \wedge \mathsf{guard}_e(\overline{x}') \wedge t < \varepsilon.$$

SEH-PILoT generates the following constraints (simplification with SLFQ is crucial as before simplification the formulae are very long); we left out the parts of the conjunction which are negations of conditions on the parameters.

$$\begin{aligned}
\Gamma_{e_{14}^1}, \Gamma_{e_{14}^2} : \ &L_{\mathsf{safe}} < \varepsilon_a - ((\mathsf{dmax} - \mathsf{dmin}) * \varepsilon) \vee e_{\mathsf{safe}} < (\varepsilon_a - (\mathsf{dmax} - \mathsf{dmin}) * \varepsilon) \vee \\
&L_{\mathsf{safe}} < (\varepsilon_a - \delta_a * \varepsilon) \vee \varepsilon_{\mathsf{safe}} < (\varepsilon_a - \delta_a * \varepsilon) \\
\Gamma_{e_{12}} : \quad &L_{\mathsf{safe}} < (L_f - (2 * \mathsf{dmax}) * \varepsilon).
\end{aligned}$$

5   Conclusions

In this paper we presented a survey of our work in the verification of parametric systems using hierarchical reasoning, and on deriving constraints on parameters and invariant strengthening using hierarchical symbol elimination, as well as an ongoing implementation. In future work we plan to extend the implementation of the hierarchical symbol elimination with further optimizations proposed in [38], use these optimizations for making invariant strenghtening more efficient, and compare the results with the orthogonal results proposed in [30]. We would like to analyze the applicability of these ideas to the verification of systems consisting of a parametric number of similar, interacting components, in the context of modular approaches to verification – for instance approaches we considered in [16] and [11], either using locality of logical theories for establishing a small model property, or using methods proposed in [41, 43].

Acknowledgments. We thank the reviewers for their helpful comments.

References

1. Alur, R., Henzinger, T.A., Ho, P.: Automatic symbolic verification of embedded systems. IEEE Trans. Softw. Eng. **22**(3), 181–201 (1996)
2. Barrett, C., et al.: CVC4. In: Gopalakrishnan, G., Qadeer, S. (eds.) CAV 2011. LNCS, vol. 6806, pp. 171–177. Springer, Heidelberg (2011). https://doi.org/10.1007/978-3-642-22110-1_14
3. Benveniste, A., et al.: Contracts for system design. Found. Trends Electron. Des. Autom. **12**(2–3), 124–400 (2018)
4. Bjørner, N., de Moura, L., Nachmanson, L., Wintersteiger, C.M.: Programming Z3. In: Bowen, J.P., Liu, Z., Zhang, Z. (eds.) SETSS 2018. LNCS, vol. 11430, pp. 148–201. Springer, Cham (2019). https://doi.org/10.1007/978-3-030-17601-3_4
5. Bradley, A.R.: IC3 and beyond: Incremental, inductive verification. In: Madhusudan, P., Seshia, S.A. (eds.) CAV 2012. LNCS, vol. 7358, p. 4. Springer, Heidelberg (2012). https://doi.org/10.1007/978-3-642-31424-7_4
6. Bradley, A.R., Manna, Z.: Property-directed incremental invariant generation. Formal Asp. Comput. **20**(4–5), 379–405 (2008)
7. Chang, C.C., Keisler, H.J.: Model Theory. Studies in Logic and the Foundations of Mathematics, vol. 73, 3rd edn. North-Holland (1992)
8. Cimatti, A., Palopoli, L., Ramadian, Y.: Symbolic computation of schedulability regions using parametric timed automata. In: Proceedings of the 29th IEEE Real-Time Systems Symposium, RTSS 2008, pp. 80–89. IEEE Computer Society (2008)
9. Cimatti, A., Roveri, M., Tonetta, S.: Requirements validation for hybrid systems. In: Bouajjani, A., Maler, O. (eds.) CAV 2009. LNCS, vol. 5643, pp. 188–203. Springer, Heidelberg (2009). https://doi.org/10.1007/978-3-642-02658-4_17
10. Damm, W., Dierks, H., Oehlerking, J., Pnueli, A.: Towards component based design of hybrid systems: safety and stability. In: Manna, Z., Peled, D.A. (eds.) Time for Verification. LNCS, vol. 6200, pp. 96–143. Springer, Heidelberg (2010). https://doi.org/10.1007/978-3-642-13754-9_6
11. Damm, W., Horbach, M., Sofronie-Stokkermans, V.: Decidability of verification of safety properties of spatial families of linear hybrid automata. In: Lutz, C., Ranise, S. (eds.) FroCoS 2015. LNCS (LNAI), vol. 9322, pp. 186–202. Springer, Cham (2015). https://doi.org/10.1007/978-3-319-24246-0_12
12. Damm, W., Ihlemann, C., Sofronie-Stokkermans, V.: Decidability and complexity for the verification of safety properties of reasonable linear hybrid automata. In: Caccamo, M., Frazzoli, E., Grosu, R. (eds.) Proceedings of the 14th ACM International Conference on Hybrid Systems: Computation and Control, HSCC 2011, pp. 73–82. ACM (2011)
13. Damm, W., Ihlemann, C., Sofronie-Stokkermans, V.: PTIME parametric verification of safety properties for reasonable linear hybrid automata. Math. Comput. Sci. **5**(4), 469–497 (2011)
14. Dillig, I., Dillig, T., Li, B., McMillan, K.L.: Inductive invariant generation via abductive inference. In: Hosking, A.L., Eugster, P.T., Lopes, C.V. (eds.) Proceedings of the 2013 ACM SIGPLAN International Conference on Object Oriented Programming Systems Languages & Applications, OOPSLA 2013, Part of SPLASH 2013, pp. 443–456. ACM (2013)
15. Dolzmann, A., Sturm, T.: Redlog: Computer algebra meets computer logic. ACM SIGSAM Bull. **31**(2), 2–9 (1997)
16. Faber, J., Ihlemann, C., Jacobs, S., Sofronie-Stokkermans, V.: Automatic verification of parametric specifications with complex topologies. In: Méry, D., Merz,

S. (eds.) IFM 2010. LNCS, vol. 6396, pp. 152–167. Springer, Heidelberg (2010). https://doi.org/10.1007/978-3-642-16265-7_12

17. Falke, S., Kapur, D.: When is a formula a loop invariant? In: Martí-Oliet, N., Ölveczky, P.C., Talcott, C. (eds.) Logic, Rewriting, and Concurrency. LNCS, vol. 9200, pp. 264–286. Springer, Cham (2015). https://doi.org/10.1007/978-3-319-23165-5_13

18. Frehse, G., Jha, S.K., Krogh, B.H.: A counterexample-guided approach to parameter synthesis for linear hybrid automata. In: Egerstedt, M., Mishra, B. (eds.) HSCC 2008. LNCS, vol. 4981, pp. 187–200. Springer, Heidelberg (2008). https://doi.org/10.1007/978-3-540-78929-1_14

19. Ghilardi, S.: Model-theoretic methods in combined constraint satisfiability. J. Autom. Reasoning **33**(3–4), 221–249 (2004)

20. Ghilardi, S., Nicolini, E., Ranise, S., Zucchelli, D.: Combination methods for satisfiability and model-checking of infinite-state systems. In: Pfenning, F. (ed.) CADE 2007. LNCS (LNAI), vol. 4603, pp. 362–378. Springer, Heidelberg (2007). https://doi.org/10.1007/978-3-540-73595-3_25

21. Ghilardi, S., Ranise, S.: Backward reachability of array-based systems by SMT solving: Termination and invariant synthesis. Logical Methods Comput. Sci. **6**(4) (2010)

22. Hodges, W.: A Shorter Model Theory. Cambridge University Press, Cambridge (1997)

23. Hune, T., Romijn, J., Stoelinga, M., Vaandrager, F.W.: Linear parametric model checking of timed automata. J. Log. Algebr. Program. **52–53**, 183–220 (2002)

24. Ihlemann, C., Jacobs, S., Sofronie-Stokkermans, V.: On local reasoning in verification. In: Ramakrishnan, C.R., Rehof, J. (eds.) TACAS 2008. LNCS, vol. 4963, pp. 265–281. Springer, Heidelberg (2008). https://doi.org/10.1007/978-3-540-78800-3_19

25. Ihlemann, C., Sofronie-Stokkermans, V.: System description: H-PILoT. In: Schmidt, R.A. (ed.) CADE 2009. LNCS (LNAI), vol. 5663, pp. 131–139. Springer, Heidelberg (2009). https://doi.org/10.1007/978-3-642-02959-2_9

26. Ihlemann, C., Sofronie-Stokkermans, V.: On hierarchical reasoning in combinations of theories. In: Giesl, J., Hähnle, R. (eds.) IJCAR 2010. LNCS (LNAI), vol. 6173, pp. 30–45. Springer, Heidelberg (2010). https://doi.org/10.1007/978-3-642-14203-1_4

27. Jacobs, S., Kuncak, V.: Towards complete reasoning about axiomatic specifications. In: Jhala, R., Schmidt, D. (eds.) VMCAI 2011. LNCS, vol. 6538, pp. 278–293. Springer, Heidelberg (2011). https://doi.org/10.1007/978-3-642-18275-4_20

28. Jacobs, S., Sofronie-Stokkermans, V.: Applications of hierarchical reasoning in the verification of complex systems. Electr. Notes Theor. Comput. Sci. **174**(8), 39–54 (2007)

29. Kapur, D.: A quantifier-elimination based heuristic for automatically generating inductive assertions for programs. J. Syst. Sci. Complex. **19**(3), 307–330 (2006)

30. Karbyshev, A., Bjørner, N., Itzhaky, S., Rinetzky, N., Shoham, S.: Property-directed inference of universal invariants or proving their absence. J. ACM **64**(1), 7:1–7:33 (2017)

31. Kramer, B., Neurohr, C., Büker, M., Böde, E., Fränzle, M., Damm, W.: Identification and quantification of hazardous scenarios for automated driving. In: Zeller, M., Höfig, K. (eds.) IMBSA 2020. LNCS, vol. 12297, pp. 163–178. Springer, Cham (2020). https://doi.org/10.1007/978-3-030-58920-2_11

32. Mal'cev, A.: Axiomatizable classes of locally free algebras of various types. In: The Metamathematics of Algebraic Systems. Collected Papers: 1936–1967. Studies in Logic and the Foundation of Mathematics, vol. 66, chap. 23. North-Holland, Amsterdam (1971)

33. Maler, O., Nickovic, D.: Monitoring temporal properties of continuous signals. In: Lakhnech, Y., Yovine, S. (eds.) FORMATS/FTRTFT -2004. LNCS, vol. 3253, pp. 152–166. Springer, Heidelberg (2004). https://doi.org/10.1007/978-3-540-30206-3_12

34. Manna, Z., Pnueli, A.: Temporal Verification of Reactive Systems - Safety. Springer, Cham (1995)

35. Marohn, P., Sofronie-Stokkermans, V.: SEH-PILoT: A system for property-directed symbol elimination - work in progress (short paper). In: Schmidt, R.A., Wernhard, C., Zhao, Y. (eds.) Proceedings of the Second Workshop on Second-Order Quantifier Elimination and Related Topics (SOQE 2021) Associated with the 18th International Conference on Principles of Knowledge Representation and Reasoning (KR 2021), Online Event, 4 November 2021. CEUR Workshop Proceedings, vol. 3009, pp. 75–82. CEUR-WS.org (2021)

36. Padon, O., Immerman, N., Shoham, S., Karbyshev, A., Sagiv, M.: Decidability of inferring inductive invariants. In: Bodík, R., Majumdar, R. (eds.) Proceedings of the 43rd Annual ACM SIGPLAN-SIGACT Symposium on Principles of Programming Languages, POPL 2016, pp. 217–231. ACM (2016)

37. Peuter, D., Marohn, P., Sofronie-Stokkermans, V.: On the verification of parametric systems (2023). arXiv/CORR https://doi.org/10.48550/arXiv.2310.18069

38. Peuter, D., Sofronie-Stokkermans, V.: On invariant synthesis for parametric systems. In: Fontaine, P. (ed.) CADE 2019. LNCS (LNAI), vol. 11716, pp. 385–405. Springer, Cham (2019). https://doi.org/10.1007/978-3-030-29436-6_23

39. Platzer, A., Quesel, J.-D.: European train control system: A case study in formal verification. In: Breitman, K., Cavalcanti, A. (eds.) ICFEM 2009. LNCS, vol. 5885, pp. 246–265. Springer, Heidelberg (2009). https://doi.org/10.1007/978-3-642-10373-5_13

40. Sangiovanni-Vincentelli, A.L., Damm, W., Passerone, R.: Taming Dr. Frankenstein: Contract-based design for cyber-physical systems. Eur. J. Control **18**(3), 217–238 (2012)

41. Sofronie-Stokkermans, V.: Fibered structures and applications to automated theorem proving in certain classes of finitely-valued logics and to modeling interacting systems. Ph.D. thesis, Johannes Kepler University, Linz, Austria (1997)

42. Sofronie-Stokkermans, V.: Hierarchic reasoning in local theory extensions. In: Nieuwenhuis, R. (ed.) CADE 2005. LNCS (LNAI), vol. 3632, pp. 219–234. Springer, Heidelberg (2005). https://doi.org/10.1007/11532231_16

43. Sofronie-Stokkermans, V.: Sheaves and geometric logic and applications to modular verification of complex systems. In: Goubault, É. (ed.) Proceedings of the Workshops on Geometric and Topological Methods in Concurrency Theory, GETCO@DISC 2004 + GETCO@CONCUR 2005 + + GETCO 2006. Electronic Notes in Theoretical Computer Science, vol. 230, pp. 161–187. Elsevier (2006)

44. Sofronie-Stokkermans, V.: Efficient hierarchical reasoning about functions over numerical domains. In: Dengel, A.R., Berns, K., Breuel, T.M., Bomarius, F., Roth-Berghofer, T.R. (eds.) KI 2008. LNCS (LNAI), vol. 5243, pp. 135–143. Springer, Heidelberg (2008). https://doi.org/10.1007/978-3-540-85845-4_17

45. Sofronie-Stokkermans, V.: Hierarchical reasoning for the verification of parametric systems. In: Giesl, J., Hähnle, R. (eds.) IJCAR 2010. LNCS (LNAI), vol. 6173, pp.

171–187. Springer, Heidelberg (2010). https://doi.org/10.1007/978-3-642-14203-1_15

46. Sofronie-Stokkermans, V.: Hierarchical reasoning and model generation for the verification of parametric hybrid systems. In: Bonacina, M.P. (ed.) CADE 2013. LNCS (LNAI), vol. 7898, pp. 360–376. Springer, Heidelberg (2013). https://doi.org/10.1007/978-3-642-38574-2_25

47. Sofronie-Stokkermans, V.: On interpolation and symbol elimination in theory extensions. In: Olivetti, N., Tiwari, A. (eds.) IJCAR 2016. LNCS (LNAI), vol. 9706, pp. 273–289. Springer, Cham (2016). https://doi.org/10.1007/978-3-319-40229-1_19

48. Sofronie-Stokkermans, V.: On interpolation and symbol elimination in theory extensions. Log. Methods Comput. Sci. **14**(3) (2018)

49. Sofronie-Stokkermans, V.: Parametric systems: Verification and synthesis. Fundam. Informaticae **173**(2–3), 91–138 (2020)

50. Tarski, A.: A Decision Method for Elementary Algebra and Geometry, 2nd edn. University of California Press, Berkeley (1951)

51. Wang, F.: Symbolic parametric safety analysis of linear hybrid systems with BDD-like data-structures. IEEE Trans. Softw. Eng. **31**(1), 38–51 (2005)

52. Wilke., T.: Specifying timed state sequences in powerful decidable logics and timed automata. In: Langmaack, H., de Roever, W.-P., and Vytopil, J. (eds.) FTRTFT 1994. LNCS, vol. 863, pp. 694–715. Springer (1994)

What if… We Applied Model-Based AI Engineering to Safety-Critical Systems?

Jens Braband[1(✉)] and Hendrik Schäbe[2]

[1] Siemens Mobility GmbH, Ackerstr. 22, 38023 Braunschweig, Germany
jens.braband@siemens.com
[2] TÜV Rheinland Intertraffic GmbH, Am Grauen Stein, 51105 Cologne, Germany
dr.hendrik.schaebe@gmail.com

Abstract. The success of the applications of so-called "artificial intelligence" (AI) are unmistakable. AI is used in numerous fields: entertainment, art, and in many technical systems, AI-supported assistance systems are in use. Despite all the remarkable achievements, however, it has not yet been possible to provide a safety case for an AI or even to obtain approval, for cases, where the AI is solely responsible for a safety-relevant decision.

The discussion about AI-supported systems has also been raised in railway technology, especially in automated driving. In this paper we explain by a thought experiment for a particular problem from safety-related communication systems, whose solution is known, the advantages of model-based AI engineering. In the thought experiment we discard the underlying results from coding theory and treat the algorithm as a black box, like a complex AI based algorithm, and exploit only empirical and structural properties. So, we can compare such results with the mathematically correct results and explore the gap between theory and practice. We argue why such an approach may "close the gap" between AI performance and safety requirements.

Keywords: Railways · Safety Case · AI · EN 50159

1 Introduction

1.1 Standardization Background

The success of the applications of so-called "artificial intelligence" (AI) are unmistakable. AI is used in numerous fields: entertainment, art, and in many technical systems, AI-supported assistance systems are in use. A good overview of AI methods and fields of application can be found in, e. g., [1]. A particular difficulty is also that so far there is not even a harmonised definition of what AI is. E. g., in the proposed EU AI Act [2] it is acknowledged that the definition is difficult and instead the term "AI system" is proposed to mean "…*software that is developed with [specific] techniques and approaches [listed in Annex 1] and can, for a given set of human-defined objectives, generate outputs such as content, predictions, recommendations, or decisions influencing the environments they interact with*". According to this annex, the notion of AI system would refer to

M. Fränzle et al. (Eds.): Werner Damm Festschrift, LNCS 15471, pp. 222–233, 2026.
https://doi.org/10.1007/978-3-031-97537-0_13

a range of software-based technologies that encompasses machine learning, logic and knowledge-based systems, and statistical approaches. In particular, the definition of the last class is amazing as it includes classical and "Bayesian estimation, search and optimization methods". So even a simple linear regression in a spreadsheet might qualify as an AI system under the AI act.

Also, technical and political preparations are also being prepared for the use of AI in safety-relevant systems, right up to a planned "TÜV for AI". Despite all the success, however, it has not yet been possible to provide a safety case for an AI or even to obtain approval, for cases, where the AI is solely responsible for a safety-relevant decision. If used outside a test field, in both medical and automotive technology, the decision of an AI must always be able to be overruled by a human operator. Especially in automotive technology, there have nevertheless already been fatal accidents. This occurred for the following two reasons:

a) the AI was not used as intended, i.e. intentionally the full responsibility has put on the AI by the driver, although it should be used as assistance system, or.
b) the safety driver was distracted.

In addition, there have been previously unexplained incidents, such as the collisions of autonomous driving cars with stationary emergency vehicles showing flashing lights. Therefore, a certain disillusionment has occurred.

The discussion about AI-supported systems has also been raised in railway technology, especially in automated driving. Compared to other fields of application, the railway system has the advantage that driverless systems are already the norm in metros and that the conditions of use are simpler than, e. g., in automotive technology.

Other possible applications can be seen in railways, for example, in the following areas:

- Driver assistance systems such as warning systems,
- Energetically optimized routing and
- Predictive Maintenance.

In the corresponding railway standards [3–5] AI does not play any role so far, only in EN 50128 [6] AI appears as a sidekick for fault correction (Annex D.1) and is explicitly "not recommended (NR)" for all safety integrity levels (Annex A table 3.13). This falls short of forbidding the technique, if it was to be used, it must be explicitly justified.

But this entry in EN 50128 leads to interesting discussions, e. g., what is really meant by this entry, e. g., what AI fault correction really means. One example might come from the area of data communication [5], where error detection and correction techniques play an important role, not based on AI but on coding theory. As some of the reasoning in safety cases according to EN 50159 are based on statistical approaches, is implemented by software (SW) and generates outputs that lead to decision e. g., if a safety-critical message is correct or not, even such an application might qualify as an AI system.

From this starting point a simple thought experiment could take off: what if we had a complex coding algorithm that we either do not know or that we do not completely understand, what would happen if we would treat such a communication system as an AI system? Could we provide a justification to base a safety-critical decision on this AI system?

So maybe by discussing this example we may come closer to answering the question how to "close the gap" [7] between AI system performance in practice and the difficulty of building up a thorough safety argument for its use, in particular in high-risk applications.

1.2 Background on Safety-Related Communication According to EN 50159

Communication systems play an increasing role in railways and are widely used. In particular, for safety-related communication EN 50159 [5] has been a huge success and is not only used in railways, but also in industry automation or process control (see the slightly generalized IEC 62280 variant). It assumes that safety-related communication is protected by a safety code, in traditional engineering mostly CRCs, but in more modern applications also hash codes. The safety case is based on statistical arguments, in particular on the bit and telegram error probability. These are the probabilities that a bit or a telegram are distorted. In particular the probability that a telegram error is not detected, plays an important role. That means that first, the problem has to be mathematically described.

We state here only the background necessary for our thought experiment, a more comprehensive and detailed discussion can be found in [8]. Nowadays mostly Bose–Chaudhuri–Hocquenghem codes (BCH) are used, which form a very flexible class of cyclic error-detection or correcting codes. They are constructed using polynomials over a finite field. A key feature of BCH codes is that during code design they allow a precise control over the number of symbol errors that are detectable and/or correctable by the code. Another advantage is the ease of coding and decoding of BCH codes so that they are undoubtedly state of the art since the late 1970s [9]. However, without a solid mathematical background, e. g., algebra and in particular Galois fields the theory is almost incomprehensible.

This holds even more for cryptographic hash codes which are mainly used to compress data into a fingerprint so that another cryptographic application, e. g., a digital signature needs only to work on the fingerprint. But in contrast to classical codes, here different design principles are necessary as it must be practicably impossible for a third party to find another data set that leads to the same fingerprint. For further details see [9].

2 Communication Coding Seen from an AI System Perspective

2.1 Setup of Communication Coding as an AI System

Imagine that we want to transmit in real-time a user message of m bit from S to R and that only one channel T exists for communication. The channel is not error free but we assume we know the HW that is used for coding and decoding and we have some information on the transmission line. We assume that errors or disturbances occur only from physical sources or systematic errors in the devices, but we exclude intentional modification of messages, e. g., by hackers. The only thing that we don't know or don't understand sufficiently well is the particular coding algorithm C, that adds r bit of redundancy to the message, extending it to $n = m + r$ bit in total.

So, imagine we are given the generator polynomial of a code.

$$g(x) = x^8 + x^7 + x^6 + x^4 + x^0,$$

which is just another representation of the binary sequence 111010001, and that we may also represent any binary message by its corresponding polynomial $m(x)$. The encoding rule is then simple that we divide $m(x)x^k$ by $g(x)$ and append the remainder of this division $r(x)$ to the original message. This means that (written as polynomials) we transmit the telegram $t(x) = m(x)x^k + r(x)$. On the receiving side we again divide by $g(x)$ and only if $t(x)$ represents a valid codeword the remainder is zero and the user message $m(x)$ is accepted. A non-zero remainder indicates a detected error and contains also information about the possible type of error.

We can now transmit as many messages as we want, and at the receiver end we always get a decision by C, whether the message has been correctly transmitted or not. Later we can also check whether the decision was correct, or not, but not in real-time as needed for our application.

Let us first check whether C might qualify as an AI system. Firstly, it is a complex SW and we assume that we don't know how the algorithm was created (this is the weak point of our argument). Certainly, it makes decisions about the environment, which, depending on the environment, may be high risk. E. g., it is not uncommon that very small error rates are required for safety-related communication systems, e. g., 10^{-9} per hour or even less. In many applications there needs to be an update not later than 10 s or so, so it is not unlikely to have applications that demand message error rates below 10^{-12} per message or even less.

Note that according to EN 50128 the use of such an AI system is not recommended if we correct errors. In order to avoid such discussion and also to make the treatment of the example easier, we will deal only with error detection, for which EN 50128 does not provide any recommendations. However, the mathematical arguments would be similar.

In the next chapters we outline and discuss some approaches to the verification of such small error probabilities. It should be noted that by model or model-driven approach we mean here any approach that describes a system or a physical phenomenon of such a system by a set of variables and a set of equations or relations between the variables. A model is called partial if it contains only some of the relevant variables.

2.2 Exhaustive Testing

For small m it might be possible to generate and encode all possible messages, but for common parameters in the railways like $m = 100$ this is infeasible. Also, even if we generate all messages, we will also need information about all possible error patterns. Even if we would assume that the bit error rate must be low in order to ensure availability of the railway application (much less than 1 error on average in 1000 bit), we would need a model of the channel T.

In EN 50159 it is assumed that for random errors the Binary Symmetric Channel (BSC) model holds (but for arbitrary bit error rates b) and also typical systematic errors need to be checked for, e. g., stuck-at or burst error (let us call the combination BSC +). But this does not make exhaustive testing easier and we may assume that it is infeasible for our AI system.

2.3 Statistical Testing

We assume like in the preceding section that we have a statistical environment model (BSC +) and that the environment is stationary. This assumption is not true in reality, but as we have to assume by EN 50159 the worst-case bit error rate (note that this is not always 0.5, as one might assume), we may safely assume it.

So assuming this stationary environment BSC +, we now have a Bernoulli experiment for any message that is transmitted and we are interested in the parameter p in this experiment (the true error rate).

As we are in classical statistics now, we can simply recall the classical results like the normal approximation of the two-sided confidence interval

$$p_{emp} \pm z \sqrt{\frac{p_{emp}}{n}} \tag{1}$$

where p_{emp} denotes the empirical error probability and z the corresponding percentile of the limiting normal distribution for confidence level (1-α). We could also formulate this as a one-sided confidence interval, but two problems would remain:

1. However small we choose α, there is always a chance that we cannot bound the true value p by this approach
2. The length of the confidence interval decrease only with the square root of the sample size, so in order to get another magnitude of accuracy we need hundred times more data

Also, we would have not only to carry out a single experiment for a particular bit error rate, but we would have to do this for a range of values, and we would have to introduce further assumptions if we want to limit the number of experiments, e. g., continuity or monotonicity of the resulting error probabilities with respect to the bit error rates.

2.4 Laplace Model Approach

The argument starts with the observation that we have 2^m user data to be transmitted, that are encoded into 2^n possible telegrams, that are sent over the channel. As the coding is deterministic, there are also only 2^m correctly coded messages. All other messages would be discarded as not correctly coded. As we do not know how the coding is performed in this thought experiment and BCH is a complex code, the coding may appear to us almost random at a first glance.

So, we might assume as an approximation, that if an error occurs on the transmission channel, then because of the complex coding process and the BSC + channel the result may appear like a Laplace experiment to us. This means that, if an error occurs, the chance of getting a formally correct, but wrong code word, is simply a Laplace experiment. So, we have 2^m formally correct code words, but we may randomly choose among 2^n possible telegrams. By Laplace the error probability would be 2^{-r}, where r is just the length of the extra redundancy bits.

Let us take a small classic BCH code with $m = 7$, $n = 15$ and $r = 8$. The code can transmit 128 different user messages, and there are 32768 possible telegrams. The

Laplace error estimate would be 2^{-8} or about 4×10^{-3}. So, if we knew the frequency f of corrupted messages, then the undetected error rate would be $4f \, 10^{-3}$. Note that for a low demand system, e. g., with f less than once a year, this might be sufficient to claim SIL 2.

This approach depends on a couple of assumptions, e. g., statistically the code should almost appear like a random number generator, and this could be checked. E. g., take a coded message, introduce a random error in the user data and code that message, too. Then XOR the redundancy part of the two messages and test the result for randomness. In the small example above this would probably not be successful as not even all possible code combinations are produced, but for larger messages or even more complex codes the result might not differ statistically significantly from random noise. And in reality, this argument was used successfully for cryptographic hash codes and also the assumption could be justified [11].

But better results may only be expected if we do not treat the complex code as black box but try to exploit some more properties.

2.5 Structural Assumptions

Even without knowing the BCH construction principle, we might experiment with some codewords and find out that the code is linear. This means that combining two valid codewords by XOR leads to another valid codeword. Even with many tests we would not find a counterexample.

If we accept this assumption (we cannot prove it without understanding BCH codes) and the BCH + model, then we know exactly the structure of undetected errors: in order to produce one, the error must have the same structure as the code itself. So, for small codes we might now count exhaustively the number of codewords W_i with a particular Hamming weight i [9]. Simply put the Hamming weight is just the number of 1s in a codeword. Now under the BCH + model we can easily derive a formula for the undetected error probability

$$p_u = \sum_{i=1}^{n} W_i b^i (1 - b)^{n-i}$$

When counting the weights, we might also observe that small weights don't occur, e. g., in the example above $W_1 = W_2 = W_3 = W_4 = 0$. This is due to the BCH design principles for the minimal Hamming distance d, but we assume here that we do not know the BCH construction. But such structural properties could also be explored and postulated purely by heuristic arguments. Also, when plotting the weight distributions for several codes we might observe that the weight distribution follows a kind of binomial structure

$$p_u \cong \sum_{i=d}^{n} \binom{n}{i} b^i (1 - b)^{n-i}$$

So even for larger codes it seems that we have now as reasonably good estimate for the undetected error probability. And if we would go for worst case approximations, we know the binomial distribution is maximized for $b = \frac{1}{2}$. So, for the worst case we would get

$$p_u \cong 2^{-r} \tag{2}$$

which corresponds well to the Laplace model. And correspondingly, for very small bit error probabilities $b < < 1$ we can derive

$$p_u \cong W_d p^d$$

So we have now arrived at a heuristic model that allows us to generate all undetected error probabilities under the BSC + model, without any deeper understanding of the construction of the BCH codes, just from empirical observations.

3 The Ground Truth

3.1 The Real Solution from Coding Theory

In Sect. 2 we discussed how far we can go with a black box view of the message coding. But being based on algebra, coding theory has developed a deep understanding of error detection and correction with mathematically designed codes such as BCH codes.

For example, it seems quite impossible to really count the code words and the weight distribution for longer codes, e. g., a BCH code with $n = 255$ and $m = 233$. But coding theory has developed the theory of dual codes [9], and from the weight distribution of the dual code V_i the weight distribution W_i of the original code can be reconstructed, and the undetected error probability can be directly calculated from the generating function of the weight distribution by

$$p_u = B(1 - 2b)2^{-r}, \text{ with } B(z) = \sum_{i=1}^{n} V_i z^i$$

The beauty of the dual code is that message content and redundancy size are swapped, so the dual code in the example has $m = 32$ (instead of $m = 233$) and can be effectively evaluated [12]. So, the exact weight distribution for almost all relevant CRCs can be exactly calculated for the BSC + model.

This allows us to compare our black box approximations to the white box ground truth. For this purpose, it is useful to introduce the concept of proper codes: a code is called proper, if its undetected error probability is strictly monotonic with respect to the bit error probability b.

For proper codes our simple Laplace model is true as a worst case estimate for the undetected error probability, see Fig. 1 for an example. This may seem a weak result, but for many cases it allows effective dimensioning of the coding mechanisms and it eases the safety case as this is a robust bound and we do not have to argue about or monitor bit error probabilities.

The evaluation of many examples has also shown that the binomial approximation works quite well for proper codes, if the minimal Hamming distance is taken into account.

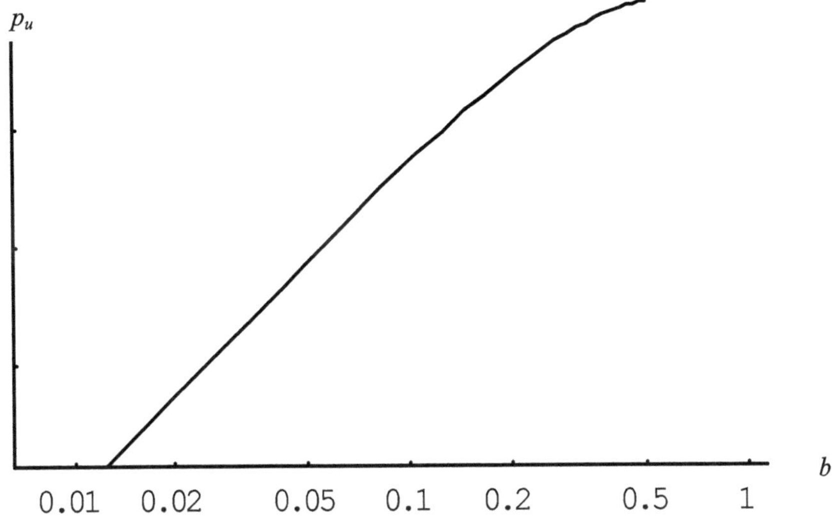

Fig. 1. Example of the undetected error probability for a proper code

And vice versa, the binomial approximation is strictly monotonic, which also suggests a good fit.

So, more or less, our simple black box approximations (2) work well for proper codes. And, for a long time, it was assumed that almost all codes were proper, though this could never be proven. But when the numerical tools became available to calculate the exact weight distributions [12], it became obvious that not all of the widely used codes were proper, see Fig. 2 for an improper code with $r = 32$.

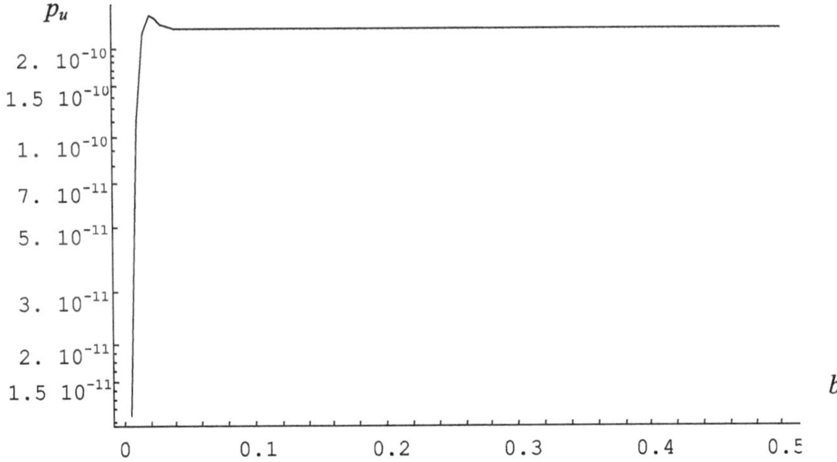

Fig. 2. Example of the undetected error probability for an improper code

3.2 Practical Implications

Unfortunately, there is no simple means to predict whether a code will be proper, or not [12]. Additionally, the construction of BCH codes for a particular set of parameters is not unique, there exist a variety of polynomials that share the same algebraic properties [9].

In EN 50159 [5] it is required either to use proper codes, or if the properness is not known, to evaluate the undetected error probability and to use the maximum of the curve for a robust estimate of the undetected error probability. So, either the Laplace approximation (2) is used (like in Fig. 1) or the curve is completely evaluated (see Fig. 2). But even today the underlying channel model, which plays an important role for the safety case, is still discussed and enhanced [14].

However, it is known that most communication systems would not work reliably at the corresponding bit error probabilities b. Also, evaluation of many improper codes has shown that the error is relatively small, in most practical cases a factor between 10 and 100. So, by taking the worst case bound (2) we are performing a practical error but in practice the bound implies overengineering by several order of magnitudes (see Fig. 2) so this error does not matter in practice.

4 Towards a Safety Case for AI Based Functions

If a safety integrity level (SIL) has been determined for a function by a risk assessment, this SIL must also be demonstrated in case this function is implemented with AI [3]. There are two conceivable ways of proving the safety of AI systems. The first one is the statistical approach, the second one would be the model-based approach. In the following, both ways will be briefly discussed.

Strictly speaking, the safety case consists of two parts that must cover two aspects. Firstly, there is a hardware on which the AI runs and a software in which it is implemented. We would therefore need to prove a safety integrity level (SIL) appropriate for the application and the required tolerable functional failure rate (TFFR) [4]. For the SIL, it should be noted that, if applicable, the tools used to determine the parameters could also be safety relevant and that the function implemented by AI must be validated in the overall system. In this context, a safety qualification period will usually also be necessary. The requirements of the SIL with regard to SW [6] should be easy to fulfil, since the generic implementation of the AI based algorithm should be easily validated, but not the parametrization, which is generated in the learning phase of the AI model.

Secondly, the AI itself is not deterministic, but it makes decisions with certain probabilities. Therefore, it makes sense to assign a part of the TFFR to the AI, i.e. the rate of dangerous failure of the technical system, see [4].

Based on the observation on the statistical approach above, e. g., based on testing or proven-in-use arguments, this seems feasible only in exceptional circumstances, e. g., for low demand systems with a low integrity requirement and a solid training data base. For higher SILs, analytical proof will usually be necessary.

Taking into account the experience from safety-related data transmission systems above, we realize that validation of AI models may be enhanced if a sensible (partial) model is found. This insight is not new, even the Turing laureate Pearl [15] has already concluded that it is necessary " formulate a model of the process that generates the data, or at least some aspects of that process". That means, at least in part, we need to understand the structure or properties of the data. Only then we can expect to be able to derive properties from the (partial) model by means of analytical results that meet the requirements for a safety case for higher SILs. Practically, this means that such a proof will be easier for data that have a certain structure, e. g., due to physical properties of the data sources, as may be the case for sensor data. On the other side, the methods side, one can expect that a safety case will be easier the more the structure of AI methods is understood. Here, some methods may be more difficult to explain than others. It could be that some AI methods as, e. g., Support Vector Machines [1] may have advantages over more complex methods like Neural Networks in establishing a safety case since they have a more explainable structure and provable properties. It is known from statistical learning theory [16] that we can enhance the statistical confidence interval to

$$p_{emp} \pm c\sqrt{\frac{S_{F,n} + \ln(1/\alpha)}{n}}$$

where $S_{F,n}$ is a constant, that depends on the complexity of the chosen function class F, which would be larger for neural networks than for support vector machines. However, the general dependency on the square root of the sample size remains. In our opinion this can only be relaxed if we can postulate stronger structural properties, comparable to the effects of the linearity assumption for the se

In any case, a "model-oriented" approach to the safety case would have the advantage that it could be conducted on the basis of today's established safety standards such as [4] or [6], and would not require any lengthy additional standardization.

5　Discussion and Conclusions

We have shown by example of the safety-relevant communication, that complex problems can be well solved (see Fig. 3) by a combination of.

1. An environment model (here the BSC +), and
2. A theory for the construction of the algorithm (coding theory), and
3. A standard that outlines the application conditions and the evaluation

The environment model contains in particular an error model, both qualitative and quantitative. As any model, it will not be correct, but it must be useful and describe the environment sufficiently well. Having said that, it is well understood that physical processes that can be described by either discrete or continuous valued time series are a good candidate for such models, e. g., models of sensors like radar or magnetic field sensors. The more complex the sensors, e. g., cameras, the harder it is to find a useful error model. And the more deterministic or foreseeable the environment is, the easier it is to formulate such a model. So, in railways or industry applications it will be much easier than in automotive.

A theory that would be as perfect as BCH coding, may be hard to find in AI applications, but we have shown that also some structural properties may suffice to move a big

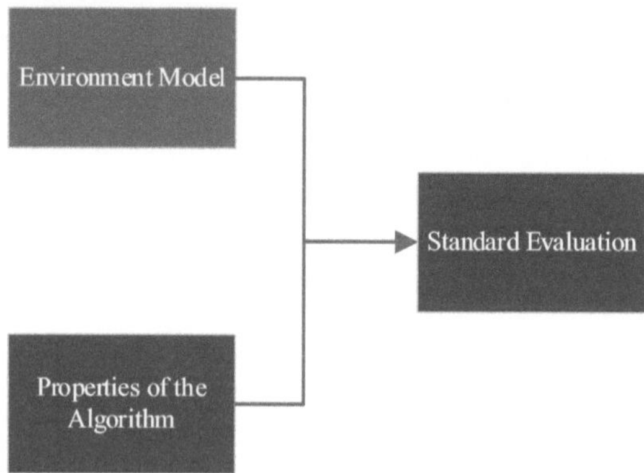

Fig. 3. Model-based AI Engineering Approach

step forward. Such properties may also be validated empirically, they would not need to be proven.

And finally, a standard is useful, as both model and structural properties will have some shortcomings or deficiencies, but publicly worked out standards, e. g., from CENELEC or IEC are worked out and voted by the majority of relevant experts in the field and represent acknowledged rules of technology. They can be also a means of organizing a level playing field, so that all users and stakeholders are bound to fulfil the same rules and may achieve a similar level of dependability.

References

1. Russell, S., Norvig, P.: Artificial Intelligence, Pearson (2012)
2. EU Commission: Proposal for a Regulation of the European Parliament and of the Council Laying Down Harmonised Rules on Artificial Intelligence (Artificial Intelligence Act) and Amending Certain Union Legislative Acts, Com/2021/206 (2021)
3. Cenelec: Railway Applications - The Specification and Demonstration of Reliability, Availability, Maintainability and Safety (RAMS), EN 50126, part 1 & 2 (2017)
4. Cenelec: Railway applications - communication, signalling and processing systems - Safety related electronic systems for signaling, EN 50129 (2018)
5. Cenelec: Railway applications communication, signalling and processing systems Safety-related communication in transmission systems, EN 50159 (2010)
6. Cenelec: Railway applications communication, signalling and processing systems: Software for railway control and protection systems, EN 50128 (2012)
7. Towards Auditable AI Systems Current status and future directions, white paper, based on the workshop "Auditing AI-Systems: From Basics to Applications", October 6th, Fraunhofer Forum, Berlin (2021)
8. Braband, J.: Using COTS Transmission Systems in Safety-critical Applications, at – Automatisierungstechnik 50, Oldenbourg Verlag, 382–388
9. MacWilliams, F.J., Sloane, N.J.A.: The Theory of Error-Correcting Codes, New York. North-Holland Publishing Company, NY (1977)
10. Schneier, B.: Applied Cryptography, Wiley (1995)

11. Braband, J.: Euroradio : Verläßliche Übertragung sicherheitsrelevanter Zugbeeinflussungsdaten über offene netzwerke. In: Brüggemann, H.H., Gerhardt-Häckl (Hrsg.), W.: Verläßliche IT-Systeme, Vieweg 1995
12. Castagnoli, G., Bräuer, S., Herrmann, M.: Optimization of cyclic redundancy-check codes with 24 and 32 parity bits. IEEE Trans. Commun. **41**, 883–892 (1993)
13. Dodunekova, R., Dodunekov, S.M.: Sufficient conditions for good and proper detecting codes. IEEE Trans. Inform. Theory **43**, 2023–2026 (1997)
14. Schiller, F., et al.: Enhancement of safe communication model – preserving the black channel concept. Automatisierungstechnik (at) **70**(1), 38–52 (2022)
15. Pearl, J., Mackenzie, D.: The Book of Why, Penguin Science (2018)
16. Vapnik, W.N.: The Nature of Statistical Learning Theory, Springer (2000)

Designing Concurrent Distributed Systems by Interface Contracts

Manfred Broy[✉]

Institut für Informatik, Technische Universität München, 80290 München, Germany
broy@in.tum.de

Abstract. The fundamental idea of Design by Contract is worked out for the specification and design of concurrent distributed interactive systems including properties of real-time behavior. We use interface predicates for specifying systems and forming system architectures by composition. We show how to begin with an interface predicate specifying a nucleus of behavioral properties and how to extend and refine it schematically by properties that hold for all implementations such as strong causality and full realizability. The goal is a modular specification and design approach for concurrent distributed interactive systems following the principles of modularity, encapsulation and information hiding.

Keywords: Contract · Interface · Concurrency · Distribution

1   Introduction

Design by contract (DbC) is a great idea (see Meyer [8]) to support the design of software systems. Originally, DbC was proposed for object orientation programming specifying the effect of method calls in terms of the attributes of classes. The popular approach "design by contract" as suggested for object orientation has its limitations less due to some weakness of this fundamental idea than due to the fact that the object oriented paradigm is typically not addressing concurrency. In the following, we describe an approach to DbC for concurrent distributed systems supporting the principles of modularity, encapsulation, and information hiding.

2   Design by Contract

Contracts have been suggested (see [8]) by Bertrand Meyer for object programming languages. There the basic idea is to specify the methods (or features, as they are called in Eiffel) of classes by preconditions, postconditions and invariants.

However, method calls in object oriented programs may not only change the local, encapsulated attributes of the class the method is defined for. Execution of method bodies may lead to further calls, not only of methods of the class in which a method is defined but of methods defined for other classes changing attributes of these classes, or more precisely objects of these classes. These attributes have to be mentioned in the

M. Fränzle et al. (Eds.): Werner Damm Festschrift, LNCS 15471, pp. 234–250, 2026.
https://doi.org/10.1007/978-3-031-97537-0_14

preconditions, postconditions, and invariants which is in conflict with the principles of information hiding and encapsulation.

In Meyer's DbC, method calls, in general, are not only affecting attributes of the class in which the method is defined but further attributes by forwarded calls of methods declared in other classes and objects breaking the rule of encapsulation, and information hiding. More subtle are forwarded calls of methods affecting further objects of the same class. Assertions then have to refer to attributes of all these classes and objects, which should, in contrast, be encapsulated, and have to introduce object identifiers explicitly in specifications with all the complications implied by this. Meyer[1] explains not to be interested in these forms of feedback loops of method calls by advocating a less sophisticated software design where for method calls their effect can always be expressed in terms of their effects on the state. However, for distributed concurrent interactive systems such feedback loops have to be considered.

Design by contract does not only refer to contracts, but, obviously, fundamentally to the notion of design. In fact, we have to understand what is meant by design. *Design means the creation and specification of architecture* (see Meyer [9]). But can architecture be described by design by contract as promoted for object orientation? This leads to the question, what a software architecture is. A quite common idea of an architecture in software and system engineering is that it describes the structuring of a system into subsystems – classes in the case of object orientation – that cooperate and interact to provide a required functionality, a set of services, to the system's users (see [6]). Thus an architecture specification is supposed to describe how, if a certain service is requested, subservices are activated which in turn may activate other services that finally work together to achieve the required service.

In the following, contracts for distributed concurrent systems are specified by interface predicates. This way, design leads to hierarchical architecture given by a set of subsystems connected by channels over which streams of data are exchanged, and where the subsystems are described by interface contracts such that the interface contract of the overall system is derived from these and verified this way.

In Bertrand Meyer's DbC this is different. In the object-oriented version of DbC specifications are given for the methods, but it is not specified which method calls activate which further methods by call forwarding. However, this touches an essential aspect of the architecture of concurrent systems. In such an architecture specification it should be described how interactive input invokes interactive output. These dynamics and causality of forwarded calls is not captured in conventional DbC.

The book "Touch of Class: Learning to Program Well with Objects and Contracts" by Bertrand Meyer (see [9]) treats DbC. Meyer's definition of design is rather brief. He writes: *"Finding appropriate classes is a central part of the task of software design, devoted to organizing the essential structure, or architecture, of a program, as opposed to writing down the details, or implementation."* So far so good – but what is an architecture? Bertrand Meyer writes: *"Design, also called architecture, builds the overall structure of a software system. It is responsible, in particular, for finding the principal units, or modules, of that system, and the relations between those units."* However, relations between the units are not described in DbC.

---

[1] Private communication.

Actually, an architecture has (see also [6]).

- a *static part*: the subsystems (often called components or classes) and their relationship seen in the OO case as the methods defined within a class with the forwarded method calls (also called feature calls) that may be invoked within a class by its methods and their corresponding return messages.
- a *dynamic part*: which method calls lead to which method calls and what are their effects to the attributes of the involved objects and their return values (as sometimes illustrated for typical instances of cascades of method invocations by message sequence charts, also called interaction diagrams in UML).

But the dynamic part is not described if only the effects of method invocations for the attributes of the classes or objects involved are specified. However, this dynamic part (which input/call leads to which output/return message) is essential and in the center of design and architecture for concurrency. Another open question for conventional DbC is, how to specify methods that change (by forwarded calls) the attributes of a number of different objects of the same class. This is handled in Arnd Poetzsch-Heffter's Habilitation thesis: "Specification and verification of object-oriented programs", submitted at the Technical University of Munich 1997, which addresses this problem leading to a quite complex specification logic.

A largely unsolved issue is concurrency not only for object orientation and design by contract. What the effects on attributes are if several threads are active concurrently in a conventional object oriented software system is largely unclear. What if a method is called in a class (or more precisely in an object) while a call of another (or even the same) method is active – this autoconcurrency in the class or object leading to the difficult questions which statements are executed in which order with or without interrupts. This is due to the inherent procedural nature of the "von Neumann style" being the basis of object orientation. An extensive overview for contracts for system design and the problems involved is given in [2].

It is a challenging task to bring in concurrency into object orientation. There are several reasons for that. The most obvious are that then (in the initially sequential execution of method bodies) we have to deal with all the difficulties of synchronization, of parallel threads, and with the challenge to provide specifications in cases, there are several threads operating concurrently. This requires a more general approach for concurrent systems.

3   Designing Concurrent Distributed Interactive Systems

We specify systems in terms of their *interface behavior*. Interface behavior is described by relations between input and output streams of data. Such relations are captured logically by interface predicates and interface assertions (for details, see [4]). Thus, a system or an interface behavior predicate is given by a syntactic interface with a corresponding interface assertion. Interface predicates are a formal model of system interface behaviors.

3.1 Streams, Time, and Behavior

Throughout this paper, we use the terms *system* and *interface behavior* in a specific way. We address discrete systems, more precisely, models of discrete concurrent time-critical systems with input and output. We introduce a corresponding model of concurrent, distributed interactive systems according to FOCUS (see [2]). In this model, the interface behavior of systems is described by relations between input and output streams. This way, for us, a *system* is an entity that shows some specific interface behavior by interacting over its interface with its surrounding *operational context*. This behavior can be structured into *functional features*. A system has a *boundary*, which defines what is inside and what is outside the system. Inside the system there is an encapsulated internal structure, often consisting of states and state attributes or of sub-architectures composed of local subsystems. Such internal implementation details are hidden following the principle of *information hiding*. The set of actions and events that may occur in the interaction of the system with its operational context determines the *syntactic ("static") interface* of the system. At its interface, a system shows a specific *interface behavior*. We specify interface behavior by *interface predicates* and *interface assertions*.

There are general logical properties that we assume for interface predicates. We require that system behaviors fulfill properties such as *strong causality* and *realizability* which hold for systems implemented by Moore machines since they hold for all Moore machines (see [10]). Causality requires that output is only generated after a corresponding input has been received. Realizability requires the existence of an interaction strategy such that, for a stream of input data, the system will produce step by step an according output stream correct for the interface predicate (see [4]). However, not all interface predicates guarantee these two properties explicitly. Nevertheless, in such cases, causality or realizability can be added by schematic refinement rules resulting in refined interface predicates for which these additional properties are valid – leading to false if the predicate cannot by implemented. As an operational model we take Moore machines (see [4]).

In FOCUS, components of distributed systems run in parallel and systems, in general, which operate concurrently, evolving in a common physical time frame. We choose an interface concept to describe interface behaviors in a property-oriented way, interface behaviors that are interactive, run in parallel, and may be time-critical.

By M* we denote the set of finite sequences (also called finite streams) of elements from set M which formally can be represented by functions ($\mathbb{N}$ denotes the natural numbers including 0, $\mathbb{N}_+ = \mathbb{N}\backslash\{0\}$; $[1{:}n] \subseteq \mathbb{N}_+$ denotes finite intervals of natural numbers for $n \in \mathbb{N}$):

$$M^* = \cup_{n\in\mathbb{N}}([1:n] \to M)$$

By M^ω we denote the infinite sequences called infinite streams over set M represented by functions:

$$M^\omega = (\mathbb{N}_+ \to M)$$

The empty sequence as well as the empty stream is denoted by $\langle\rangle$. By

$$M^{*|\omega} = M^* \cup M^\omega$$

238   M. Broy

we denote the set of streams over M.

In a synchronous timed stream, every position in the stream indicates which message is transmitted in the respective time interval[2]. This may be a data element or nothing indicated by the symbol $\bullet$. We define $M_\bullet = M \cup \{\bullet\}$ (where we assume $\bullet \notin M$). By

$$(M_\bullet)^{*|\omega}$$

we denote the set of finite or infinite timed streams. A stream $x \in (M_\bullet)^{*|\omega}$ is called a *timed* stream, since we understand the set $\mathbb{N}_+$ to represent time in terms of an infinite sequence of time intervals. Each time interval is numbered by a number $t \in \mathbb{N}_+$. Then $x(t)$ with $t \in \mathbb{N}_+$ denotes the message communicated by the timed stream x in time interval t, which is empty if $x(t) = \bullet$. For $m \in M$, we denote the one element sequence and the one element time free stream by

$$\langle m \rangle$$

By $\langle \rangle$ we denote the empty stream for untimed streams. By $\langle \bullet, \ldots \rangle$ we denote the (infinite) empty timed stream. This shows an important difference between finite timed streams and finite untimed streams.

A finite timed stream always represents the result of an interaction until to a particular point of time. This is different for untimed streams. An untimed finite stream may represent what happens in a finite computation and therefore represents an approximation of a computation, but also the result of an infinite computation in cases where in this infinite time only a finite number of messages are communicated.

Note that timed and untimed streams are both streams with the only difference that in a timed stream we have in addition the entry "$\bullet$" which indicates that in the particular time slot no message is communicated. Therefore, all definitions for streams can be applied both for timed and untimed streams.

By $\#x \in \mathbb{N} \cup \{\infty\}$ we denote the number of copies of messages from M occurring in a timed stream $x \in (M_\bullet)^{*|\omega}$ or in a time free stream $x \in M^{*|\omega}$. For a timed stream $x \in (M_\bullet)^{*|\omega}$ or a time free stream $x \in M^{*|\omega}$ and a subset $D \subseteq M$ we denote by

$$D\#x$$

the number of copies of elements from set D in stream x; we write $a\#x$ for $\{a\}\#x$ and $\#x$ for $M\#x$.

Given a sequence s of elements from M (or in the case of timed streams of elements, from $M_\bullet$) and a stream $x \in (M_\bullet)^{*|\omega}$ or $x \in M^{*|\omega}$ we denote by:

$$s \,\hat{}\, x$$

the concatenation of sequence s with stream x.

For a timed stream $x \in (M_\bullet)^{*|\omega}$ we denote by $|x|$ its length and by $\bar{x} \in (M)^{*|\omega}$ its time abstraction defined for $m \in M$ by

---

[2] In [4] we use a slightly different model of timed streams, where streams carry in each time interval a finite sequence of messages.

$$\overline{\langle\rangle} = \langle\rangle$$
$$\overline{\langle m\rangle\hat{\ }x} = \langle m\rangle\hat{\ }\overline{x}$$
$$\overline{\langle\bullet\rangle\hat{\ }x} = \overline{x}$$

For $x \in (M_\bullet)^{*|\omega}$ or $x \in (M)^{*|\omega}$ and $t \in \mathbb{N}$ we denote a *time cut* of maximal length t as follows:

$$x \downarrow t \in ([1 : n] \to (M_\bullet)^*)$$

where $n = \min(t, |x|)$ and for all $j \in \mathbb{N}_+$

$$(x \downarrow t)(j) = x(j) \Leftarrow 1 \le j \le n$$

The stream $x \downarrow t$ denotes the prefix of the first t elements in an infinite timed stream x, representing the messages transmitted in the first t time intervals.

Streams are sent over channels that connect systems. Given a set X of typed channel names c_j with types T_j:

$$X = \{c_1 : T_1, \ldots, c_m : T_m\}$$

The set X is called a *channel signature*. By $\overrightarrow{X}$ we denote channel histories given by families of timed streams, one timed stream for each channel:

$$\overrightarrow{X} = (X \to (M_\bullet)^\omega)$$

where for $x \in \overrightarrow{X}$, the timed stream $x(c_k)$, $1 \le k \le m$, for the channel $c_k \in X$ carries messages of type T_k. All introduced notations, such as time abstraction $\overline{x}$, concatenation $s\hat{\ }x$, time slice $x \downarrow t$, and prefix order carry over to channel histories $x \in \overrightarrow{X}$, as well.

Let T_i and S_i be data types. Given two channel signatures:

$$X = \{x_1 : T_1, \ldots, x_n : T_n\}$$
$$Y = \{y_1 : S_1, \ldots, y_m : S_m\}$$

we denote a syntactic interface of a system by $(X \blacktriangleright Y)$ where X denotes the set of input channels and Y denotes the set of output channels of the system interface. This way, in syntactic interfaces each channel is identified by an identifier with a specified data type indicating which types of data are communicated.

The elements of $\overrightarrow{X}$ are called *histories*. Formally, a history $x \in \overrightarrow{X}$ is a mapping from the set of channels X onto streams and therefore just a family of timed streams named by the channel identifiers. Given a history $x \in \overrightarrow{X}$ and a subset $C \subseteq X$ we denote by $x|C \in \overrightarrow{C}$ the history which is the restriction of x to the channels in channel signature C.

For history $x \in \overrightarrow{X}$ and $y \in \overrightarrow{Y}$, where $X \cap Y = \emptyset$, the history $x \oplus y \in \overrightarrow{Z}$ where $Z = X \cup Y$ is defined for $c \in C$ by:

$$(x \oplus y)(c) = x(c) \Leftarrow c \in X$$
$$(x \oplus y)(c) = y(c) \Leftarrow c \in Y$$

We specify the interface behavior of a system by a syntactic interface $(X \blacktriangleright Y)$ and a predicate Q

$$Q : \vec{X} \times \vec{Y} \to \mathbb{B}$$

called an *interface predicate* for the channel histories over X and Y. We write $Q::(X \blacktriangleright Y)$ to express that Q is an interface predicate over the syntactic interface $(X \blacktriangleright Y)$. An interface predicate $R::(X \blacktriangleright Y)$ is called a *refinement* of interface predicate $Q::(X \blacktriangleright Y)$ if $R \Rightarrow Q$.

In concrete specifications, we describe interface predicates $Q::(X \blacktriangleright Y)$, for better readability, by logical assertions A in higher order predicate logic with channels from X and Y as free variables for timed streams of the respective type. The logical formula A is called *interface assertion* over the syntactic interface $(X \blacktriangleright Y)$. It defines an interface predicate Q uniquely. We write $Q = (X \blacktriangleright Y): A$.

In the following, we show interface specifications written in the form of templates. We introduce composition and refinement, and illustrate these concepts by examples.

3.2 Contracts: Specification Syntax

We may represent interface specifications graphically by data flow nodes (see Fig. 1) and use the following form of templates for interface specifications as demonstrated by the following brief example.

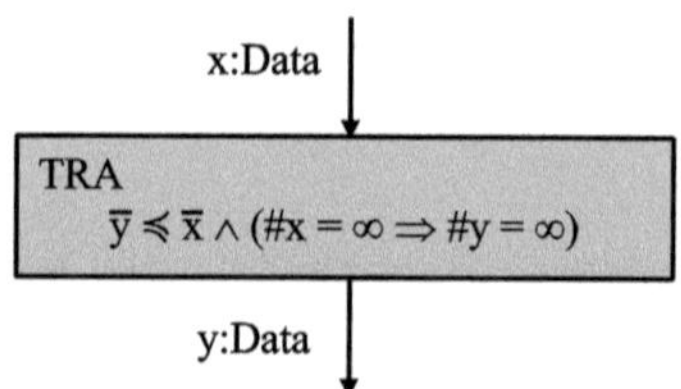

Fig. 1. System TRA as a data flow node

Two partial orderings on streams x, y are used ($\hat{ }$ denotes concatenation):

Prefix: $x \sqsubseteq y = \exists z: x\hat{ }z = y$

Substream: $x \preccurlyeq y = (x = \langle \rangle \vee rest(x) \preccurlyeq y \vee (first(x) = first(y) \wedge rest(x) \preccurlyeq rest(y)))$

We use first and rest as function on streams x with first(x) delivering the first element of stream x (if stream x is not empty) and $\bot$ otherwise and with rest(x) the stream without the first element or the empty stream if x is the empty stream.

TRA

in x: Data
out y: Data
$\overline{y} \preccurlyeq \overline{x} \wedge (\#x = \infty \Rightarrow \#y = \infty)$

The above template gives a name (in our example, TRA) to an interface predicate. It describes the syntactic interface (in our example $(\{x: \text{Data}\} \blacktriangleright \{y: \text{Data}\})$)and the predicate by an interface assertion (in our example $\overline{y} \preccurlyeq \overline{x} \wedge (\#x = \infty \Rightarrow \#y = \infty)$) specifying the interface behavior (note for all timed streams x: $\#x = \#\overline{x}$). We derive

$$\text{TRA} = (x: \text{Data} \blacktriangleright y: M): \overline{y} \preccurlyeq \overline{x} \wedge (\#x = \infty \Rightarrow \#y = \infty)$$

from the template which defines the predicate TRA. Using [4] we may deduce further properties from the specifying assertion.

In the interface assertion

$$\overline{y} \preccurlyeq \overline{x} \wedge (\#x = \infty \Rightarrow \#y = \infty)$$

x and y denote timed streams. System specifications of this form are called *untimed*, since according to the interface assertion the output data stream y does not depend on the timing of the input stream x. Hence, TRA is called *non-time-critical*. We write TRA(a, b) to denote the application of predicate TRA to streams a and b.

We consider as a more complex example the well-known alternating bit protocol (ABP) with the interface predicates SEND and RECE specifying the subsystems for the sender and the receiver resp.

SEND = (x: Data, r: Bit $\blacktriangleright$ u: (Data, Bit)):
$$\text{ab}(\overline{u}) \wedge \text{data}(\overline{u}) \sqsubseteq \overline{x} \wedge \#\text{data}(\overline{u}) = \min(\#x, 1+\#\text{alt}(\overline{r})) \wedge (\#\text{alt}(\overline{r}) < \#x \Rightarrow \#u = \infty)$$
RECE = (z: (Data, Bit) $\blacktriangleright$ y: Data, c Bit): $\overline{y} = \text{data}(\overline{z}) \wedge \overline{c} = \text{bits}(\overline{z})$

where ab is a predicate on streams $u \in (\text{Data} \times \text{Bit})^{*|\omega}$ defined by:

$$\text{ab}(u) = \forall\, b \in \text{Bit}, d, d' \in \text{Data}, a \in (\text{Data}\times\text{Bit})^*: a^\smallfrown\langle(b,d)\rangle^\smallfrown\langle(b,d')\rangle \sqsubseteq u \Rightarrow d = d'$$

and we use the following auxiliary prefix monotonic functions on untimed streams:

$$\text{bits}(\langle\rangle) = \langle\rangle$$
$$\text{bits}(\langle(d, b)\rangle\char`^u) = \langle b\rangle\char`^\text{bits}(u)$$
$$\text{alt}(\langle\rangle) = \langle\rangle$$
$$\text{alt}(\langle b\rangle\char`^r) = \langle b\rangle\char`^\text{alt}(\text{del}(b, r))$$
$$\text{alt}(\langle(d, b)\rangle\char`^u) = \langle b\rangle\char`^\text{alt}(\text{del}((d, b), u))$$
$$\text{data}(\langle\rangle) = \langle\rangle$$
$$\text{data}(\langle(d, b)\rangle\char`^u) = \langle d\rangle\char`^\text{data}(\text{del}((d, b), u))$$

and for an element e of type M and a stream u of type $M^{*|\omega}$ we specify.

$$e\#u = 0 \Rightarrow \text{del}(e, u) = \langle\rangle$$
$$\text{del}(e, \langle e\rangle\char`^u) = \text{del}(e, u)$$
$$\text{del}(e, \langle b\rangle\char`^u) = \langle b\rangle\char`^u \Leftarrow b \neq e$$

Note that in the definition of the predicate SEND the assertion $(\#\text{alt}(\bar{r}) < \#x \Rightarrow \#u = \infty)$ permits that $\#u = \infty$ in the case $\#\text{alt}(\bar{r}) \geq \#x < \infty$ indicating that only a finite number of data are to be sent and all sent data are acknowledged. The predicate SEND may be refined to $(\#\text{alt}(\bar{r}) \geq \#x \wedge \#x < \infty) \Rightarrow \#u < \infty$ that indicates that in the case $\#\text{alt}(\bar{r}) = \#x < \infty$ transmission eventually stops.

The example ABP will be completed in the following sections.

3.3 Contracts: Semantics – Strong Causality and Full Realizability

Given channel signatures X and Y, a function on histories (the definition applies as well to functions on streams):

$$f : \vec{X} \rightarrow \vec{Y}$$

is called *strongly causal* if:

$$\forall\, t \in \mathbb{N}, x, z \in \vec{X}: x{\downarrow}t = z{\downarrow}t \Rightarrow f(x){\downarrow}t+1 = f(z){\downarrow}t+1$$

Then we write SC[f].

An interface predicate Q for the syntactic interface $(X \blacktriangleright Y)$ is called *strongly causal*. if for all $t \in \mathbb{N}, x, z \in \vec{X}, y \in \vec{Y}$:

$$Q(x, y) \wedge x{\downarrow}t = z{\downarrow}t \Rightarrow (\exists\, y' \in \vec{Y}: Q(z, y') \wedge y{\downarrow}t+1 = y'{\downarrow}t+1)$$

Then we write SC[Q].

The weakest predicate $Q^{\copyright}$ that is strongly causal and a refinement of Q can also be expressed explicitly as follows (see [4]).

$$Q^{\copyright}(x, y) = (Q(x, y) \wedge \forall\, t \in \mathbb{N}, z \in \vec{X}: \exists\, y' \in \vec{Y}: Q((x{\downarrow}t)\char`^z, (y{\downarrow}(t+1))\char`^y'))$$

Predicate $Q^\copyright$ is the weakest refinement of Q that is strongly causal. Assuming that Q specifies a system implemented by a Moore machine then the system fulfils predicate $Q^\copyright$ since behaviors of Moore machines are strongly causal (for details see [4]).

We refine the interface assertion specifying TRA into a strongly causal one

$$\forall\, t \in \mathbb{N}: \overline{y \downarrow (t{+}1)} \leqslant \overline{x \downarrow t} \wedge \overline{y} \leqslant \overline{x} \wedge (\#x = \infty \Rightarrow \#y = \infty)$$

An interface predicate Q::(X ▶Y) is called *realizable*, if the there exists a strongly causal function f: $\vec{X} \to \vec{Y}$ such that:

$$\forall x \in \vec{X} : Q(x,\ f(x))$$

Then the function f is called *realization* of Q. Predicate Q is called *fully realizable* if it is realizable and

$$Q(x,\ y) = (\exists f : \vec{X} \to \vec{Y} : y = f(x) \wedge Q[f]$$

where Q[f] holds if f is a realization for Q:

$$Q[f] = (SC[f] \wedge \forall x \in \vec{X} : Q(x,\ f(x)))$$

Then for all $x \in \vec{X}$ and $y \in \vec{Y}$ for which Q(x, y) holds, there exists a realization f for predicate Q such that x = f(y). Every realization

$$f : \vec{X} \to \vec{X}$$

has exactly one unique fixpoint which can be proved by Banach's fixpoint theorem since f is a contractive function (see [1]).

A fully realizable interface predicate is always strongly causal. An interface predicate Q for the syntactic interface (X ▶Y) is refined into predicate $Q^\circledR$ (see [4]):

$$Q^\circledR(x, y) = (\exists\ f: \vec{X} \to \vec{Y}: y = f(x) \wedge Q[f])$$

Either predicate $Q^\circledR$ = false and there does not exist a refinement of Q that is fully realizable or $Q^\circledR$ the weakest predicate that is a refinement of Q and fully realizable. If we assume that predicate Q specifies a system that is implemented by a Moore machine then the system fulfils predicate $Q^\circledR$ since the behaviors of Moore machines are fully realizable, provided $Q^\circledR \neq$ false (for details see [4]). The proposition $Q^\circledR$ = false implies that an implementation by a Moore machine does not exist. We assume that systems are implemented (realized) by generalized Moore machines which serve as operational semantics. Then a system implemented by a Moore machine and specified by a predicate Q fulfills both the predicates $Q^\copyright$ and $Q^\circledR$ (see [4]).

A fully realizable interface predicate defines all interface properties that hold for all its implementations by Moore machines. Therefore, a calculus can be defined that comprises the rule to refine interface predicates into fully realizable ones is sound and relative complete (see [5]). In most of the practical cases, we get

$$Q^{\circledR} = Q^{\copyright}$$

In these cases, the weakest strongly causal refinement $Q^{\copyright}$ is already sufficient to prove all properties that hold for all Moore machines that realize Q.

3.4 A Somewhat Artificial Illustrative Example

Let us look at an illustrative example. Consider the predicate

$$IE = (x: Nat \blacktriangleright y, z: Nat): (\#x < \infty \Leftrightarrow \#z = \infty) \wedge \#y = \#x$$

The predicate IE is neither strongly causal nor fully realizable. We get

$$IE^{\copyright} = (x: Nat \blacktriangleright y, z: Nat): (\#x < \infty \Leftrightarrow \#z = \infty)$$

$$\wedge \#y = \#x \wedge \forall\, t \in \mathbb{N}: \#(y{\downarrow}t+1) \leq \#(x{\downarrow}t)$$

Given input x, the assertions $\#x < \infty$, $\#z = \infty$ and $\#y = \#x$ are liveness conditions for output y, z while the assertion $\#(y \downarrow t + 1) \leq \#(x \downarrow t)$ is a safety condition. We get

$$IE^{\circledR} = (x: Nat \blacktriangleright y, z: Nat): false$$

To prove $IE^{\circledR} = false$ construct a contradiction to the assumption that $IE^{\circledR}$ is realizable.

3.5 Hiding

Given an interface predicate $Q::(X \blacktriangleright Y)$ and a subset $Z \subseteq Y$ of the output channels in Y we hide the channels in Z and denote the resulting predicate by

$$\textbf{(hide } Z \textbf{ in } Q)::(x \blacktriangleright Y\backslash Z)$$

defined for $x \in \vec{X}$, $y \in \overrightarrow{YZ}$ by

$$\textbf{(hide } Z \textbf{ in } Q)(x, y) = \exists\, z \in \vec{Z}: Q(x, y{\oplus}z)$$

We get a local scope this way for the output channels in Z in which these are the local variables.

3.6 Concurrent Composition with Interaction via Feedback Loops

To define concurrent composition, systems are composed by putting them side by side introducing feedback loops for mutual communication via their channels that fit together. Systems are composed as shown in Fig. 1. There the channels z_1 and z_2 serve as feedback channels carrying streams for the communication between the two systems S_1 and S_2. These streams z_1 and z_2 are defined by fixpoints.

Interface predicates $Q_k::(X_k \blacktriangleright Y_k)$ with $k = 1, 2$ are called *composable*, if their channel types are consistent and $Y_1 \cap Y_2 = \emptyset$. Moreover, we assume that for $k = 1, 2$ the channel sets X_k and Y_k are disjoint $X_k \cap Y_k = \emptyset$. We define concurrent composition with feedback of system specified by interface predicates by logical conjunction of their interface predicates.

Consider predicates $Q_k::(X_k \blacktriangleright Y_k)$ for $k = 1, 2$, that are composable. Define

$$X = (X_1 \cup X_2) \backslash Z \qquad \textit{input channels for the composed system}$$
$$Y = Y_1 \cup Y_2 \qquad \textit{output channels for the composed system}$$
$$Z = Y \cap (X_1 \cup X_2) \qquad \textit{feedback channels}$$

The channel set Z denotes the set of feedback channels in the composition.

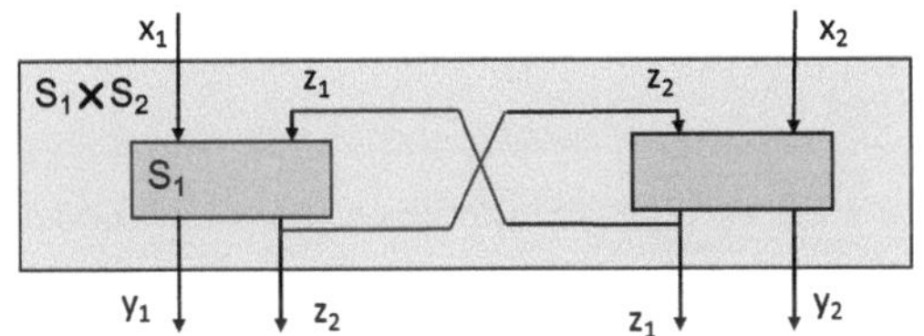

Fig. 2. Composition of systems S_1 and S_2 with feedback channels z_1 and z_2

We specify for the composite system $S_1 \times S_2$ the interface predicate $Q_1 \times Q_2$ where we assume that the predicates Q_k are given by assertions A_k with $Q_k = (X_k \blacktriangleright Y_k): A_k$; then their composition *including feedback channels as output* is given by

$$(Q_1 \times Q_2) = (X \blacktriangleright Y): A_1 \wedge A_2$$

This is a very nice and logically simple way to define composition. If both predicates Q_k, for $k = 1, 2$, are fully realizable so is predicate $(Q_1 \times Q_2)$ (see [4]). In this case, fixpoints for feedback loops are precisely captured (see [4]). For every refinement R_k of Q_k for $k = 1, 2$, we get

$$((R_1 \Rightarrow Q_1) \wedge (R_2 \Rightarrow Q_2)) \Rightarrow ((R_1 \times R_2) \Rightarrow (Q_1 \times Q_2))$$

If Q_1 and Q_2 are strongly causal so is $Q_1 \times Q_2$; if Q_1 and Q_2 are realizable so is $Q_1 \times Q_2$; if Q_1 and Q_2 are fully realizable so is $Q_1 \times Q_2$ (see [4]).

The feedback channels in Z can be hidden by

$$\textbf{hide } Z \textbf{ in } (Q_1 \times Q_2)$$

We get (see Fig. 3) for the design of the alternating bit protocol the definition

$$ABP = \textbf{hide } u, z, c, r \textbf{ in} (SEND(x, r, u) \times TRA(u, z) \times RECE(z, y, c) \times TRA(c, r))$$

This shows a design described by a concurrent composition of interface predicates.

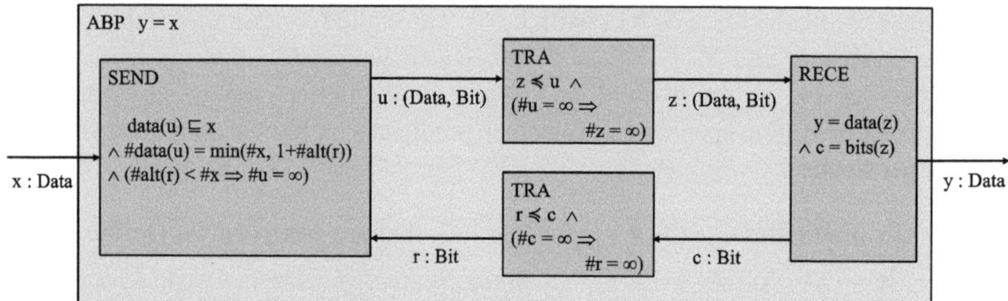

Fig. 3. Alternating Bit Protocol

3.7 Modularity and Refinement

We define *modularity* as the following property of a specification technique:

- systems are specified by predicates describing their interface behavior by which details of their implementation are *encapsulated* and hidden, following the principles of *information hiding*,
- composite systems are the result of the composition of subsystems leading to interface behaviors specified by interface predicates derived by composing the interface predicates specifying their subsystems.

A key issue is the relationship between layered architectures and key modularity concepts such as components, encapsulation, and information hiding.

3.8 Interface Contracts for Physical Systems

A challenging approach is the development of cyber-physical systems which means that we develop systems composed of subsystems that are digital and subsystems that are physical. In the end, we have software and hardware, on one hand, digital hardware and computing devices as well as communication devices. On the other hand, we have physical systems that are concrete mechanical machines and mechanical devices which are usually controlled by the software.

Here it is important to understand that the functionality of those systems is a result of the interaction of the software with physical devices. One way to model this is to work it out in all its technical details. Then, one has to study the complex technicalities of software and hardware, often called software/hardware co-design, and the details of the physical devices. Such an approach is very difficult to explain and to document. In contrast, it might be more interesting to come up with a more abstract description of embedded physical devices. We describe the functionality of the software controllers, and, as well, the functional properties of the physical devices at an abstract digital level. We model the behavior of physical devices by interface predicates that describe the relation between the input streams of signals for the actuators and the output streams generated by the sensors. We introduce a state machine representation for physical systems. To illustrate the idea, we give a short simple example of a gate in the following. The gate is modelled by a Mealy machine, a state machine with input and output. The state space is used to model the state of the gate. The set of states is defined as follows:

$$\text{State} = \{\text{isopened, isclosed, opening, closing, error}\}$$

The control input to the system and its output are given by the following two sets defined as follows:

$$\text{Input} = \{\text{open, close, interrupt, reset}\}$$
$$\text{Output} = \{\text{started, opened, closed, alarm}\}$$

The state transition function is defined as follows:

$$\Delta: \text{State} \times \text{Input}_\bullet \to \wp(\text{State} \times \text{Output}_\bullet)$$

It is defined by Table 1 that represents a formula that specifies the state transition function.

Table 1. State Transitions of the Gate

state	input	next state	output
	interrupt	error	alarm
error	reset	isclosed	closed
error	$\neq$ reset	error	$\bullet$
isopened	close	closing	started
closing	$\bullet$	closed	closed
		closing	$\bullet$
closing	$\neq \bullet$	error	alarm
$\neq$ isopened	close	error	alarm
isclosed	open	opening	started
opening	$\bullet$	opening	$\bullet$
		isopened	opened
opening	$\neq \bullet$	error	alarm
$\neq$ isclosed	open	error	alarm

The table Table 1 defines the state transition relation by a disjunctive formula. Every line in the table defines an assertion. For instance, the following line

closing	$\neq \bullet$	error	alarm

represents the conjunctive formula:

$$(\text{state} = \text{closing}) \wedge (\text{input} \neq \bullet) \wedge \left(\text{state}' = \text{error}\right) \wedge (\text{output} = \text{alarm})$$

These conjunctive formulas represented by the lines of the table are connected by disjunction to deduce the formula that specifies the state transition relation for the gate.

Note that the state machine is highly nondeterministic since when opening or closing the time needed is not fixed – it may need an arbitrary even an infinite amount of time. However, if it takes too long we may always send an interrupt. Another idea for the specification would be to specify as a refinement a maximal number of steps with input of the message $\bullet$ in the cases the gate is opening or closing until the gate is opened or closed.

Note that this way we define a Mealy machine and not a Moore machine for the gate. However, this is not a problem. Either we change the specification to a Moore machine by delaying the output by one step and specify some initial output (the element $\bullet$) or work with a Mealy machine here which defines a weakly causal behavior which composed with a strongly causal and fully realizable behavior of the controller still guarantees unique fixpoints (see [4]).

In fact, it is straightforward to abstract the behavior of this state machine into an interface assertion between a timed input and a timed output stream. This way the state is abstracted away – we no longer have the information how the input and output streams are related to certain states of the physical system.

The function mapping states onto interface predicates

$$\varphi: \text{State} \to (\,(\text{Stream Input} \times \text{Stream Output}) \to \mathbb{B}\,)$$

describes the behavior of the physical device which is derived from the state transition function as follows

$$\varphi(\sigma)(\langle e \rangle\hat{}\,a, \langle r \rangle\hat{}\,b) = \exists\, \sigma' \in \text{State}:\ (\sigma', r) \in \Delta(\sigma, e) \wedge \varphi(\sigma')(a, b)$$

This example shows that we can model physical systems by state machines with input and output. Actually, it means that these state machines have input that at a technical level corresponds to signals sent to actuators, and that the physical systems produce output which – viewed from the software systems – are signals and messages that are generated by the sensors of the systems. As a result, we have a view onto a physical device as a system that produces streams of sensor signals given a stream of actuator messages. By the states of the state machines we can model the reaction of the physical devices at the physical level such that we can see what the effects of these input and output streams are.

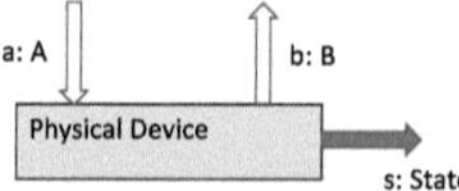

Fig. 4. Physical Device, Showing the Stream of States

To make states observable we may enrich the model by the stream of states. We define the function

$$\phi: \text{State} \to (\,(\text{Stream Input} \times \text{Stream Output} \times \text{Stream State}) \to \mathbb{B}\,)$$

that describes the specification of the behavior of the physical device. It is derived from the state transition function as follows

$$\phi(\sigma)(\langle e \rangle\hat{}\,a, \langle r \rangle\hat{}\,b, \langle \sigma' \rangle\hat{}\,s) = ((\sigma', r) \in \Delta(\sigma, e) \wedge \phi(\sigma')(a, b, s))$$

This way we get a stream of states that shows the modes and the positions of the physical system, in our example the automatic window, depending on the input stream a.In addition, we may introduce a software system which controls the physical device, in our example the gate, and form an architecture (see Fig. 5). We can prove properties both about the gate and about the system that consists of the controller which may get input from some user interfaces and controls the gate. We may compose the physical device modelled by the interface predicate φ (σ) with initial state σ with the control layer described by interface predicate CL and get predicate CL $\times$ φ (σ) defined by the assertion

$$CL(x,\ b,\ y,\ a) \wedge \varphi(\sigma)(a,\ b)$$

We get the interface assertion for the system in Fig. 5 where the behavior of the physical device is captured by the interface predicate ϕ (σ_0) where σ_0 is the initial state of the physical device as follows

$$\textbf{hide}\ a,\ b\colon CL(x, b, y, a) \wedge \phi(\sigma_0)(a, b, s)$$

This way we specify the dependency between input x and the states of the physical device. We end up with a model based documentation of physical subsystems as they are part of cyber-physical systems.

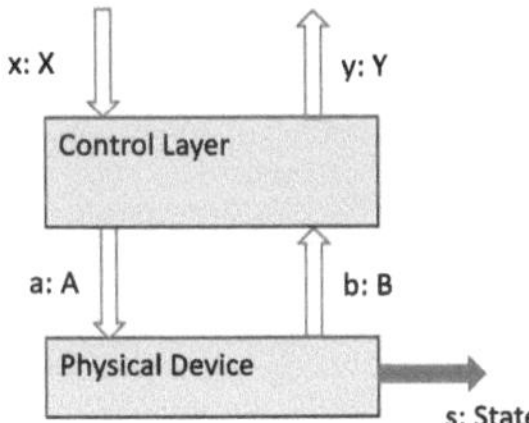

Fig. 5. Control Layer Composed with a Physical Device

3.9   From Abstract Ideas to Design and Implementation

What we have introduced allows us to design hierarchical architectures where by hiding we may introduce scopes containing the hidden channels as local variables for timed streams. In the approach introduced above we are able to design hierarchical architectures. In fact, we can describe infinite architectures (see [5]). The neat thing is that each subsystem, in the case it has its own local architecture has an interface specification which is the result of composing (conjoining) the interface specifications of its subsystems by conjunction.

The model we have introduced is fully concurrent. The concurrency is explicitly part of the model. The fact that we use timed streams has two advantages: First, it allows to formulate the concept of strong causality and based on this a concept realizability. Second, it enables the specification of real time properties. Nevertheless, we are free to formulate specification that do not refer to timing.

4   Conclusion

The introduced specification approach fulfills all the requirements of a design by contract for concurrent distributed systems. It is based on the idea of strong causality and full realizability which is a property of generalized Moore machines which form the operational semantics of our approach. What is neat is the fact that we can write a logical nucleus in form of an interface predicate to start with and then add strong causality and full realizability to include details about possible implementations.

It is interesting that we use time in the approach for two substantially different reasons: One is to express causality and realizability, the other is to express timing properties of systems. As a result, we get a very compact and powerful specification technique supporting design by contract where the logic involved can be captured by a calculus which is shown to be sound and relatively complete (see [5]). In [2] a rich set of specification approaches to interface specification for concurrent systems are referenced. There, however, the key idea of the approach presented here namely how to identify computationally feasible fixpoints is only treated at the level of Moore machines. Note the difference between the FOCUS approach (see [2]) as presented here, which aims at functional specification of system behavior by interface predicates and the design of hierarchical architectures of distributed systems including real time and approaches to model concurrency between statements (also called actions) which address, in particular, the specification of concurrent algorithms such as in Lamport's TLA (see [7]).

References

1. Banach, S.: Sur les opérations dans les ensembles abstraits et leur application aux équations intégrales. Fundam. Math. **3**, 133–181 (1922)
2. Benveniste, A., et al.: Contracts for System Design, p. 296. Now Publishers Inc., Norwell, 2018
3. Broy, M., Stølen, K.: Specification and Development of Interactive Systems: Focus on Streams, Interfaces, and Refinement. Springer, Berlin, 2001
4. Broy, M.: Specification and Verification of Concurrent Systems by Causality and Realizability. Accepted for publication in Theoretical Computer Science
5. Broy, M.: A Calculus for the Specification, Design, and Verification of Distributed Concurrent Systems. Submitted
6. Clements, P.C., et al.: Documenting Software Architectures: views and beyond. Addison Wesley, Boston, 2003
7. Lamport, L.: Specifying Systems: The TLA+ Language and Tools for Hardware and Software Engineers. Addison-Wesley, Boston (2003)
8. Meyer, B.: Applying design by contract. Computer **25**, 10, 40–51, Oct. 199
9. Meyer, B.: Touch of Class: Learning to Program Well with Object and Contracts. Springer (2009)
10. Moore, E.F.: Gedanken-experiments on Sequential Machines. Automata Studies, Annals of Mathematical Studies, no. 34, pp. 129–153. Princeton University Press, Princeton, N.J. (1956)

Information-Flow Interfaces and Security Lattices

Ezio Bartocci[1], Thomas A. Henzinger[2], Dejan Nickovic[3],
and Ana Oliveira da Costa[2(✉)]

[1] Technische Universität Wien, Vienna, Austria
`ezio.bartocci@tuwien.ac.at`
[2] IST Austria, Klosterneuburg, Austria
`{tah,ana.costa}@ist.ac.at`
[3] AIT Austrian Institute of Technology, Vienna, Austria
`dejan.nickovic@ait.ac.at`

Abstract. Information-flow interfaces is a formalism recently proposed
for specifying, composing, and refining system-wide security requirements. In this work, we show how the widely used concept of security
lattices provides a natural semantic interpretation for information-flow
interfaces.

1 Introduction

Modern information and communication technologies are reaching unprece-
dented size and complexity, exposing them more and more to a wide variety of
cyber-attacks. The security-by-design engineering approach addresses this problem by enforcing security requirements throughout all phases of the design cycle,
including early design stages.

Information-flow interfaces [3,4] is a recently introduced security-by-design
formalism for the compositional development of systems that are guaranteed to implement specified information-flow policies. This formalism defines
information-flow requirements using *no-flow* relations. A no-flow relation specifies a forbidden exchange of information from one system variable to another.
This framework allows the engineer to combine *top-down* and *bottom-up* design
activities: a top-down step consists of decomposing the system-wide requirements and mapping them to sub-systems and components, while a bottom-up
step consists of assembling the system by combining available elements. To support the compositional design of systems, information-flow interfaces distinguish
between *assumptions* about the component's environment and *guarantees* that
the component provides when it operates in a proper environment. In addition,
information-flow interfaces satisfy the two main properties of an interface theory: *incremental design* and *independent implementability*. Incremental design
enables composing sub-systems and components without knowing the complete
design context. Independent implementability allows the development of subsystems by different design teams with system integration that guarantees the
preservation of system-level requirements.

© The Author(s), under exclusive license to Springer Nature Switzerland AG 2026
M. Fränzle et al. (Eds.): Werner Damm Festschrift, LNCS 15471, pp. 251–263, 2026.
https://doi.org/10.1007/978-3-031-97537-0_15

252   E. Bartocci et al.

We illustrate the top-down design using information-flow interfaces with an automotive example – a simplified version of a shared communication infrastructure connecting a wheel sensor and distance warners to the braking system and the odometer [3]. The shared communication infrastructure consists of (1) two (front and back) distance warners, which are sensors that estimate the proximity of the vehicle to other objects; (2) the wheel sensor, which senses the wheel rotations; (3) the odometer, a component that uses the wheel sensor data to measure the distance travelled by a vehicle; (4) a braking system, which takes the decision when to brake based on the information from the distance warners and (5) a shared bus that enables the communication between the other components. In this example, the braking system has a safety-critical function; consequently, communication with the distance warners has high-integrity. In contrast, the odometer and wheel sensor communication has low-integrity requirement. The main system-level requirement for this system is to guarantee the integrity of the communication channel when performing the safety-critical function. In other words, the design forbids any flow of information from the wheel sensor to the braking system.

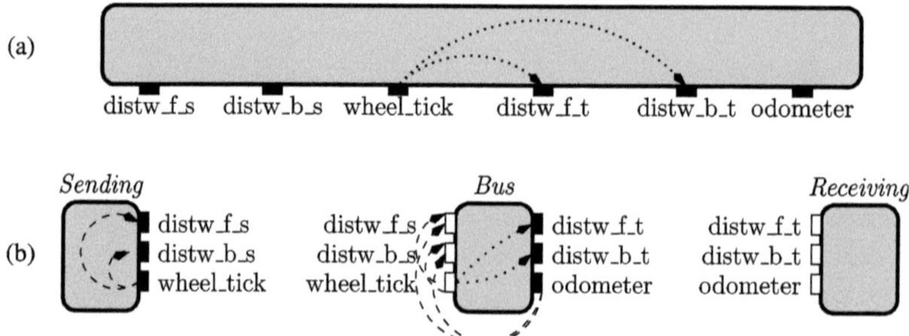

Fig. 1. An example of an automotive shared communication infrastructure specified with information-flow interfaces.

Figure 1 shows (a) the system-level interface specifying the above information-flow requirements and (b) one decomposition and refinement step. We depict an interface with a gray rounded-edged box. The white and black rectangles attached to the interface denote its input and output ports, respectively. We use dashed arrows to specify no-flow relations. A no-flow relation is an assumption if its target is an input port and a guarantee otherwise. The system-level security requirement from Fig. 1 (a) consists of two no-flow relations that forbid the exchange of information between the wheel sensor and the front and back distance warner targets. We observe that this interface specifies the properties of a closed system without defining its components and their interaction.

In the refinement step, shown in Fig. 1 (b), the interface is decomposed into three sub-systems: sending, receiving and bus. The two system-level requirements are naturally transferred to the bus sub-system. However, mapping these two properties from the system-level interface to the bus interface is not sufficient to enforce the original requirements. Without additional restrictions, there would be for instance an allowed flow in the decomposed system from the wheel sensor to the odometer, from the odometer to the distance wheeler sources, and from the sources to the distance wheeler targets, resulting in the violation of the system level requirements. To guarantee the enforcement of the system-level requirement after integration of sub-systems, we need to add additional assumptions (no-flow relations from the odometer and the wheel sensor to the distance wheelers in the bus sub-system) and guarantees (no-flow relations from the wheel sensor to the distance wheelers). In the subsequent design step, these three interfaces can be either further decomposed into smaller components, or assigned to different design teams for implementation.

Information-flow interfaces follow a declarative style of specification with the focus on what are the forbidden flows of information between system variables, rather then on how to guarantee the absence of the specified flows. This is a design choice that gives a syntactic flavor to the interface theory and its operations – composition of interfaces and their refinement consist of manipulating no-flow relations only. The advantage of this approach is that it enables simple incremental development of secure system architectures, while allowing to postpone semantic and implementation choices for the components to the later design stages. Information-flow interfaces have been equipped with a semantic interpretation in terms of *contracts* defined as pairs of assumptions and guarantees being sets of *flow relations* [4].

In this paper, we propose *security lattices* as an alternative semantic interpretation for information-flow interfaces. Equipping this interface theory with semantics based on security lattices has two major advantages. First, security lattices are commonly used in the security community for specifying information-flow and other security-related policies [1,2,7,9,10]. Therefore, their integration into the interface theory facilitates the security engineers' adoption of the compositional design framework. Second, the information-flow theory's compositional nature can help improve the efficiency of verification mechanisms whose performance depends on the shape of the lattice (e.g., number of labels, number of interactions between labels, ...). By using the information-flow interfaces equipped with security lattices, we can help to design more efficient lattices by (i) decomposing the system design into sub-systems and removing unnecessary labels and (ii) providing a list of lattices to choose from.

2 Background

This section provides the necessary background for equipping the information-flow interface theory with a semantic interpretation that uses security lattices. We first recall information-flow contracts as the original semantic interpretation for information-flow interfaces [4]. We then provide an overview of security lattices [6] for defining security policies. Throughout this section, we use the shared communication infrastructure example from Fig. 1 to illustrate the semantic interpretation of interfaces using information-flow contracts and sketch the idea of replacing contracts with more common security lattices.

Information-Flow Contracts. The information-flow interfaces use the syntax described in the shared communication infrastructure example and illustrated in Fig. 1 to specify security requirements in terms of no-flow relations. *Information-flow contracts* [4] provide a semantic interpretation to the interface theory. They define assumptions on the environment and guarantees on implementations in terms of *flow relations*. A relation $\mathcal{M} \subseteq (U \cup V) \times V$ is a flow relation iff it is a transitive relation over $U \cup V$, and reflexive over V. Let X and Y be disjoint sets of input and output variables, respectively, with the set of all variables defined as $Z = X \cup Y$. An *information-flow contract* is a tuple $(X, Y, A_{\text{flow}}, G_{\text{flow}})$ where $A_{\text{flow}} \subseteq 2^{Z \times X}$ is a set of flow relations to input variables, called *(contract) assumption*; and $G_{\text{flow}} \subseteq 2^{Z \times Y}$ is a set of flow relations to output variables, called *(contract) guarantee*.

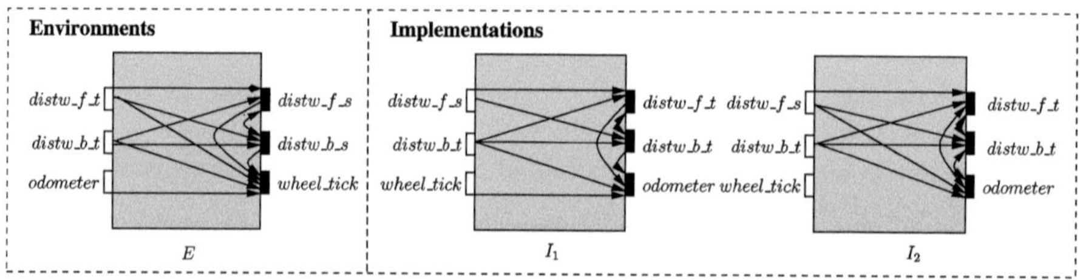

Fig. 2. Information-flow contract for the shared communication infrastructure interface, where E is the maximal permissible environment, while I_1 and I_2 are maximal implementations that satisfy the interface specification.

Example 1. We refer to the bus sub-system in the shared communication infrastructure from Fig. 1 to illustrate its semantic interpretation with information-flow contracts. Figure 2 depicts the information-flow contract for this interface. We use sets of components to graphically portray contract assumption and guarantee, where we depict a component as a gray box decorated with input and output variables, with solid arrows representing allowed flows. The assumption contains the set of all permissible environments, while the guarantee contains the set of all correct implementations. In this example, we depict only the maximal components, i.e., the contract assumption and guarantee contain all components that refine these components. In particular, there is one maximal environment E and two maximal implementations I_1 and I_2.

Security Lattices. Information-flow policies are usually defined for a set of *(security) labels*, also referred to in the literature as security classes. Labels and objects of a system (for example, variables or ports) usually define distinct sets of objects, with a label categorizing the role of an object within an information-flow policy.

A policy is specified with a *label can-flow* relation $\sqsubseteq$ over a set of security labels SC. Then, $L \sqsubseteq L'$, for labels $L \in$ SC and $L' \in$ SC, means that information on entities assigned to the label L is allowed to flow to entities in L'. Note that flow relations are defined for system's objects while can-flow relations are defined over security labels. Additionally, a policy provides a joint operator $\oplus \subseteq$ SC $\times$ SC specifying how to combine entities assigned with different labels. This operator can be used to dynamically assign a security label to objects that had no prior assignment. A policy is fully specified by a tuple $(SC, \sqsubseteq, \oplus)$. In [6], Denning proposed the following axioms for security policies $(SC, \sqsubseteq, \oplus)$:

Finiteness the set of security classes SC is finite;
Order $(SC, \sqsubseteq)$ defines a partial order;
Public Label there exists an unique lower bound in SC with respect to $\sqsubseteq$;
Totally of Label Combining the joint operator $\oplus$ is a least upper bound operator defined for every pair of security labels.

A security policy that satisfies these assumptions, defines a bounded lattice [6]. Then, $\sqsubseteq$ is a reflexive, transitive, and antisymmetric relation over SC with a unique upper bound (represented by $\top$) and a unique lower bound (represented by $\bot$). From now on, for simplicity of presentation, we denote security policies by $(SC, \sqsubseteq)$ only. Whenever $(SC, \sqsubseteq)$ defines a lattice, the definition of $\oplus$ over a pair of labels in SC is given by their least upper bound in $(SC, \sqsubseteq)$.

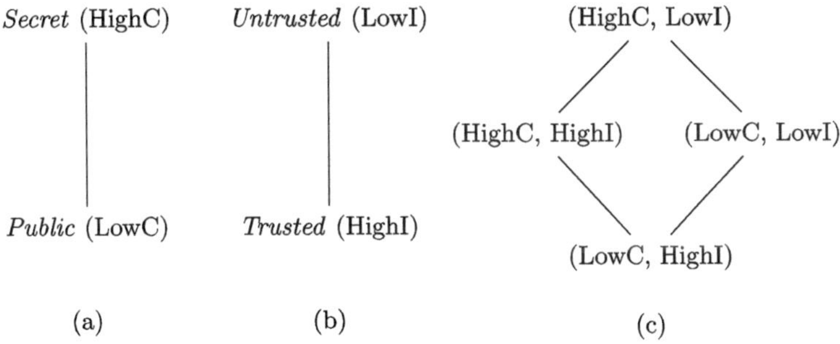

Fig. 3. Security lattices for (a) confidentiality, (b) integrity and (c) both combined.

Example 2. In Fig. 3, we have three examples of security lattices represented as Hasse diagrams. Reflexive and transitive arrows are omitted in the diagrams. The two Hasse diagrams to the left depict two linear lattices: one for confidentiality and one for integrity. As illustrated by the lattices, confidentiality and integrity adopt inverse views on the flow from high to low labelled variables. The lattice on the right side depicts the result of combining the confidentiality and integrity lattices. Figure 4 below depicts a security lattice that provides an elegant and succinct representation of the maximal implementation I_1 from Fig. 2.

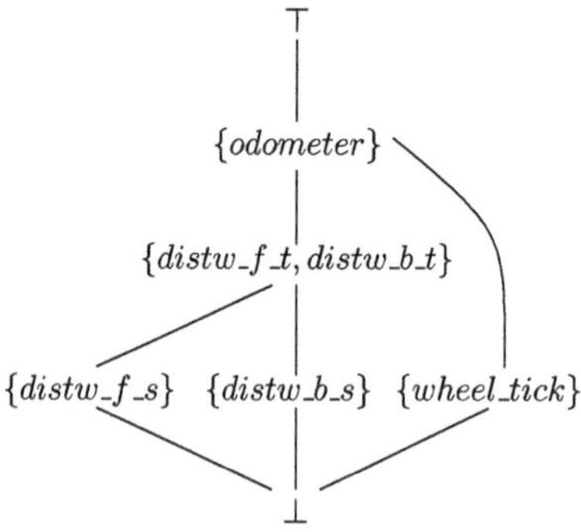

Fig. 4. Security lattice representation of the maximal implementation I_1.

3   Security Lattices For Information-Flow Interfaces

In this section, we introduce security lattice contracts and show how to translate from flow relations to security lattices and vice-versa. As for information-flow contracts, we define security lattice contracts over disjoint sets of input and output variables, denoted X and Y respectively, with the set of all variables defined as $Z = X \cup Y$. Note that while our atomic objects are variables, security labels are defined by sets of variables. For a set of variables Z, we define the set with all possible security labels over Z as $\mathrm{SC}^Z = 2^Z \cup \{\top, \bot\}$. A variable x is labelled with $L \in \mathrm{SC}$ iff $x \in L$.

Definition 1. *A security lattice contract is a tuple* (X, Y, A, G) *where* X *and* Y *are disjoint sets of variables and* $Z = X \cup Y$; $A = (\mathrm{SC}_A, \sqsubseteq_A)$ *is a security lattice with* $\mathrm{SC}_A \subseteq \mathrm{SC}^Z$, *called contract* assumption; *and* $G = (\mathrm{SC}_G, \sqsubseteq_G)$ *is a security lattice with* $\mathrm{SC}_G \subseteq \mathrm{SC}^Z$, *called contract* guarantee.

From Flow Relations to Security Lattices and Back. While security lattices specify requirements on security labels that are later assigned to objects of our design (for example, variables), flow relations specify flow requirements directly between the objects of our design. For this reason, flow relations should not be anti-symmetric, as we want to keep the design objects distinct even if they are indistinguishable to this relation.

We split the translation from a flow relation $\mathcal{M}$ to security lattice in three steps: (1) defining the set of security labels; (2) defining a partial-order over that set preserving the flow requirements from $\mathcal{M}$; and (3) extending the set of security labels and the can-flow relation to guarantee there is a least-upper bound for each pair of labels. The algorithm for this translation is presented in Algorithm 1. We explain next the steps in detail.

Algorithm 1. From flow relation $\mathcal{M}$ to security lattice – $[\![\mathcal{M}]\!]_{\text{lat}}$

Input: *Flow relation* $\mathcal{M} \subseteq (U \cup V) \times V$
Output: *Security Lattice* $(SC, \sqsubseteq)$

```
 1: SC ← {⊤, ⊥}
 2: ⊑ ← {}
 3: for all sequences (z₀, z₁) ∘ · · · ∘ (zₙ, z₀) defined by pairs in M do
 4:     L₁ ← {z₀, . . . , zₙ}
 5:     isToInclude ← true
 6:     for all L₂ ∈ SC do
 7:         if L₁ ⊆ L₂ then
 8:             isToInclude ← false        ▷ We do not add L₁ to SC as L₂ subsumes it
 9:         else if L₂ ⊆ L₁ then
10:             SC ← SC \ {L₂}             ▷ Remove labels from SC subsumed by L₁
11:         end if
12:     end for
13:     if isToInclude then SC ← SC ∪ {L₁}
14:     end if
15: end for
16: for all L₁ ∈ SC do
17:     ⊑ ← ⊑ ∪{(⊥, L₁), (L₁, ⊤), (L₁, L₁)}     ▷ Label relates with ⊤, ⊥ and itself
18:     for all L₂ ∈ SC do
19:         if L₁ × L₂ ⊆ M then
20:             ⊑ ← ⊑ ∪{(L₁, L₂)}  ▷ Add pair of labels to can-flow, if they satisfy M
21:         end if
22:     end for
23: end for
24: if (SC, ⊑) is not a lattice then
25:     (SC, ⊑) ← addLeastUpperLabels((SC, ⊑))                    ▷ Complete order
26: end if
27: return SC, ⊑)
```

Security Labels. In our first step, we collect all sets of variables in which all variables are in a loop defined using pairs in the given flow relation (c.f., line 3 in Algorithm 1). Formally, for a flow relation $\mathcal{M}$ and for a loop defined by a sequence $(z_0, z_1) \circ \cdots \circ (z_n, z_0)$, where $(z_i, z_{i+1}) \in \mathcal{M}$ for $0 \leq i < n$, then $\{z_0, \ldots, z_n\}$ defines a security label. Note that as flow relations are reflexive, for each variable z in the domain of $\mathcal{M}$, the singleton set $\{z\}$ defines a security label that satisfies the loop condition. We observe, additionally, that due to transitivity of flow relations all subsets of labels defined as explained above will also be security labels according to this definition (i.e., with transitivity we can skip steps in the sequence and find a loop that supports the label with fewer variables). To remove redundant labels, we keep only the largest set of variables necessary to define the loop (c.f., lines 7 to 13 in Algorithm 1).

Can-flow Order. After creating the labels, we proceed to defining the can-flow relation as a partial order over the set of labels derived in the previous step. This

---

Algorithm 2. Add least upper bound labels – addLeastUpperLabels

---

Input: $(SC, \sqsubseteq)$
Output: Updated $(SC, \sqsubseteq)$
```
 1: ToProcess ← {(L₁, L₂) | {L₁, L₂} ⊆ SC}                    ▷ Labels to process
 2: for all (L₁, L₂) ∈ ToProcess do
 3:     if L₁ and L₂ has no least upper bound then
 4:         L' = {newVar(SC)}              ▷ Singleton label with new variable name
 5:         ToProcess ← ToProcess ∪ {(L', L) | L ∈ SC}
 6:         SC ← SC ∪ {L'}
 7:         ⊑←⊑ ∪{(⊥, L'), (L', ⊤), (L', L'), (L₁, L'), (L₂, L')}
 8:         for all L ∈ SC do
 9:             if {(L₁, L), (L₂, L)} ⊆⊑ then
10:                 ⊑←⊑ ∪{(L', L)}        ▷ Connect L' to all common upper bounds
11:             end if
12:         end for
13:     end if
14:     ToProcess ← ToProcess \ {(L₁, L₂)}                ▷ Remove processed labels
15: end for
16: return (SC, ⊑)
```

---

is straightforward, as the flow relation is already reflexive and transitive. A pair of security labels (L, L') is in the can-flow relation iff all the variables in them are related by the flow relation (c.f., lines 16 to 23 in Algorithm 1). Formally, given a flow relation $\mathcal{M}$, for all $z \in L$ and $z' \in L'$, we require $(z, z') \in \mathcal{M}$. Additionally, we make sure that all labels are appropriately connected to the top and bottom elements of the lattice and to themselves to guarantee reflexivity (c.f., line 17 in Algorithm 1).

Example 3. In Fig. 5, we depict an implementation of the shared communication infrastructure, I_3. To the right of that implementation, we depict the partial order $(SC, \sqsubseteq)$ derived by line 23 of the Algorithm 1. Note that the pair of labels $(\{distw\_f\_s\}, \{distw\_b\_s\})$ do not have a lower upper bound, because $\{distw\_f\_t\}$ and $\{distw\_b\_t\}$ are uncomparable for the relation $\sqsubseteq$. When this is the case, we proceed in Algorithm 1 to line 25 and add the missing labels.

Least Upper Bounds. As illustrated in the example above, the set of labels together with the can-flow relation by line 23 in Algorithm 1 may not define a lattice. In Algorithm 2, we present a possible way to extend the set of security labels and the can-flow relation to guarantee that each pair of labels has a unique least upper bound. There are multiple ways one could define and implement this extension. In this work, we focus on presenting a simple and intuitive extension for demonstration purposes.

Intuitively, for a flow relation over $(U \cup V) \times V$, we need to create intermediate labels connecting labels with U variables to labels with V variables. Note that there are no labels that mix variables of V and U because there is no loop containing both types of variables in the flow relation $\mathcal{M} \subseteq (U \cup V) \times V$. To

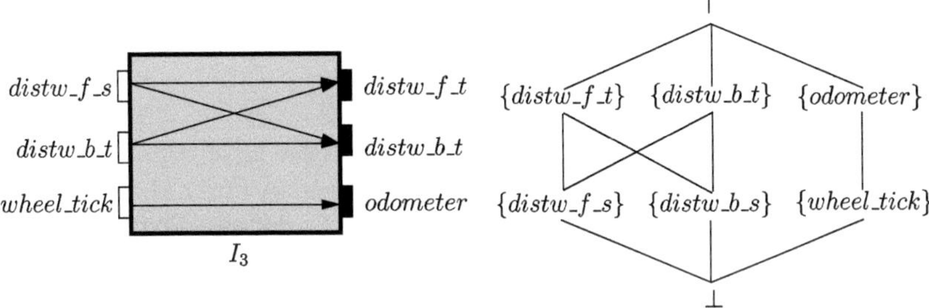

Fig. 5. An implementation for the shared communication I_3 with the partial order derived from its can-flow by line 23 in Algorithm 1. Note that labels $\{distw\_f\_s\}$ and $\{distw\_b\_s\}$ do not have a lower upper bound, which will be added in the next steps of the Algorithm.

define the intermediate labels, we extend the domain of security labels with sets over fresh variables (obtained with the function newVar that returns a variable name not yet in SC). The new labels are internal to a contract implementation (i.e., irrelevant to the system's design) and characterize the expected security label when we combine information from different sources.

Example 4. Algorithm 1 outputs lattices in Fig. 4 and Fig. 6 when the input are the flow relations in implementations I_1 and I_3, respectively.

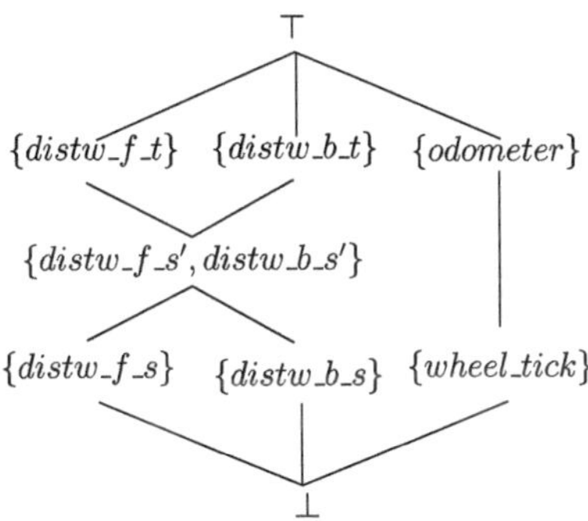

Fig. 6. Security lattice returned from Algorithm 1 for the flow relation in I_3.

We prove below that the relation output by Algorithm 1 is indeed a lattice and is equivalent to the flow restrictions of the input $\mathcal{M}$. For a flow relation $\mathcal{M} \subseteq Z \times V$, a structure $(SC, \sqsubseteq)$ *is equivalent to it,* denoted $(SC, \sqsubseteq) \equiv_{Z \times V} \mathcal{M}$ iff for all pairs of labels $(L_1, L_2) \in 2^Z \times 2^V$, they are in the can-flow relation, $(L_1, L_2) \in \sqsubseteq$, iff, for all $(z_1, z_2) \in L_1 \times L_2$ we have $(z_1, z_2) \in \mathcal{M}$.

Theorem 1. *Let $\mathcal{M} \subseteq Z \times V$ be a flow relation and $[\![\mathcal{M}]\!]_{\mathrm{lat}} = (SC, \sqsubseteq)$. Then, $(SC, \sqsubseteq)$ defines a security lattice that is equivalent to the can-flow restrictions of $\mathcal{M}$ over $Z \times V$, i.e., $(SC, \sqsubseteq) \equiv_{Z \times V} \mathcal{M}$.*

Proof. We start by proving that $(SC, \sqsubseteq)$ defined at line 23 of Algorithm 1 (before we call addLeastUpperLabels procedure) (i) defines an order that is equivalent to $\mathcal{M}$ over its domain, $(SC, \sqsubseteq) \equiv_{Z \times V} \mathcal{M}$, and (ii) $\sqsubseteq$ defines a partial order over the set of labels SC.

We observe that in Algorithm 1 (a) by line 16, we only have labels with variables in the domain of $\mathcal{M}$ (i.e., $SC \subseteq 2^{U \cup V} \cup \{\top, \bot\}$); (b) only in lines 17 and 20 we add pairs of labels to $\sqsubseteq$; and (c) in line 17 we add the reflexive pairs and the connections to $\top$ and $\bot$, while, by the condition at line 19, in line 20 we only add the pair (L_1, L_2) to the can-flow relation when all their elements $z_1 \in L_1$ and $z_2 \in L_2$ are pairs in the flow relation, $(z_1, z_2) \in \mathcal{M}$. By these three observations, it follows that by line 23 of Algorithm 1 $(SC, \sqsubseteq) \equiv_{Z \times V} \mathcal{M}$ holds. We prove now (ii). After line 17 in Algorithm 1, $\sqsubseteq$ is a reflexive relation. By transitivity of flow relations and the point (i) proved above, then $\sqsubseteq$ is also a transitive relation. Finally, by removal of the loops in lines 3 to 15, it follows that $\sqsubseteq$ is antisymmetric.

Now, we are only missing to prove that by line 27 in Algorithm 1 the returned $(SC, \sqsubseteq)$ (iii) is equivalent to $\mathcal{M}$, $(SC, \sqsubseteq) \equiv_{Z \times V} \mathcal{M}$, and (iv) defines a lattice. First, we observe that all pairs of labels added to $\sqsubseteq$ in lines 7 and 10 of Algorithm 2 do not have variables from Z. Then, by (i) proved previously, (iii) holds. If $(SC, \sqsubseteq)$ is a lattice by line 24 Algorithm 1, then $(SC, \sqsubseteq)$ is not changed so (iv) holds. We consider now the case that $(SC, \sqsubseteq)$ does not define a lattice by line 24. Due to the elements added to $\sqsubseteq$ in line 7 of Algorithm 2 and (ii) proved above, it follows that $\sqsubseteq$ is reflexive relation. Moreover, after the same line and from line 17 of Algorithm 1, it follows that $\bot$ and $\top$ are the lower and upper bound of $(SC, \sqsubseteq)$, respectively. Transitivity of $\sqsubseteq$ follows from the pairs added in line 7 and 10 in Algorithm 1 and (ii) proved above. By lines 1 and 5 in Algorithm 2, we have the guarantee that all pair of labels added to SC will be checked for having a least upper bound. If they do not have it, a new label is created that is connected to all their upper bounds (c.f., line 10). As all the upper bounds to any of pair of labels processed are either in the algorithm input structure or are added only during the loop iteration we add a new label, then we have the guarantee that no new upper bounds are added after a pair of labels is processed. Then, at each iteration the new label L' is the least upper bound for the labels L_1 and L_2. Hence the returned $(SC, \sqsubseteq)$ is a lattice. □

The translation from a security lattice to a flow relation is straightforward, as a lattice is by definition already reflexive and transitively closed. Formally, given a security lattice $(SC, \sqsubseteq)$ we define its derived flow relation over a domain $Z \times V$ as follows:

$$[\![(SC, \sqsubseteq), Z \times V]\!]_{\mathrm{flow}} = \{(z, z') \mid (L, L') \in\ \sqsubseteq,\ z \in (L \cap Z) \text{ and } z' \in (L' \cap V)\}.$$

We prove below that after we translate a flow relation to a security lattice, using the operator defined by Algorithm 1, we can use the operator defined above to get back the original flow relation.

Theorem 2. *Let* $\mathcal{M} \subseteq (U \cup V) \times V$ *be a flow relation with* $Z = U \cup V$. *Then:*

$$[\![[\![\mathcal{M}]\!]_{\mathrm{lat}}, Z \times V]\!]_{\mathrm{flow}} = \mathcal{M}.$$

Proof. By Theorem 1, $[\![\mathcal{M}]\!]_{\mathrm{lat}} \equiv_{Z \times V} \mathcal{M}$, i.e., for $(SC, \sqsubseteq) = [\![\mathcal{M}]\!]_{\mathrm{lat}}$ and for all pairs of labels $(L_1, L_2) \in 2^Z \times 2^V$, they are in the can-flow relation, $(L_1, L_2) \in \sqsubseteq$, iff, for all $(z_1, z_2) \in L_1 \times L_2$, we have $(z_1, z_2) \in \mathcal{M}$. Hence, by definition of $[\![.]\!]_{\mathrm{flow}}$, $[\![[\![\mathcal{M}]\!]_{\mathrm{lat}}, Z \times V]\!]_{\mathrm{flow}} = \mathcal{M}$ holds. $\qquad\qquad\square$

We show in the theorem below that, under reasonable constraints, when given a security lattice, if we translate it to a flow relation followed by a translation to a lattice, then at the end, we get back the original lattice. The first constraint, specifying that variables in U can only define singleton labels, is motivated by the observation that for a flow relation $\mathcal{M} \subseteq (U \cup V) \times V$, there are only outgoing flows from U. Then, there cannot be loops with variables in U, and in our translation from flow relations to lattices, we defined that loops are the only way to get non-singleton labels. The second constraint is derived from our implementation decision in Algorithm 1 only to keep maximal sets.

Theorem 3. *Let* $SC \subseteq \{\{u\} \mid u \in U\} \cup 2^V \cup \{\top, \bot\}$ *be a set of maximal sets of variables, i.e., for all* $\{L_1, L_2\} \subseteq SC$ *then* $L_1 \nsubseteq L_2$. *For all relations* $\sqsubseteq \subseteq SC \times SC$ *s.t.* $(SC, \sqsubseteq)$ *defines a lattice, then:*

$$[\![[\![(SC, \sqsubseteq), (U \cup V) \times V]\!]_{\mathrm{flow}}]\!]_{\mathrm{lat}} = (SC, \sqsubseteq).$$

Proof. Let $\mathcal{M} = [\![(SC, \sqsubseteq), (U \cup V) \times V]\!]_{\mathrm{flow}}$. Note that, by $\sqsubseteq$ being a partial order over SC and by definition of $[\![.]\!]_{\mathrm{flow}}$, the only loops in the flow relation output by $[\![(SC, \sqsubseteq), (U \cup V) \times V]\!]_{\mathrm{flow}}$ are between variables in the same label in SC. Let SC_{15} be the set of security labels by line 15 in Algorithm 1. By SC having only maximal sets, then $SC_{15} = SC$. Now, let $\sqsubseteq_{23}$ be the can-flow relation defined by line 23 of Algorithm 1 and SC_{23} the set of labels (note that $SC_{15} = SC_{23} = SC$). By a reasoning analogous to the proof of Theorem 1, we know that $(SC_{23}, \sqsubseteq_{23}) \equiv_{Z \times V} \mathcal{M}$. As $\mathcal{M}$ is derived from a lattice, then $(SC_{23}, \sqsubseteq_{23})$ is also a lattice and $[\![[\![(SC, \sqsubseteq), (U \cup V) \times V]\!]_{\mathrm{flow}}]\!]_{\mathrm{lat}} = (SC_{23}, \sqsubseteq_{23})$. Then, by $(SC_{23}, \sqsubseteq_{23}) \equiv_{Z \times V} \mathcal{M}$, it follows that $\sqsubseteq_{23} = \sqsubseteq$. Hence $[\![[\![(SC, \sqsubseteq), (U \cup V) \times V]\!]_{\mathrm{flow}}]\!]_{\mathrm{lat}} = (SC_{23}, \sqsubseteq_{23}) = (SC, \sqsubseteq)$ $\qquad\square$

Information-Flow and Security Lattice Contracts. Given an information-flow contract $C_{\mathrm{flow}} = (X, Y, A_{\mathrm{flow}}, G_{\mathrm{flow}})$, we define the derived security lattice contract as: $[\![C_{\mathrm{flow}}]\!] = (X, Y, [\![A_{\mathrm{flow}}]\!]_{\mathrm{lat}}, [\![G_{\mathrm{flow}}]\!]_{\mathrm{lat}})$. Information-flow contracts were introduced in [4] along with a composition operator and a refinement relation over such contracts. Using the translations presented above, we define composition and refinement over security lattice contracts directly from their counterpart in information-flow contracts (we use the same definitions and translate back and forth between flows and lattices).

In [4], we proved that the composition and refinement of information-flow contracts adequately capture the semantics of their counterpart in information-flow interfaces. These results show that information-flow contracts, just as for information-flow interfaces, satisfy incremental design and independent implementability [4]. Then, using Theorem 2, all the results mentioned above for information-flow contracts also hold for security lattices contracts.

4 Conclusions

In this paper, we proposed security lattices as an alternative semantics for the information-flow interface theory. Security lattices are the backbone of many successful, readily available techniques to verify and enforce security policies. Examples range from type systems [7] to static analysis [8] or runtime enforcement using secure multi-execution [2,5], to name a few.

Equipping interfaces with a semantic interpretation based on a commonly used formalism for defining and enforcing security policies facilitates the bridge between the security practitioners and the security-by-design paradigm.

Acknowledgements. This project was funded in part by the Austrian Science Fund (FWF) SFB project SpyCoDe F8502 and by the ERC-2020-AdG 101020093.

References

1. Algehed, M., Flanagan, C.: Transparent IFC enforcement: possibility and (in)efficiency results. In: Proceedings of Computer Security Foundations Symposium (CSF), pp. 65–78 (2020). https://doi.org/10.1109/CSF49147.2020.00013
2. Austin, T.H., Flanagan, C.: Multiple facets for dynamic information flow. In: Proceedings of ACM SIGPLAN-SIGACT Symposium on Principles of Programming Languages (POPL), pp. 165–178 (2012)
3. Bartocci, E., Ferrère, T., Henzinger, T.A., Nickovic, D., da Costa, A.O.: Information-flow interfaces. In: Proceedings of Fundamental Approaches to Software Engineering (FASE), pp. 3–22 (2022). https://doi.org/10.1007/978-3-030-99429-7_1
4. Bartocci, E., Ferrère, T., Henzinger, T.A., Nickovic, D., da Costa, A.O.: Information-flow interfaces (2020). https://arxiv.org/abs/2002.06465
5. De Groef, W., Devriese, D., Nikiforakis, N., Piessens, F.: FlowFox: a web browser with flexible and precise information flow control. In: Proceedings of the 2012 ACM Conference on Computer and Communications Security (CCS), pp. 748–759 (2012). https://doi.org/10.1145/2382196.2382275
6. Denning, D.E.: A lattice model of secure information flow. Commun. ACM **19**(5), 236–243 (1976). https://doi.org/10.1145/360051.360056
7. Focardi, R., Maffei, M.: Types for security protocols. Formal Model. Tech. Anal. Secur. Protoc. **5**, 143–181 (2011). https://doi.org/10.3233/978-1-60750-714-7-143

8. Huang, Y.W., Yu, F., Hang, C., Tsai, C.H., Lee, D.T., Kuo, S.Y.: Securing web application code by static analysis and runtime protection. In: Proceedings of the 13th International Conference on World Wide Web, pp. 40–52. WWW '04 (2004). https://doi.org/10.1145/988672.988679
9. Sabelfeld, A., Myers, A.C.: Language-based information-flow security. IEEE J. Sel. Areas Commun. **21**(1), 5–19 (2003). https://doi.org/10.1109/JSAC.2002.806121
10. Sandhu, R.: Lattice-based access control models. Computer **26**(11), 9–19 (1993). https://doi.org/10.1109/2.241422

Identification of Classification Clusters in Convolutional Neural Networks

Felix Brüning[1] , Felix Höfer[2] , Wen-ling Huang[1] , and Jan Peleska[1(✉)]

[1] University of Bremen, Department of Mathematics and Computer Science,
Bibliothekstrasse 1, 28359 Bremen, Germany
`{fbrning,huang,peleska}@uni-bremen.de`
[2] Princeton University, Department of Operations Research and Financial
Engineering, Princeton, NJ 08540, USA
`fhoefer@princeton.edu`

Abstract. In this paper, we evaluate the first of two parts of a novel approach for the assessment of residual error probabilities in trained convolutional neural networks (CNN). We consider CNNs for camera image classification, as needed in a safety-critical context, for example in autonomous road vehicles or trains, for the purpose of obstacle detection. The objective of the strategy's first part is to identify so-called classification clusters of CNNs: these are subsets of the input space, whose elements are all mapped to the same classification result. To this end, we apply a new technique based on mathematical analysis.

Keywords: Convolutional neural networks · differentiable manifolds · singular Riemannian metrics · null curves · mathematical analysis

1 Introduction

1.1 Objectives

This paper has been created in the context of research project HiDyVe (Highly Dynamic Virtual and Hybrid Validation and Verification), in which existing and future challenges for the verification and validation (V&V) of complex – potentially autonomous – transportation systems are investigated. The work presented here has been significantly inspired by Werner Damm, who founded the working group *"Virtuelle Absicherung"* (engl. *Virtual Assurance [of safety-critical autonomous systems]*) as part of the SafeTRANS[1] initiative.

In this paper, we present a novel method to identify the *classification clusters* of convolutional neural networks (CNN) [2] used for the identification of objects in camera images. Following the terminology introduced in the new safety standard ANSI/UL 4600 for the evaluation of autonomous products [29], such a

---

[1] https://www.safetrans-de.org/en/index.php.

F. Brüning, W. Huang and J. Peleska—Are funded by the German Ministry of Economics, Grant Agreement 20X1908E.

M. Fränzle et al. (Eds.): Werner Damm Festschrift, LNCS 15471, pp. 264–287, 2026.
https://doi.org/10.1007/978-3-031-97537-0_16

cluster is a subset of the CNN input space, whose elements are all mapped to the same classification result. The method presented here is one building block of a verification strategy for CNNs used for image classification in a safety-critical context, such as the obstacle detection on railway tracks (application domain *autonomous trains*) and on roads (application domain *autonomous road vehicles*). The overall strategy has been described in [14]: with a complete set of classification clusters at hand, it is possible to determine the residual misclassification probability of a trained CNN with a given confidence level. Intuitively speaking, statistical tests are used to determine the discrete probability distribution that predicts whether an image will be a member of classification cluster c_i, where $i \in \{1, \ldots, \ell\}$, and ℓ is the number of clusters that have been identified for the trained CNN with the method described here. The analysis of the Coupon Collector's Problem [12] can be applied to this empirical distribution to estimate the residual probability that a cluster $c_{\ell+1}$ has been overlooked during the training phase. For the cluster identification, a new method proposed by Benfenati and Marta [6,7] is evaluated here for the first time with respect to suitability for classification CNNs.

1.2   Contributions

We present a new approach to re-model the layers of a trained CNN by means of subsets of $\mathbb{R}^m$ with layer-dependent dimension m. The inter-layer transformations are modelled precisely as piecewise differentiable mappings between these subsets. This new approach is inspired by original work of Benfenati and Marta [6,7], but has been considerably revised and extended:

- The more complex method of these authors based on singular Riemannian differentiable manifolds and metric pullbacks is simplified in a considerable way.
- We show that Benfenati's and Marta's requirement that the real inter-layer mappings should always be approximated by smooth mappings can be dropped, so that the original (often only piecewise differentiable) mappings of the CNN can be directly used in the mathematical analysis.

Instead of pre-determining all classification clusters *before* the statistical evaluation starts, we elaborate an on-the-fly cluster identification technique that allows to start the statistical evaluation immediately after the training and initial validation and optimisation phases.

To evaluate the new analysis approach, a trained image classification CNN which is based on the well-known MNIST data set[2] is used. While this still results in fairly small CNN models, it suffices to show that the method can cope with the complexity of high-dimensional image input spaces and with the inter-layer mappings typically used in trained CNNs. A "real-world" application to image data bases representing obstacles on railway tracks will be performed in the future, after more data sets of this kind have become available.[3]

---

[2] http://yann.lecun.com/exdb/mnist/.

[3] As of today, a publicly available image database for railway obstacles does not exist yet.

1.3   Background and Related Work

Sensing and perception are the initial steps of any *autonomy pipeline* [29] controlling an autonomous system. These two steps are essential for creating *situation awareness*, that is, for updating the internal system state space used by the subsequent pipeline steps (planning, prediction, control, actuation) with data regarding the current environment state. For safety-critical autonomous systems, sensing and perception need to be sufficiently trustworthy, because an erroneous representation of the environment state (e.g. a false negative indicating "no obstacle present" while there is an obstacle on the railway track) can result in catastrophic consequences. There is a common understanding that sensing and perception need to be based on a fusion of different redundant sensor technologies and perception techniques [13,21]. Moreover, it is important to evaluate the trustworthiness of the fused sub-system during runtime, in order to perform safe system degradations (e.g. remote or manual control of a train whose obstacle detection sub-system has failed) if automated sensing/perception can no longer be trusted [8,13,21]. Each of the sensing and perception methods applied, however, need to be verified and validated for type certification, to show that the method and its associated technical design can guarantee that the *safety of the intended functionality* [17] is ensured with a sufficiently small residual failure probability: otherwise it would be impossible to draw conclusions about the residual risk of the fused sensor/perceptor sub-system.

Due to their complex transformation functions whose weight and bias parameters are determined during their training phase, deep neural networks represent a considerable verification challenge. In particular, the correctness of the network's software implementation is an insufficient indicator for its performance: the training data applied to optimise the network parameters and the validation data used to fine-tune them influence the resulting safety of the intended functionality in an essential way. It is impossible to prove in a formal way that the training and validation data used are sufficient for the network to handle *every* input that might occur in the operational design domain (ODD)[4] correctly.

The training and validation-related root causes for insufficient performance of a trained neural network have been identified in a fairly comprehensive way. (1) *Overfitting* occurs when the training and validation sets have been too small in comparison to the degrees of freedom given by the number of weights and bias parameters of the network. As a consequence, the trained network performs perfectly on the training and validation data, but fails frequently in the real world (or for a verification set that has only few similarities to the training and validation data) [2]. (2) *Brittleness* [29] occurs when a trained network works correctly for a certain input value v, but fails for values v' that are very close to v. Here, *closeness* is interpreted in the sense of the intended functionality. (3) *Explanation errors* occur when a CNN obtains a correct classification result, but "for the wrong reasons". Infamous examples for this type of error have occurred in situations where irrelevant image information (e.g. watermarks or photo studio

---

[4] The ODD is a well-defined restriction of the real world where an autonomous system is expected to operate with an acceptable quantified risk [29].

names) have been considered by a trained network as necessary and sufficient for the classification result, due to unfortunate choices of the training data [27].

During the last decade, considerable progress has been made in the field of neural network verification. Today, the avoidance and detection of overfitting is well understood [2,23], and both model agnostic and model sensitive methods have been designed to effectively provide explanations for the classification results achieved [4,10,27]. Moreover, the detection of brittleness can be achieved with effective coverage-driven testing methods [28].

The remaining main challenge for the use of image classification CNNs in safety-critical systems is to (a) justify at type certification time that the trained CNN comes with an acceptable residual misclassification risk, and (b) determine at runtime whether a CNN-based perceptor provides trustworthy results, or should be excluded from the configuration of fused sensors and perceptors. For Problem (b), a new approach has been proposed by Gruteser et al. [15].

Our current research focus is on providing a comprehensive approach to the solution of Problem (a) [14]. There have been several attempts to determine the residual risk for misclassification in trained CNNs [24–26]. These, however, were mostly based on statistical analyses alone, and did not take the internal structure of the CNN model into account. Our approach to this problem regards the identification of classification clusters in trained CNNs as an essential prerequisite to speed up the statistical analyses and to justify convincingly that the residual probability that certain images will be classified in the wrong way since no appropriate cluster has been created during training is acceptably small. In this paper, we only described the method for the identification of these clusters. The overall verification approach involving a specific statistical testing strategy has been described by Gleirscher et al. [14].

Braband et al. [9] argue that for freight trains in the railway domain, the safety integrity level SIL-3 is sufficient. This corresponds to a tolerable hazard rate (THR) of 10^{-7} obstacle detection failures per hour. According to the publicly available literature [3,22], CNN-based obstacle detection is unlikely to perform better than with a classification error probability of $p_f = 0.02$. Assuming an obstacle occurrence rate of $2/24h^{-1}$ or higher, this probability does not fulfil the THR requirement of SIL-3. We have shown, however, that a fusion of three or more sensor/perceptor components with $p_f = 0.04$ or better, the THR requirement can be fulfilled [14].

1.4 Overview

In Sect. 2, we review basic facts about CNNs and summarise the work of Benfenati and Marta [6,7] as theoretical background. In Sect. 3, our revised theory for the analytic modelling of CNNs is presented. In Sect. 4, a new technique for identifying classification clusters on-the-fly, while the statistical verification tests are performed, is presented. A practical evaluation exampled is discussed in Sect. 5. In Sect. 6, we analyse threats to validity. Section 7 contains a conclusion, and future work is discussed.

2 Theoretical Foundations

2.1 Convolutional Neural Network Structure

We consider CNNs with the usual layers and inter-layer transformations, such as convolutions with kernels of varying sizes, max pooling transformations, flattening maps and dense maps [2]. All differentiable activation functions are admissible. The activation function $\texttt{ReLU}(x) = \max(0, x)$, though not differentiable in 0, is admissible as well. Likewise, max pooling transformations are not differentiable everywhere, but still admissible. Convolutions, flattening maps, and dense maps are differentiable. They are, however, often combined with the $\texttt{ReLU}$ activation function applied to the elements of the result matrices or result vectors produced by the differentiable maps.

We consider camera image-based obstacle detection functions implemented by trained CNNs. The CNN $\mathcal{N} : [0,1]^{L \times B \times d} \longrightarrow [0,1]^k$ maps input images[5] of size $L \times B \times d$ to a k-dimensional output tuple $\boldsymbol{p} = (p_1, \ldots, p_k)$ satisfying $\sum_{i=1}^{k} p^i = 1$, so that $\boldsymbol{p}$ represents a probability distribution calculated by $\mathcal{N}$. The probabilistic interpretation is ensured by applying the *softmax classifier* (multinomial logistic regression)

$$\texttt{Softmax} : \mathbb{R}^k \longrightarrow [0,1]^k; \quad (v_1, \ldots, v_k) \mapsto \boldsymbol{p} = (e^{v_1}, \ldots, e^{v_k}) / \sum_{i=1}^{k} e^{v_i} \qquad (1)$$

to the k-dimensional result vector of the last dense transformation [2, Section 2.3.3]. For $i = 1, \ldots, (k-1)$, vector component p_i of $\boldsymbol{p}$ represents the probability that the image contains an obstacle of type i. The last vector component p_k is the probability that *no* obstacle is contained in the image. The decision *'no obstacle present'* is made if

$$p_k = \max\{p_1, \ldots, p_k\}. \qquad (2)$$

Likewise, $p_j = \max\{p_1, \ldots, p_k\}$ for $j < k$ indicates that an obstacle of type j is present. For the monitoring of sensor/perceptor trustworthiness at runtime [13], the result tuple $\mathcal{N}(x)$ will typically be analysed, since the "uncertainty" in a classification result $\mathcal{N}(x)$ can be detected, for example, by all p_i being approximately of the same size. To make the decision *'obstacle/no obstacle'*, an activation function for the tuple $\mathcal{N}(x)$ is induced by the observation that

$$p_k = \max\{p_1, \ldots, p_k\} \text{ if and only if } \sum_{i=1}^{k-1} \texttt{ReLU}(p_i - p_k) = 0. \qquad (3)$$

Therefore, we define

$$\Omega_k : [0,1]^k \longrightarrow [0,1]; \quad (p_1, \ldots, p_k) \mapsto \sum_{i=1}^{k-1} \texttt{ReLU}(p_i - p_k) \qquad (4)$$

---

[5] Typically, each grey-scale or colour pixel value is normalised to range $[0, 1]$.

and the final obstacle detection function

$$\Lambda^k = \Omega_k \circ \mathcal{N} : \mathbb{R}^{L \times B \times d} \longrightarrow [0, 1].$$

An image x is mapped by Λ^k to value zero, if and only if x does not contain an obstacle according to the trained CNN $\mathcal{N}$. Conversely, $\Lambda^k(x) > 0$ if and only if the CNN $\mathcal{N}$ has identified an obstacle in x.

Similarly, activation functions for the evaluation whether '*NO obstacle of type j is present*' are defined by

$$\Omega_j : [0, 1]^k \longrightarrow [0, 1]; \quad (p_1, \ldots, p_k) \mapsto \sum_{i=1}^{k} \text{ReLU}(p_i - p_j). \tag{5}$$

The function value $\Omega_j(\boldsymbol{p}) = 0$ indicates '*an obstacle of type j is present*', $\Omega_j(\boldsymbol{p}) > 0$ states that no obstacle of this type has been detected. For $j = 1, \ldots, k - 1$, we define $\Lambda^j(x) = \Omega_j \circ \mathcal{N}(x)$, so $\Lambda^j(x) = 0$ indicates that image x contains an obstacle of type j.

2.2   A Differential Geometric Approach to DNN Analysis

Our calculus-based approach to CNN analysis described in Sect. 3 has been inspired by Benfenati and Marta [6,7], who proposed a differential geometric interpretation of deep neural networks. Their main application focus was on DNNs modelling solutions to physical problems (thermodynamics) and low-dimensional classification problems (point classes in $\mathbb{R}^2$ separated by mathematical functions). The authors expressed the expectation that a generalisation to more complex image classification problems involving CNNs should be possible.

Benfenati and Marta consider the layers of a deep neural network as differentiable manifolds[6] $M_0, \ldots, M_n$, where manifold M_0 represents the input layer, $M_1, \ldots, M_{n-1}$ the intermediate "hidden" layers, and M_n the output layer with its classification results. Between each pair of manifolds, differentiable mappings

$$M_0 \xrightarrow{\Lambda_1} M_1 \xrightarrow{\Lambda_2} M_2 \ldots M_{n-1} \xrightarrow{\Lambda_n} M_n$$

are defined, each mapping Λ_i a differentiable approximation of the true inter-layer mapping applied in the CNN model. The $\text{ReLU}(x)$ activation function, for example, which is not differentiable in $x = 0$, can be approximated by the softplus function which is defined as

$$\text{Softplus}(x) = \frac{\ln(1 + e^{kx})}{k},$$

which results in increasingly precise approximations of $\text{ReLU}(x)$ with growing values $k \geq 1$.

---

[6] For an introduction to differentiable manifolds, we recommend Kupeli [19] and the definitions and explanations given in [6].

If M_n represents classification results as points on a real interval, this manifold can be equipped with a Riemannian metric by simply choosing the Euclidean distance $|a - b|$ between points a, b on the real axis. On the higher dimensional manifolds $M_0, \ldots, M_{n-1}$, this induces Riemannian metrics δ_i on each M_i, $i = 0, \ldots, n-1$ by defining

$$\delta_i(x, y) = |\Lambda_n \circ \Lambda_{n-1} \circ \cdots \circ \Lambda_{i+1}(x) - \Lambda_n \circ \Lambda_{n-1} \circ \cdots \circ \Lambda_{i+1}(y)| \quad \text{for } x, y \in M_i. \quad (6)$$

This metric on M_0, however, is *singular*: this means that different points in M_0 can have distance zero. This happens exactly if both points are mapped to the same classification result in M_n. Given any point $p_0 \in M_0$, all other points p with the same classification as p (that is, the *classification cluster of p*) can now be formally specified by

$$\texttt{cluster}(p) = \{p \in M_0 \mid \delta_0(p_0, p) = 0\} \quad (7)$$

A main result of Benfenati's and Marta's work consists in the insight that distance-zero points in the vicinity of some point $p_0 \in M_0$ can be determined locally in an analytic way. To this end, one proceeds as follows.

1. A metric on the vector spaces of the tangent bundle TM_i is introduced[7] by means of the pullback of the Riemannian metric on M_n through the interlayer mappings $\Lambda_n \circ \Lambda_{n-1} \circ \cdots \circ \Lambda_i$.[8]
2. The length of a differentiable curve connecting two points in M_i is defined as the integral over the lengths of its tangent vectors.
3. A second metric $\delta'(p_1, p_2)$ between points $p_1, p_2 \in M_i$ is defined as the infimum over the lengths of all differentiable curves connecting p_1 and p_2.

As a result of the pullback construction for defining metrics on TM_i, the distance $\delta'_0(p_1, p_2)$ between two points $p_1, p_2 \in M_0$ coincides *locally* with the distance $\delta_0(p_1, p_2) = |\Lambda(p_1) - \Lambda(p_2)|$: distance value $\delta'_0(p_1, p_2) = 0$ occurs exactly, if the minimal length of differentiable curves

$$\gamma : I \longrightarrow M_0 \text{ with open interval } I \text{ containing } [0, 1] \text{ and } \gamma(0) = p_0 \text{ and } \gamma(1) = p_1$$

connecting p_0 and p_1 is zero. Such a *null curve* γ of M_0 has (non-null) tangent vectors that always have length zero in the pullback metric. Length-zero tangent vectors in this metric imply that $\Lambda(\gamma(t))$ remains constant for all $t \in I$. Consequently,

$$\Lambda(p_0) = \Lambda(\gamma(0)) = \Lambda(\gamma(1)) = \Lambda(p_1),$$

so $|\Lambda(p_0) - \Lambda(p_1)| = \delta_0(p_0, p_1) = 0 = \delta'_0(p_0, p_1)$.

The singular metric δ'_0 induces an equivalence relation on M_0: two points $p_0, p_1 \in M_0$ are equivalent if and only if they are connected by a null curve. Using fundamental concepts of Riemannian geometry, Benfenati and Marta show

---

[7] Each point of a manifold M_i is associated with such a vector space.

[8] Readers who are not familiar with these differential geometric terms need not despair: we will give a simpler calculus-based explanation of the underlying theory in Sect. 3.

that each equivalence class $[p_0] \subseteq M_0$ forms a maximal integral manifold of the vertical bundle $\mathcal{V}M_0$. As a consequence, each classification cluster of a deep classification network can be represented as a union over maximal integral manifolds $[p_i]$, that is,

$$\texttt{cluster}(p) = \{p \in M_0 \mid \delta_0(p_0, p) = 0\} = [p] \cup \bigcup_{i=1}^{q} [p_i] \tag{8}$$

for suitable equivalence points $p_1, \ldots, p_q$ fulfilling $\delta_0(p, p_1) = 0$ for $i = 1, \ldots, q$.

3   Revised Mathematical Theory

An in-depth analysis of Benfenati's and Marta's differential geometric approach [6,7] shows that the theory can be presented in a simpler form, based on mathematical analysis alone, as described, for example, in the text book by Apostol [5]. This is elaborated in the current section, and we add several results that are useful for practical application of the theory to CNNs.

3.1   Inter-Layer Mappings as Piecewise Differentiable Functions

To construct an explicit mathematical function representation of a trained CNN for image classification,

$$\Lambda^k = \Omega_k \circ \mathcal{N} : [0, 1]^{L \times B \times d} \longrightarrow [0, \infty),$$

we decompose Λ^k into its inter-layer mappings as described in Sect. 2.2, that is, $\Lambda^k = \Lambda_n \circ \cdots \circ \Lambda_1$ with

$$M_0 \xrightarrow{\Lambda_1} M_1 \xrightarrow{\Lambda_2} M_2 \ldots M_{n-1} \xrightarrow{\Lambda_n} M_n, \tag{9}$$

where $M_0 = [0, 1]^{L \times B \times d}$ is the input image space, and $M_n = [0, 1]$ is the classification output, $\Lambda^k(x) = 0$ meaning "*no obstacle detected*". The input image space has the usual tensor encoding as a $L \times B$ pixel matrix, where every pixel is represented by $d = 3$ RGB channels. As described in Sect. 2.1, we assume further that $M_{n-1} = [0, 1]^k$ for a final classification into $(k - 1)$ features and a further indicator p_k giving the probability that none of the $(k - 1)$ features are present.

The intermediate layers $M_1, \ldots, M_{n-2}$ have varying dimensions with varying sub-ranges of $\mathbb{R}$, depending on the CNN model chosen. A concrete example is given below in Sect. 5.

Convolutional Maps. Typically, the first inter-layer mapping of a CNN is a convolution with an $a \times a$ matrix as filter. The convolution reduces the dimension of the original image, but quite often the original size is kept by means of padding the original image matrix with extra columns and extra rows containing zeroes only. Specialising on 28×28 greyscale images and 3×3 kernels $(z_{ij})_{i,j=1,2,3}$, as used for the evaluation example in Sect. 5, the inter-layer mapping from M_0 to M_1 can be explicitly represented as

$$\Lambda_1 : [0,1]^{28 \times 28} \longrightarrow [0,1]^{28 \times 28}$$

$$\left(m_{ij}\right)_{i,j=1,\ldots,28} \mapsto \left(b + \sum_{p,q=1,2,3} z_{pq} \cdot m_{i+p-2,j+q-2}\right)_{i,j=1,\ldots,28}, \tag{10}$$

where $m_{i,j} = 0$ for $i \vee j \in \{0, 29\}$. The bias b and the kernel $(z_{ij})_{i,j=1,2,3}$ are determined during the training phase. Obviously, $\Lambda_1 = (\Lambda_1^{k\ell})_{k,\ell=1,\ldots,28}$ is differentiable with respect to all partial derivatives

$$\mathbf{D}_{ij}\Lambda_1^{k\ell} = \frac{\partial \Lambda_1^{k\ell}}{\partial m_{ij}}$$

We observe that convolutional maps are *affine transformations*: these consist of a linear transformation (the sum over matrix elements multiplied by kernel weights z_{pq} in Eq. (10)) followed by a *translation* (value b is added to each element of the image matrix resulting from the linear transformation). Affine transformations preserve straight lines and parallelism.

Non-linear Activation Functions. Affine transformations are not capable of separating input data (i.e. image matrices) from different classification clusters lying on the same straight line of the input space, since they preserve straight lines [2, Section 1.5.1]. As a consequence, affine inter-layer transformations are frequently followed by *non-linear activation functions* $\mathbb{R} \longrightarrow \mathbb{R}$. These are applied to every element of an affine transformation's image, so that the dimension of this transformation's image space remains unchanged.

Important activation functions are

- the *rectified linear unit activation function* $\texttt{ReLU}(x) = \max(0, x)$, which is non-differentiable at 0,
- the differentiable *softplus* function $\texttt{Softplus}(x) = \frac{\ln(1+e^{kx})}{k}$ with $k \geq 1$;

these were already discussed in Sect. 2.2. Further activation functions (sigmoid, tanh and others) that are used in CNNs like *LeNet, AlexNet, VGG16, GoogLeNet* are discussed, for example, by Aggarwal [2]. All of them can be used in inter-layer transformations conforming to the approach discussed in this paper.

Maxpooling Maps. The $\texttt{MaxPooling}$ map transforms a matrix to a smaller one by building the maximum over square sub-matrices. The variant used in the evaluation in Sect. 5 uses 2×2 sub-matrices and is defined as

$$\texttt{MaxPooling} : [0,1]^{28 \times 28} \longrightarrow [0,1]^{14 \times 14} \tag{11}$$

$$\left(m_{ij}\right)_{i,j=1,\ldots,28} \mapsto \left(\max\{m_{1+2i,1+2j}, m_{2+2i,1+2j}, m_{1+2i,2+2j}, m_{2+2i,2+2j}\}\right)_{i,j=0,\ldots,13},$$

As shown in the proof of Corollary A2, each image element of the $\texttt{MaxPooling}$ transformation can be considered as an expression over $\texttt{ReLU}$ functions, so it is piecewise differentiable.

Flattening Transformation. The `Flatten` mapping transforms an $n \times m$-matrix into a vector of length $n \cdot m$ by concatenating the matrix rows.

Dense Transformation. Dense transformations are affine transformations mapping input vectors to (usually shorter) vectors, using weight elements $z_{i,j}$ in the linear transformation part $(x_1, \ldots, x_n)^\mathsf{T} \mapsto \left(\sum_{j=1}^{n} z_{1,j} \cdot x_j, \ldots, \sum_{j=1}^{n} z_{m,j} \cdot x_j \right)^\mathsf{T}$ and a vector $(b_1, \ldots, b_m)^\mathsf{T}$ of bias elements to be added to the result of the linear transformation.

The Softmax Classifier. The differentiable `Softmax` transformation already defined in Eq. (1) is used to map a preliminary result vector $v \in \mathbb{R}^n$ (in our evaluation example, $n = 128$) to a shorter vector $p \in \mathbb{R}^m$, whose elements indicate the probabilities for features being present in the classified image. We use $m = 4$ in the evaluation example.

The Obstacle Activation Function. As described in Sect. 2.1, the final 'obstacle/no obstacle' result aggregated from the probability vector $p \in [0,1]^k$ satisfying $\sum_{i=1}^{k} p_i = 1$, is calculated by activation function Ω_k specified in Eq. (4). Again, Ω_k is an expression over `ReLU` function applications, so it is piecewise differentiable.

Observe that all piecewise differentiable transformation used in a CNN are "good-natured" in the sense that the subsets of input data where the function is non differentiable are single points or ℓ-dimensional hyperplanes in $\mathbb{R}^m$, $\ell \in \{1, \ldots, m-1\}$.

3.2   Gradient and Jacobian Matrix

For any differentiable function $f : \mathbb{R}^n \longrightarrow \mathbb{R}$, its *gradient vector* $\nabla f(x)$ at $x \in \mathbb{R}^n$ is defined by

$$\nabla f(x) = (\mathbf{D}_1 f(x), \ldots, \mathbf{D}_n f(x))^\mathsf{T}, \text{ with partial derivative } \mathbf{D}_i f(x) = \frac{\partial f}{\partial x_i}(x).$$

Extending this concept to differentiable mappings

$$f : \mathbb{R}^n \longrightarrow \mathbb{R}^m; \ x \mapsto \left(f_1(x), \ldots, f_m(x) \right)^\mathsf{T},$$

the *Jacobian matrix* $\mathbf{J}_f(x)$ at $x \in \mathbb{R}^n$ is defined by

$$\mathbf{J}_f(x) = \begin{bmatrix} \nabla f_1(x)^\mathsf{T} \\ \cdot \\ \cdot \\ \cdot \\ \nabla f_m(x)^\mathsf{T} \end{bmatrix} = \begin{pmatrix} \mathbf{D}_1 f_1(x) & \ldots & \mathbf{D}_n f_1(x) \\ \mathbf{D}_1 f_2(x) & \ldots & \mathbf{D}_n f_2(x) \\ \cdot & \cdot & \cdot \\ \cdot & \cdot & \cdot \\ \cdot & \cdot & \cdot \\ \mathbf{D}_1 f_m(x) & \ldots & \mathbf{D}_n f_m(x) \end{pmatrix}$$

If $f : \mathbb{R}^{m_0} \longrightarrow \mathbb{R}$ is expressed as a chain $f = h_n \circ h_{n-1} \circ \cdots \circ h_1$ of n differentiable maps like the neural network function Λ^k described in Eq. (9), it is useful to be able to calculate ∇f by means of these intermediate functions and their Jacobians. Suppose that

$$h_i : \mathbb{R}^{m_{i-1}} \longrightarrow \mathbb{R}^{m_i} \quad \text{for} \quad i = 1, \dots, n-1, \quad \text{and} \quad h_n : \mathbb{R}^{m_{n-1}} \longrightarrow \mathbb{R}.$$

Setting

$$\boldsymbol{x}_0 \in \mathbb{R}^{m_0}, \; \boldsymbol{x}_i = (h_i \circ h_{i-1} \circ \cdots \circ h_1)(\boldsymbol{x}_0), \; i = 1, \dots, n-1,$$

the chain rule for Jacobian matrices [5, 12.10] implies that

$$\nabla f(\boldsymbol{x}_0)^\mathsf{T} = \nabla h_n(\boldsymbol{x}_{n-1})^\mathsf{T} \mathbf{J}_{h_{n-1}}(\boldsymbol{x}_{n-2}) \cdots \mathbf{J}_{h_1}(\boldsymbol{x}_0), \tag{12}$$

so the gradient of f can be calculated by means of the matrix product of the Jacobian matrices associated with $h_1, \dots, h_{n-1}$, respectively, multiplied with the gradient of the final function chain element h_n.

3.3   Null Spaces and Null Curves

Given any matrix $\boldsymbol{m} \in \mathcal{M}_{m \times n}(\mathbb{R})$, its *null space* $\mathtt{Null}(\boldsymbol{m})$ (also called the *kernel* of $\boldsymbol{m}$) is the vector space spanned by a basis $\{\boldsymbol{n}_1, \dots, \boldsymbol{n}_k\} \subseteq \mathbb{R}^n$, such that $\boldsymbol{m} \cdot \boldsymbol{v}^t = \boldsymbol{0}$ for any linear combination $\boldsymbol{v} = a_1 \boldsymbol{n}_1 + \cdots + a_k \boldsymbol{n}_k$. Given a differentiable mapping $f : \mathbb{R}^n \longrightarrow \mathbb{R}^m$, the null space of the Jacobian $\mathbf{J}_f(\boldsymbol{x})$ at some point $\boldsymbol{x} \in \mathbb{R}^n$ has the intuitive interpretation that $f(\boldsymbol{x} + h\boldsymbol{v})$ "remains constant for changes of $\boldsymbol{x}$ by a null vector $h\boldsymbol{v}$ of infinitesimal length, $h \to 0$". More formally, a differentiable curve $\gamma : [-1, 1] \longrightarrow \mathbb{R}^n$ is a *null curve of f through* $\boldsymbol{x} \in \mathbb{R}^n$, if and only if

1. $\gamma(0) = \boldsymbol{x}$,
2. $\dot{\gamma}(t) \in \mathtt{Null}(\mathbf{J}_f(\gamma(t)))$ for $t \in (-\delta, \delta) \subset [-1, 1]$ with a suitable $\delta > 0$.

Here, $\dot{\gamma}(t)$ denotes the *tangent vector* of γ in t: if $\gamma(t) = (g_1(t), \dots, g_n(t))^\mathsf{T}$ for differentiable functions $g_i : [-1, 1] \longrightarrow \mathbb{R}$, then

$$\dot{\gamma}(t) = \left(\frac{dg_1}{dt}(t), \dots, \frac{dg_n}{dt}(t) \right)^\mathsf{T}.$$

These two properties ensure that $f(\gamma(t))$ remains constant with value $f(\boldsymbol{x})$ for $t \in (-\delta, \delta)$.[9] Considering the gradient $\nabla f(\boldsymbol{x})$ of a function $f : \mathbb{R}^n \longrightarrow \mathbb{R}$ at some

---

[9] Considering $\mathbb{R}^n$ as a differentiable manifold, the Jacobians $\mathbf{J}_f(\boldsymbol{x})$, $\boldsymbol{x} \in \mathbb{R}^n$ span a fibre bundle E on $\mathbb{R}^n$. The sub-bundle $\mathtt{Null}(\mathbf{J}_f(\boldsymbol{x}))$, $\boldsymbol{x} \in \mathbb{R}^n$ is called the vertical bundle $\mathcal{V}E$ of E already mentioned in Sect. 2.2, the complementary space the horizontal bundle $\mathcal{H}E$, satisfying $E = \mathcal{V}E \oplus \mathcal{H}E$. Null curves have tangent vectors in the vertical bundle, so f remains constant along these curves. In contrast to that, f changes along curves whose tangent vectors are contained in the horizontal bundle, and then the change is proportional to the length of the tangent vectors.

point $x \in \mathbb{R}^n$ as an $n \times 1$ matrix, its null space $\mathtt{Null}(\nabla f(x))$ has dimension n if $\nabla f(x)$ is the null vector; otherwise $\mathtt{Null}(\nabla f(x))$ has dimension $n - 1$. The geometric interpretation is that $\mathtt{Null}(\nabla f(x))$ contains the vectors that are perpendicular to the gradient $\nabla f(x)$. In any case, $f(\gamma(t))$ keeps its constant value $f(x)$ along a null curve γ through x. Summarising, we have found a simpler approach to construct the null curves specified in the singular Riemannian metric approach of Benfenati and Marta [6,7] (see Sect. 2.2): instead of using differentiable manifolds, metrics and their pullbacks, we can simply consider mappings between subsets of $\mathbb{R}^n$ and $\mathbb{R}^m$, and evaluate Jacobians, gradients, and their null spaces.

3.4 Closed Interval Subsets of $\mathbb{R}^k$

Typically, an image input space is a Cartesian product of *closed* intervals. Consequently, the Jacobi matrix and the gradient, respectively, do not exist on boundary points of the input space, where at least one tuple component lies on an interval boundary. To extend the Jacobi matrix to boundary points in a well-defined way, we apply *directional derivatives* as follows. Let $I = [a, b] \subset \mathbb{R}$ be a closed interval and

$$f : I^n \longrightarrow \mathbb{R}^m; \;\; (x_1, \ldots, x_n) \mapsto (f_1(x_1, \ldots, x_n), \ldots, f_m(x_1, \ldots, x_n))$$

a CNN transformation $f = \Lambda_q$ from layer $(q-1)$ to layer q. Define the *one-sided directional derivatives* of f_i for (i, j), $i = 1, \ldots, m$, $j = 1, \ldots n$ by

$$\mathbf{D}_j^+ f_i(x_1, \ldots, x_n) = \lim_{h \to 0, h > 0} \frac{f_i(x_1, \ldots, x_{j-1}, x_j + h, x_{j+1}, \ldots x_n) - f_i(x_1, \ldots, x_n)}{h}$$

$$\mathbf{D}_j^- f_i(x_1, \ldots, x_n) = \lim_{h \to 0, h < 0} \frac{f_i(x_1, \ldots, x_{j-1}, x_j + h, x_{j+1}, \ldots x_n) - f_i(x_1, \ldots, x_n)}{h}$$

Note that all functions involved in CNN inter-layer mappings have well-defined (potentially differing) directional derivatives in all points $x = (x_1, \ldots, x_n)^\mathsf{T}$, even if they are not differentiable in x. For example, $\mathtt{ReLU}(x) = \max(0, x)$ is not differentiable in $x = 0$ with $\mathbf{D}^+\mathtt{ReLU}(0) = 1$, and $\mathbf{D}^-\mathtt{ReLU}(0) = 0$.

3.5 Dealing With Non-differentiabilities

The use of activation functions such as $\mathtt{ReLU}$ or $\mathtt{MaxPooling}$ leads to points of non-differentiability in $\Lambda^k = \Omega_k \circ \mathcal{N}$. However, in the search for piecewise smooth curves along which Λ^k is constant, it turns out that, with a suitable notion of the Jacobian of Λ^k, the relation $\mathbf{J}_{\Lambda^k}(\gamma(t)) \cdot \dot{\gamma}(t) = 0$ for all t still provides a sufficient condition. More precisely, we first consider the case where the only points of non-differentiability stem from the presence of $\mathtt{ReLU}$ functions in Λ^k. By defining $\mathtt{ReLU}'(0) := 0$ we proceed *formally* to define generalised partial derivatives of Λ^k by the chain rule, using $\mathtt{ReLU}'(0) = 0$ whenever needed. The generalised Jacobian

$\widehat{\mathbf{J}}_{\Lambda^k}(x)$ of Λ^k is then defined entrywise by the generalised partial derivatives. It is proven in Appendix A that with this definition, $\Lambda^k \circ \gamma$ is indeed constant whenever γ is a piecewise smooth curve and $\widehat{\mathbf{J}}_{\mathcal{N}}(\gamma(t)) \cdot \dot{\gamma}(t) = 0$ holds for all t. This result is extended to the use of `MaxPooling` and to the activation functions Ω_i specified in Sect. 2.1 by expressing them as the composition of affine maps and `ReLU` functions. Proofs and details can be found in Appendix A.

4   Identification of Classification Clusters

Given an equivalence class $[p_i]$ contributing to a classification cluster, each pair of points $p, p' \in [p_i]$ can be connected by a piecewise differentiable null curve: since p, p' are each connected to p_i by null curve (this is the requirement for p, p' to be contained in $[p_i]$), the concatenation of null curves $p \longrightarrow p_i$ and $p_i \longrightarrow p'$ results in a null curve from p to p' that may be non-differentiable in p_i. We now restrict this definition of equivalence classes further by defining

$$[p_i]' = \{p \mid p \text{ is reachable from } p_i \text{ by a polygonal chain of null line segments}\} \tag{13}$$

Trivially, the vertex points of the polygonal chain are also members of $[p_i]'$, since they are end points of straight null segments. Obviously, $[p_i]' \subseteq [p_i]$. Since arbitrary null curves can be approximated by polygonal chains, $[p_i]'$ is actually an acceptable approximation of $[p_i]$. The identification of equivalence classes $[p_i]'$ is now performed in two phases as follows.

Initial Setup. In the first phase, an initial set of equivalence classes, each contributing to a specific classification cluster, is identified from the training and validation data as follows.

ALGORITHM 1.

1. **Input.** Labelled training and validation images v_i^j with $j \in \{1, \ldots, k\}$ representing obstacle types for $j < k$ and 'no obstacle' images for $j = k$. Index i denotes the ith image of type j.
2. **Output.** A mapping τ from training and validation images v to class representatives r, so that $\tau(v) = r$, if and only if $v \in [r]'$.
3. **Algorithm.**
   (a) Initialise $\tau := \{\}$ (the empty map);
   (b) For each type $j \in \{1, \ldots, k\}$
       i. Extend τ by setting $\tau := \tau \oplus \{v_1^j \mapsto v_1^j\}$;
       ii. For all images v^j of type j that are not yet contained in $\mathrm{dom}\ \tau$
           A. Find $v' \in \mathrm{dom}\ \tau$ such that v^j and v' are connected by a null segment.[10] If such a v' exists, it must belong to the same classification type j as v^j, so add v^j to the class of v' by setting $\tau := \tau \oplus \{v^j \mapsto \tau(v')\}$;

B. If v' cannot be found in Step A, but v^j is on the boundary of $V_j = \{v \mid \Lambda^j(v) = 0\}$, find a new inner point of V_j by setting $v'' = v^j - \delta \cdot \nabla \Lambda^j(v^j)$ with a small value $\delta > 0$.[11] Then v^j and v'' are connected by a null segment by construction, since $-\nabla \Lambda^j(v^j)$ points from v^j into the interior of V_j. If a $v' \in \mathsf{dom}\ \tau$ exists, such that also v'' and v' are connected by a null segment, extend τ by setting $\tau := \tau \oplus \{v^j \mapsto \tau(v'), v'' \mapsto \tau(v')\}$.

C. Otherwise, if v' could not be found in Step A or Step B, select v^j as a new class representative by setting $\tau := \tau \oplus \{v^j \mapsto v^j\}$;

Note that ALGORITHM 1 under-approximates $[r]'$, because at a later time, another image v' might be added to $[r]'$, from where v^j is directly reachable on a null segment. Our experiments have shown, however, that this rarely happens in practice, so that an unnecessary creation of a new class $[v^j]'$ does not occur very often. Therefore, the simplicity of ALGORITHM 1 outweighs the risk to create a superfluous new class.

We observe that the statistical approach [14] to the estimation of a residual classification error probability does not require to determine the *maximal* equivalence classes. Instead more fine-grained ones can be chosen, without affecting the estimate for the error probability.[12]

Verification Phase. In the second phase, the verification of the CNN is started with new images that were not contained in the training and validation set, while the collection of identified equivalence classes (map τ) is incrementally extended as necessary for each new verification image. This is performed by ALGORITHM 2 which is executed for each new verification image.

ALGORITHM 2.

1. **Input.** A new verification image v with its correct classification label $\underline{j} \in \{1, \ldots, k\}$, and the current version of τ.
2. **Output.** FAIL, v is misclassified in an unacceptable way; otherwise PASS and an updated version of the map τ.

---

[10] The check whether two images v_j and v' are connected by a null segment can be simply performed by checking that $\Lambda^j\big((1 - t) \cdot v_j + t \cdot v'\big) = 0$ for all $t \in [0, 1]$.

[11] The gradient is calculated by means of directional derivatives approaching the boundary point v^j from the outside of V_j. Then boundary points v^j are characterised by $\Lambda^j(v^j) = 0 \wedge \nabla \Lambda^j(v^j) \neq \mathbf{0}$.

[12] Using finer equivalence classes still affects the number of statistical tests to be performed, but we consider this as acceptable, since we benefit from the simplicity of ALGORITHM 1. It is interesting to note that a refinement of input equivalence classes in software testing also does not affect the test strength, but only the size of the generated test suites [16].

3. **Algorithm.**
   If $\Lambda^{\underline{j}}(v) > 0$ and $\mathcal{N}(v)$ does not indicate any uncertainty[13], return FAIL and terminate. Otherwise,
   (a) Extend τ in analogy to ALGORITHM 1, Steps A, B, C, with input v taking the role of v^j in ALGORITHM 1.
   (b) Return PASS and the updated version of τ.

If ALGORITHM 2 returns FAIL, the CNN needs to be improved. Otherwise, as long as new equivalence classes $[v]'$ are found, the verification tests need to be continued. While verification tests do not detect new classes, the probability for a new image v to be associated with an existing class $[r_\ell^j]'$ can be estimated using τ, following the test strategy described in [14]. Finally the probability p_f of a new image to be associated with a *missing* class for which the CNN has not been trained can also be estimated [14]. As soon as the estimate for p_f is small enough and has a sufficiently high confidence value,[14] the verification tests are terminated.

The fairly easy method to identify equivalence classes is possible, because inner points of an image set $V_j = \{v \mid \Lambda^j(v) = 0\}$ have gradient $\nabla\Lambda^j(v) = \mathbf{0}$. Consequently, inner points of V_j always possess convex ε environments that have the same dimension $(L \times B \times d)$ as the input image space. Moreover, an inner straight line segment connecting two inner points of V_j is automatically a null curve. Only boundary points of V are associated with a gradient (calculated by directional derivatives) whose null space has dimension $(L \times B \times d) - 1$ (in our evaluation setting, this is dimension $(28 \times 28) - 1$). Two boundary points v, v' of V can be connected by two null line segments in a simple way: the first connects v to a suitable inner point v'', the second connects v'' to v'. If v'' is chosen in such a way that the two connecting segments are completely contained in V, they are automatically null curves.

5   Evaluation

The MNIST data set [20] was used to train, validate and test the convolutional neural network. This data set consists of nearly evenly distributed handwritten digits from zero to nine and is divided into 60000 images as a training set and 10000 images as a test or validation set. Each image has a shape of 28×28 pixels,

---

[13] Uncertainty can be detected by finding one or more other components in the probability vector $\boldsymbol{p} = \mathcal{N}(v)$ having an only marginally larger value than $p_{\underline{j}}$. In such a case, the obstacle detection system would mark the classification result as not trustworthy and use more reliable ones of the sensor fusion system or fall to the safe side, deciding for 'obstacle present'.

[14] As elaborated in [14], a failure probability $p_f < 0.04$ suffices to apply sensor fusion and obtain a redundant sensor/perceptor system whose hazard rate is smaller than the tolerable hazard according to SIL-3. The required confidence value still needs to be defined; we expect that this well be performed in an associated future standard for safety-critical autonomous systems.

where each (grey-scale) pixel value is in range $[0, 255]$. As frequently applied in image classification problems, a normalisation to pixel value range $[0, 1]$ was applied. We modified the set of labels in the sense that labels of digits 0, 1, 2 remain unchanged and labels of digits 3 to 9 are changed to label class 3. Images 0, 1, 2 are interpreted in our experimental setting as three different classes of "obstacles", while the other images are interpreted as "no obstacle present". This data set is appropriate for our purpose, since it is fairly simple and meaningful at the same time as an artificial example of the obstacle classification problem.

We used TensorFlow [1] and Keras [11] to design and train a convolutional neural network. Our CNN consists of the following layers: The first layer is a convolutional layer that applies one 3×3 filter over the input-image of shape $28 \times 28 \times 1$ with stride 1×1. The padding of the image is kept the same and the activation function is `ReLU`. The second layer applies `MaxPooling` by down-sampling the output of the convolutional layer along its spatial dimensions by taking the maximum value of the pooling window of size 2×2 with stride 2×2. Flattening is applied afterwards to transform the 14×14-matrix returned by the `MaxPooling` transformation into a vector of length 196.

The last two layers are densely-connected layers: The first dense layer has 128 units and `ReLU` as activation function to calculate the outputs. The outputs of the second dense-layer are calculated by the `Softmax` function to output a 4-vector as a probability distribution over detected label classes $[0, 1, 2, 3]$.

We trained the model in 50 epochs and reached 98.99% accuracy and 7.95% loss. Accuracy is the rate of correctly classified images from the validation set, while loss is the quantified difference between two probability distributions: the predicted result from the CNN and the correct classification of an image. As a loss-metric we used Keras sparse categorical cross entropy which is best suited for more than two label classes and a probability distribution over label classes as outputs. We used the stochastic gradient descent optimiser Adam [18] for training.

Learnt parameters were extracted from the trained CNN model to re-model the CNN behaviour with Mathematica[15]. By using the chain rule described in Sect. 3, the gradients of the classification functions Λ^j, $j = 1, 2, 3, 4$, could be effectively calculated with help of the Jacobians of each individual inter-layer transformation. ALGORITHM 2 defined in Sect. 4 requires a few 100ms per check and can be significantly accelerated by means of parallelisation on several CPU cores.

During our evaluation of ALGORITHM 1, we identified five equivalence classes in total, distributed across four classification clusters. Within cluster 0, all images belong to a single equivalence class. Cluster 1, on the other hand, is divided into two equivalence classes, except for one image, which was erroneously classified as false negative *'no obstacle present'*. Most of the images in cluster 2 are associated with a single equivalence class, with five images being erroneously classified as false negatives *'no obstacle present'* and one image being misclassified with an incorrect *'obstacle present'* label. In the case of cluster 3, most

---

[15] https://www.wolfram.com/mathematica/.

of the images denoting 'no obstacle present' belong to a single equivalence class. However, five images were false positives, leading to a misclassification 'obstacle present' (Table 1).

Table 1. The first row describes the number of equivalence classes found for each cluster. The false positives are images that are labelled as 'no obstacle present' but were classified as 'obstacle present'. False negatives are images that were classified as 'no obstacle present' but labelled as 'obstacle present'. The last row shows the number of images that are labelled as 'obstacle present' but were classified with wrong 'obstacle present' label.

	'obstacle present'			'no obstacle present'
	Cluster 0	Cluster 1	Cluster 2	Cluster 3
Equivalence classes	1	2	1	1
False positives	0	0	0	5
False negatives	0	1	5	0
Wrong 'obstacle' cluster	0	0	1	0

From this fairly trivial MNIST dataset we can already conclude that the calculus-based approach in ALGORITHM 1 for the identification of classification clusters is therefore meaningful and effective to find equivalence classes in a large set of training images.

The implementation of ALGORITHM 1 and ALGORITHM 2 consists of approx. 950 lines of Python code. We executed the program in parallel for each classification cluster separate on our kubernetes cluster with 1 AMD EPYC 7702 CPU core and 16 GiB of RAM allocated for each docker container[16]. We bundled four docker containers in one kubernetes pod[17], where one docker container was responsible to execute the implementation of ALGORITHM 1 for one classification cluster. We used `tensorflow:2.14.0` as a base image. The runtime of the program and the number of images in the respective dataset is shown in Table 2 for ALGORITHM 1.

Table 2. Recorded runtime values for ALGORITHM 1 for each classification cluster.

Cluster	#images in training set	runtime
Cluster-0	5923	370.81 s
Cluster-1	6742	1998.12 s
Cluster-2	5958	1751.82 s
Cluster-3	41377	12203.33 s

---

[16] https://www.docker.com/.

[17] https://kubernetes.io/docs/concepts/workloads/pods/.

As shown in Table 2, the runtime on the cluster 0 is much lower than on the runtime on cluster 2 although the number of images in the datasets is almost the same. This is quite likely attributed to the fact that the algorithm found a suitable image that has a class closer to the beginning of the dictionary of images already analysed. The much higher runtime for the cluster 3 is due to the much higher number of images in that cluster.

The evaluation of ALGORITHM 2 yielded the following results: The runtime was 21.23 s, 327 images of the test set were analysed and two images were found that got classified with a wrong label of type *'obstacle present'*. The program returned FAIL due to a fatal misclassification: an image with label *'obstacle present'* was classified as *'no obstacle present'* with high probability.

Overall, in ALGORITHM 1, we identified 6 out of the 60,000 images in the training set as fatal misclassifications (classified by the CNN as'no obstacle present,' but actually'obstacle present'). This results to a rate of 0.1%. In the entire test set (without terminating ALGORITHM 2), 52 out of 10,000 test images were identified as critical errors, resulting in an error rate of 0.52%. Lastly, our findings demonstrated the suitability of the trained CNN model for integration into a sensor fusion system for an autonomous freight train.

6   Threats to Validity

The use of the `Softmax` function for the final transformation of the last hidden layer into the output layer (see Sect. 2.1) is appropriate for *multi-class classification* problems, in which an image contains exactly one object of some class $1, \ldots, k - 1$ or none of these objects at all. This is appropriate for the simple MNIST data set we have used for the experimental evaluation described in Sect. 5, since each image only contains one digit or no digit at all (we have added white-noise images to the original MNIST data set). For a "real-world obstacle detection function", it is of course possible that more than one obstacle type is present in an image (e.g. a motor cycle standing in front of a truck, both located on the railway track at a level crossing). This is a *multi-label classification* problem, where the `Sigmoid` function

$$\text{Sigmoid} : \mathbb{R}^k \longrightarrow (0,1)^k; \quad (v_1, \ldots, v_k) \mapsto \left(\frac{1}{1 + e^{-v_1}}, \ldots, \frac{1}{1 + e^{-v_k}} \right)$$

is typically applied, since each result vector component $1/(1 + e^{-v_j})$ is independent of the other component values $v_i, i \neq j$. An obstacle of type $j \in \{1, \ldots, k-1\}$ is considered to be present in an image if $1/(1 + e^{-v_j})$ is greater or equal to 0.5 [2, Section 2.2.3] and the *"no obstacle present"* result vector component k has a value less than 0.5. Uncertainty is expressed here by result vectors, where either all components are less than 0.5 (so neither an obstacle has been detected, nor the *"no obstacle present"* result seems to be trustworthy), or one or more obstacle classes *and* the *"no obstacle present"* result vector component have a value greater or equal to 0.5. In analogy to the functions $\Omega_j, j = 1, \ldots, k$

defined in Eq. (4) and Eq. (5), the `Sigmoid` function induces

$$\Omega_j' : [0,1]^k \longrightarrow [0,1); \quad (y_1, \ldots, y_k) \mapsto \text{ReLU}(0.5 - y_j) \text{ for } j = 1, \ldots, k. \tag{14}$$

Again, $\Omega_k'(\boldsymbol{y}) = 0$ indicates that 'NO obstacle is present', and $\Omega_j'(\boldsymbol{y}) = 0$ for $j = 1, \ldots, k-1$ indicates that 'obstacle of type j is present'. These considerations show that the usage of `Sigmoid` instead of `Softmax` does not introduce any new problems that could not be handled with our mathematical analysis approach, since again, only differentiable expressions or `ReLU` are involved.

The only real increase in terms of mathematical complexity to occur when using "real-world" obstacle data sets is that these would be colour images, so the dimension of the input space is increased from $[0,1]^{L \times B \times 1}$ used for the MNIST greyscale images to $[0,1]^{L \times B \times 3}$ needed to encode RGB values of colour pixels. Since the performance observed during the evaluation of the CNN trained with the MNIST data set was very good as reported in Sect. 5, we expect that the increase of dimensions for considering colour images will still allow for equivalence classes to be identified with acceptable performance.

7   Conclusion

We have presented a novel approach to the identification of classification clusters of trained convolutional neural networks by means of techniques from mathematical analysis, building a precise representation of the CNN's inter-layer mappings. The evaluation based on the simple MNIST data set shows that the approach scales well and covers all mappings typically needed for CNN modelling. The next evaluation step will be to use more realistic colour images representing real obstacles, as discussed in Sect. 6. Moreover, the analytic methods that have become possible by means of the precise mathematical representation can also be used to detect unwanted occurrences of brittleness in the CNN. This will also be investigated in the near future.

Acknowledgements. The authors are grateful to Roland Höfer for his valuable advice on special problems of mathematical analysis and the effective use of Mathematica. We also thank Peter Maaß for a stimulating discussion about numerical problems related to this work.

A   Non-differentiabilities

Consider a generic trained neural network $\mathcal{N}$ of the following form:

$$\mathcal{N} = A^L \circ \sigma^{L-1} \circ A^{L-1} \circ \cdots \circ \sigma^1 \circ A^1$$

for affine maps

$$A^\ell(x) := W^\ell x + b^\ell \qquad\qquad (x \in \mathbb{R}^{n_{\ell-1}})$$

$$A^\ell(x) := W^\ell x + b^\ell \qquad\qquad (x \in \mathbb{R}^{n_{\ell-1}})$$

where $W^\ell \in \mathbb{R}^{n_\ell \times n_{\ell-1}}$, $b \in \mathbb{R}^{n_\ell}$, $1 \leq \ell \leq L$, and for generic activation functions $\sigma^\ell = (\sigma_1^\ell, \ldots, \sigma_\ell^\ell)$ where $\sigma_j^\ell : \mathbb{R} \to \mathbb{R}$, $1 \leq j, \ell \leq L - 1$. Here $n_0 \in \mathbb{N}$ denotes the input dimension of $\mathcal{N}$ and $n_\ell \in \mathbb{N}$ is the width of the ℓ.th layer, $1 \leq \ell \leq L$.

For now, assume that σ_j^ℓ is either smooth or equal $\texttt{ReLU}$ for all $1 \leq j, \ell \leq L-1$. As explained in the main text, we define Jacobian matrices formally by the chain rule using $(\sigma_j^\ell)'(0) = 0$ if $\sigma_j^\ell = \texttt{ReLU}$. In particular, the generalised Jacobian matrix $\widehat{\mathbf{J}}_{\mathcal{N}}(x) \in \mathbb{R}^{n_L \times n_0}$ is defined. Further, note that the function $\Lambda^k = \Omega_k \circ \mathcal{N}$ from Sect. 2.1 can be represented in the above form, see below for an explicit construction.

Proposition A1 *Let $\gamma : [0, 1] \to \mathbb{R}^{n_0}$ be a piecewise smooth curve and $D \subset [0, 1]$ an open set such that γ is differentiable on D and $\bar{D} = [0, 1]$. If*

$$\forall t \in D : \quad \widehat{\mathbf{J}}_{\mathcal{N}}(\gamma(t))\gamma'(t) = 0,$$

then $\mathcal{N} \circ \gamma$ is a constant on $[0, 1]$.

Proof. Let $\texttt{ReLU}_1, \ldots, \texttt{ReLU}_k : \mathbb{R} \to \mathbb{R}$ be any enumeration of the $\{\sigma_j^\ell\}_{1 \leq j, \ell \leq L-1}$ which are a $\texttt{ReLU}$ function. For $1 \leq i \leq k$ define

$$\Gamma_i := \{t \in [0, 1] \, | \, (A^\ell \circ \sigma^{\ell-1} \circ \cdots \circ A^1 \circ \gamma)_j(t) = 0\}$$

whenever $\texttt{ReLU}_i = \sigma_j^\ell$ for some $1 \leq j, \ell \leq L - 1$. Then every Γ_i is closed since it is the preimage of $\{0\}$ under a continuous function. Define

$$B := \bigcup_{i=1}^{k} \partial\Gamma_i, \text{ and } I := [0, 1] \setminus B.$$

Claim #1. $\mathcal{N} \circ \gamma$ is constant on the closure of any connected component C of I.

Note that, for all i, $C \cap \Gamma_i \in \{\emptyset, C\}$: Suppose there are $c_1, c_2 \in C$ such that $c_1 \in \Gamma_i$, $c_2 \notin \Gamma_i$. Then there exists some $b \in \partial\Gamma_i \subset B$ with $c_1 < b < c_2$, or $c_2 < b < c_1$. Since C is connected, $b \in C$. This leads to the contradiction $C \subset I$ and $I \cap B = \emptyset$. We now define a new neural network $\widetilde{\mathcal{N}}$ as follows: Start with a copy of $\mathcal{N}$. For $i = 1, \ldots, k$ do the following: If $C \cap \Gamma_i = C$, then replace the activation function in $\mathcal{N}$ corresponding to $\texttt{ReLU}_i$ by the zero function. Then $\widetilde{\mathcal{N}} \circ \gamma = \mathcal{N} \circ \gamma$ on C and $\widetilde{\mathcal{N}} \circ \gamma$ is differentiable in $t \in \mathring{C} \cap D$. Since

$$\forall t \in \mathring{C} \cap D : \quad \frac{d}{dt}(\widetilde{\mathcal{N}} \circ \gamma)(t) = \mathbf{J}_{\widetilde{\mathcal{N}}}(\gamma(t))\gamma'(t) = \widehat{\mathbf{J}}_{\mathcal{N}}(\gamma(t))\gamma'(t) = 0,$$

$\widetilde{\mathcal{N}} \circ \gamma$ is constant on $\mathring{C} \cap D$ and, by continuity of $\widetilde{\mathcal{N}} \circ \gamma$, also on the closure of C. The first claim follows from recalling that $\widetilde{\mathcal{N}} \circ \gamma = \mathcal{N} \circ \gamma$ on C, and by continuity, on $\bar{C}$ as well.

Claim #2. $\mathcal{N} \circ \gamma$ is constant on $[0,1]$.

The definition of the topological boundary implies that B does not contain an open set. Therefore, for any point $t \in B$ we can find a sequence $t_n \in I$ converging to t. Continuity together with Claim #1 implies the second claim. $\square$

Corollary A2 *Assume that $\mathcal{N}$ is a composition of smooth mappings, ReLU, or* MaxPooling. *Define Jacobian matrices formally by the chain rule using* ReLU$'(0) = 0$. *Then the Jacobian matrix $\widehat{\mathbf{J}}_{\mathcal{N}}(x) \in \mathbb{R}^{n_L \times n_0}$ is defined and Proposition A1 holds true.*

Proof Since $\max(x,y) = y + \text{ReLU}(x - y)$,

$$
\begin{aligned}
\max(x_1, \ldots, x_n) &= \max(\max(x_1, \ldots, x_{n-1}), x_n) \\
&= x_n + \text{ReLU}(\max(x_1, \ldots, x_{n-1}) - x_n) \\
&= x_n + \text{ReLU}(x_{n-1} + \text{ReLU}(\max(x_1, \ldots, x_{n-2}) - x_{n-1}) - x_n) \\
&= x_n + \text{ReLU}(x_{n-1} + \text{ReLU}(x_{n-2} + \cdots + \\
&\qquad\qquad \text{ReLU}(x_2 + \text{ReLU}(x_1 - x_2) \ldots) - x_{n-1}) - x_n) \\
&= x_n + \text{ReLU}(x_{n-1} - x_n + \text{ReLU}(x_{n-2} - x_{n-1} + \\
&\qquad\qquad \text{ReLU}(\cdots + \text{ReLU}(x_1 - x_2))))
\end{aligned}
$$

can be formulated as a composition of ReLU operations. $\square$

The neural network

$$
\Lambda^k = \Omega_k \circ \mathcal{N} : [0,1]^{L \times B \times d} \longrightarrow [0,1].
$$

with

$$
\Omega_k : [0,1]^k \longrightarrow [0,1]; \quad (p_1, \ldots, p_k) \mapsto \sum_{i=1}^{k-1} \text{ReLU}(p_i - p_k)
$$

as defined in Sect. 2.1 can be re-written equivalently as

$$
\Lambda^k = B \circ \sigma \circ A \circ \mathcal{N},
$$

where $\sigma = (\sigma_1, \ldots, \sigma_{k-1})$, $\sigma_1 = \cdots = \sigma_{k-1} = \text{ReLU}$, $A(p) = \begin{pmatrix} 1 & 0 & 0 & \cdots & 0 & -1 \\ 0 & 1 & 0 & \cdots & 0 & -1 \\ & & \cdots & & & \\ 0 & 0 & 0 & \cdots & 1 & -1 \end{pmatrix} \begin{pmatrix} p_1 \\ \vdots \\ p_k \end{pmatrix}$, and $B(x) = (1 \cdots 1) \begin{pmatrix} x_1 \\ \vdots \\ x_{k-1} \end{pmatrix}$. Therefore, Λ^k is still a chain of affine maps and ReLU applications, so that Proposition A1 can be applied.

References

1. Abadi, M., et al.: TensorFlow: Large-scale machine learning on heterogeneous systems (2015). https://www.tensorflow.org/, software available from tensorflow.org
2. Aggarwal, C.C.: Neural Networks and Deep Learning. Springer Nature Switzerland AG, Cham, Switzerland (2018)
3. Akai, N., Hirayama, T., Murase, H.: Experimental stability analysis of neural networks in classification problems with confidence sets for persistence diagrams. Neural Netw. **143**, 42–51 (2021). https://doi.org/10.1016/j.neunet.2021.05.007, https://doi.org/10.1016/j.neunet.2021.05.007
4. Anders, C.J., Weber, L., Neumann, D., Samek, W., Müller, K.R., Lapuschkin, S.: Finding and removing clever Hans: using explanation methods to debug and improve deep models. Inform. Fusion **77**, 261–295 (2022). https://doi.org/10.1016/j.inffus.2021.07.015
5. Apostol, T.M.: Mathematical Analysis. Addison-Wesley Publishing Company, reading, Massachusetts, second edn. (1974)
6. Benfenati, A., Marta, A.: A singular Riemannian geometry approach to deep neural networks i. theoretical foundations. Neural Netw. **158**, 331–343 (2023). https://doi.org/10.1016/j.neunet.2022.11.022
7. Benfenati, A., Marta, A.: A singular Riemannian geometry approach to deep neural networks ii. reconstruction of 1-d equivalence classes. Neural Netw. **158**, 344–358 (2023). https://doi.org/10.1016/j.neunet.2022.11.026
8. Bhardwaj, N., Liggesmeyer, P.: A runtime risk assessment concept for safe reconfiguration in open adaptive systems. In: Tonetta, S., Schoitsch, E., Bitsch, F. (eds.) SAFECOMP 2017. LNCS, vol. 10489, pp. 309–316. Springer, Cham (2017). https://doi.org/10.1007/978-3-319-66284-8_26
9. Braband, J., Lindner, L., Rexin, F.: Risk analyses for obstacle detection in automatic driving. SIGNAL+DRAHT **115**(3), 12–20 (2023)
10. Chockler, H., Kroening, D., Sun, Y.: Compositional Explanations for Image Classifiers. arXiv:2103.03622 [cs] (2021). http://arxiv.org/abs/2103.03622, arXiv: 2103.03622
11. Chollet, F., et al.: Keras. https://keras.io (2015)
12. Flajolet, P., Gardy, D., Thimonier, L.: Birthday paradox, coupon collectors, caching algorithms and self-organizing search. Discret. Appl. Math. **39**(3), 207–229 (1992). https://doi.org/10.1016/0166-218X(92)90177-C
13. Flammini, F., Marrone, S., Nardone, R., Caporuscio, M., D'Angelo, M.: Safety integrity through self-adaptation for multi-sensor event detection: methodology and case-study. Futur. Gener. Comput. Syst. **112**, 965–981 (2020). https://doi.org/10.1016/j.future.2020.06.036
14. Gleirscher, M., Haxthausen, A.E., Peleska, J.: Probabilistic risk assessment of an obstacle detection system for goa 4 freight trains. In: Proceedings of the 9th ACM SIGPLAN International Workshop on Formal Techniques for Safety-Critical Systems, pp. 26–36. FTSCS 2023, Association for Computing Machinery, New York, NY, USA (2023). https://doi.org/10.1145/3623503.3623533
15. Gruteser, J., Geleßus, D., Leuschel, M., Roßbach, J., Vu, F.: A formal model of train control with AI-based obstacle detection. In: Milius, B., Dutilleul, S.C., Lecomte, T. (eds.) Reliability, Safety, and Security of Railway Systems. Modelling, Analysis, Verification, and Certification - 5th International Conference, RSSRail 2023, Berlin, Germany, October 10-12, 2023, Proceedings. Lecture Notes in Computer Science, vol. 14198, pp. 128–145. Springer (2023). https://doi.org/10.1007/978-3-031-43366-5_8

16. Huang, W., Peleska, J.: Complete model-based equivalence class testing. Int. J. Softw. Tools Technol. Transfer **18**(3), 265–283 (2014). https://doi.org/10.1007/s10009-014-0356-8

17. ISO: ISO/DIS 21448: Road vehicles — Safety of the intended functionality. European Committee for Electronic Standardization (2021). iCS: 43.040.10, Draft International Standard

18. Kingma, D.P., Ba, J.: Adam: a method for stochastic optimization. In: Bengio, Y., LeCun, Y. (eds.) 3rd International Conference on Learning Representations, ICLR 2015, San Diego, CA, USA, May 7-9, 2015, Conference Track Proceedings (2015). http://arxiv.org/abs/1412.6980

19. Kupeli, D.N.: Singular Semi-Riemannian Geometry. Springer Science+Business Media Dordrecht (1996). https://doi.org/10.1007/978-94-015-8761-7

20. LeCun, Y., Cortes, C., Burges, C.: MNIST handwritten digit database. ATT Labs. http://yann.lecun.com/exdb/mnist **2** (2010)

21. Peleska, J., Haxthausen, A.E., Lecomte, T.: Standardisation considerations for autonomous train control. In: Margaria, T., Steffen, B. (eds.) Leveraging Applications of Formal Methods, Verification and Validation. Practice - 11th International Symposium, ISoLA 2022, Rhodes, Greece, October 22-30, 2022, Proceedings, Part IV. Lecture Notes in Computer Science, vol. 13704, pp. 286–307. Springer (2022). https://doi.org/10.1007/978-3-031-19762-8_22

22. Ristić-Durrant, D., Franke, M., Michels, K.: A review of vision-based on-board obstacle detection and distance estimation in railways. Sensors (Basel, Switzerland) **21**(10), 3452 (2021). https://doi.org/10.3390/s21103452, https://www.ncbi.nlm.nih.gov/pmc/articles/PMC8156009/

23. Santos, C.F.G.D., Papa, J.P.: Avoiding overfitting: a survey on regularization methods for convolutional neural network. ACM Comput. Surv. **54**(10s), 213:1–213:25 (2022). https://doi.org/10.1145/3510413

24. Sensoy, M., Saleki, M., Julier, S., Aydogan, R., Reid, J.: Misclassification risk and uncertainty quantification in deep classifiers. In: IEEE Winter Conference on Applications of Computer Vision, WACV 2021, Waikoloa, HI, USA, January 3-8, 2021, pp. 2483–2491. IEEE (2021). https://doi.org/10.1109/WACV48630.2021.00253

25. Sharma, N., Jain, V., Mishra, A.: An analysis of convolutional neural networks for image classification. Procedia Comput. Sci. **132**, 377–384 (2018). https://doi.org/10.1016/j.procs.2018.05.198

26. Shi, W., Alawieh, M.B., Li, X., Yu, H., Aréchiga, N., Tomatsu, N.: Efficient statistical validation of machine learning systems for autonomous driving. In: Liu, F. (ed.) Proceedings of the 35th International Conference on Computer-Aided Design, ICCAD 2016, Austin, TX, USA, November 7-10, 2016, p. 36. ACM (2016). https://doi.org/10.1145/2966986.2980077

27. Sun, Y., Chockler, H., Huang, X., Kroening, D.: Explaining image classifiers using statistical fault localization. In: Vedaldi, A., Bischof, H., Brox, T., Frahm, J. (eds.) Computer Vision - ECCV 2020 - 16th European Conference, Glasgow, UK, August 23-28, 2020, Proceedings, Part XXVIII. Lecture Notes in Computer Science, vol. 12373, pp. 391–406. Springer (2020). https://doi.org/10.1007/978-3-030-58604-1_24

28. Sun, Y., Wu, M., Ruan, W., Huang, X., Kwiatkowska, M., Kroening, D.: Concolic testing for deep neural networks. In: Huchard, M., Kästner, C., Fraser, G. (eds.) Proceedings of the 33rd ACM/IEEE International Conference on Automated Software Engineering, ASE 2018, Montpellier, France, September 3-7, 2018, pp. 109–119. ACM (2018). https://doi.org/10.1145/3238147.3238172
29. Underwriters Laboratories Inc.: ANSI/UL 4600-2020 Standard for Evaluation of Autonomous Products – First Edition. Underwriters Laboratories Inc., 333 Pfingsten Road, Northbrook, Illinois 60062-2096, 847.272.8800 (2020)

ResNets, NeuralODEs and CT-RNNs are Particular Neural Regulatory Networks

Radu Grosu[(✉)]

Fakultät für Informatik, Technische Universität Wien, Vienna, Austria
`radu.grosu@tuwien.ac.at`

Abstract. This paper shows that ResNets, NeuralODEs, and CT-RNNs, are particular neural regulatory networks (NRNs), a biophysical model for the nonspiking neurons encountered in small species, such as the *C.elegans* nematode, and in the retina of large species. Compared to ResNets, NeuralODEs and CT-RNNs, NRNs have an additional multiplicative term in their synaptic computation, allowing them to adapt to each particular input. This additional flexibility makes NRNs M times more succinct than NeuralODEs and CT-RNNs, where M is proportional to the size of the training set. Moreover, as NeuralODEs and CT-RNNs are N times more succinct than ResNets, where N is the number of integration steps required to compute the output $F(x)$ for a given input x, NRNs are in total $M{\cdot}N$ more succinct than ResNets. For a given approximation task, this considerable succinctness allows to learn a very small and therefore understandable NRN, whose behavior can be explained in terms of well-established architectural motifs, that NRNs share with gene regulatory networks, such as activation, inhibition, sequentialization, mutual exclusion, and synchronization.

1 Introduction

Artificial neurons (ANs) were inspired by spiking neurons [2,9]. However, ANs are nonspiking and from a biophysical point of view, wrong. This precluded their adoption in neuroscience, from which they drifted apart. The main mistake of ANs was to combine in one unit the linear transformation of the inputs of a neuron, with a nonlinear transformation, inspired by spiking. However, the nonlinear transformation itself is smooth, that is, it is nonspiking.

Despite of this, the way in which an artificial neural network (ANN) puts the neurons together, corrects the mistake. In ANNs it is irrelevant what is the exact meaning of a neuron, and what is that of a synapse. What matters, is the mathematical expression of the network itself. This was best exemplified by ResNets, which were forced for technical reasons, to separate the linear transformation from the nonlinear one, and introduce new state variables, which are the outputs of the summation, and not of the nonlinear transformation [10,11].

This separation, although not recognized as such in ResNets, allows us to reconcile ResNets with a biophysical model for nonspiking neurons, we call neural regulatory networks (NRNs), as they share many architectural motifs with

M. Fränzle et al. (Eds.): Werner Damm Festschrift, LNCS 15471, pp. 288–297, 2026.
https://doi.org/10.1007/978-3-031-97537-0_17

gene regulatory networks. In particular, important binary motifs are activation, inhibition, sequentialization, mutual exclusion, and synchronization. As NRNs have a liquid time constant, we refer to them as LTCs in [14]. NRNs capture neural behavior of small species, such as the C.elegans nematode [15], and of nonspiking neurons in the retina of large species [13]. In this model the neuron is a capacitor, and its rate of change is the sum of a leaking current, and of synaptic currents. The conductance of synapses varies in a smooth nonlinear fashion with the potential of the presynaptic neuron, and this is multiplied with a difference of potential of the postsynaptic neuron, to produce the synaptic current. Hence, the smooth nonlinear transformation is the one that synapses perform, which is indeed the case in nature, and not the one of neurons.

The main difference between NRNs and ResNets, (augmented) NeuralODEs, and CT-RNNs [3,6,8], is that NRNs multiply the conductance with a difference of potential. This is dictated by physics, as one needs to obtain a current. This constraint turns out to be very useful, as it allows NRNs to adapt to each particular input x. In other words, for each input, the discrete-time unfolding of an NRN is a ResNet, even though, there may be no ResNet that would properly work for all inputs. If the training set has size is S, this flexibility allows NRNs to be M times more succinct than (augmented) NeuralODEs and CT-RNNs, where M is proportional to S. Moreover, since (augmented) NeuralODEs and CT-RNNs are N times more succinct than ResNets, where N is the number of integration steps necessary to compute the output y for a given input x, NRNs are in total $M \cdot N$ more succinct than ResNets. For a given approximation task, this considerable succinctness allows to learn a very small and understandable NRN, which can be explained in terms of the architectural motifs.

Another important difference of CT-RNNs and NRNs, compared to ResNets and (augmented) NeuralODEs, is the leaking current, which can be seen as a regularization term. If a neuron is not excited, the leaking current forces it to return to its equilibrium state, that is, it makes it stable. This regularization imposes beneficial restrictions when learning the parameters of an NRN.

The rest of the paper is structured as follows. In Sect. 2 we shortly review ANNs, and discuss their misconceptions. We then discuss ResNets in Sect. 3 and show how they can provide a fresh view to ANNs. In Sect. 4 we review how one can smoothly transition from ResNets to NeuralODEs, and how approximation becomes a control problem. In Sect. 5 we show that CT-RNNs are a slight extension of NeuralODEs. In Sect. 6 we show that all previous models are in fact a simplification of NRNs, and finally in Sect. 7 we draw our conclusions and discuss future work.

2   Artificial Neural Networks

A main claim of ANNs is that ANs were inspired by their biological counterpart [2]. The story goes as follows: an AN receives one or more inputs (from other ANs), sums them up in a linear fashion, and passes the result through a non-linear thresholding function, known as an activation function. The latter frequently has a sigmoidal shape. The threshold itself is a condition for the neuron to fire. Mathematically, this is expressed as follows:

$$y_i^{t+1} = \sigma(\sum_{j=1}^{n} w_{ji}^t\, y_j^t, \mu_i^{t+1}) \qquad \sigma(x,\mu) = \frac{1}{1 + e^{-(x-\mu)}} \tag{1}$$

where, as shown in Fig. 1, y_i^{t+1} is the output of neuron i at layer $t+1$, y_j^t, for $j = 1:n$, is the output of neuron j at layer t, w_{ji}^t is the weight associated to the synapse between neuron j at layer t and neuron i at layer $t+1$, μ_i^{t+1} is the threshold of neuron i at layer $t+1$, and σ is the activation function, such as the logistic function above. A network with one input layer, one output layer, and $N \geq 2$ hidden layers, is called a deep neural network (DNN) [9].

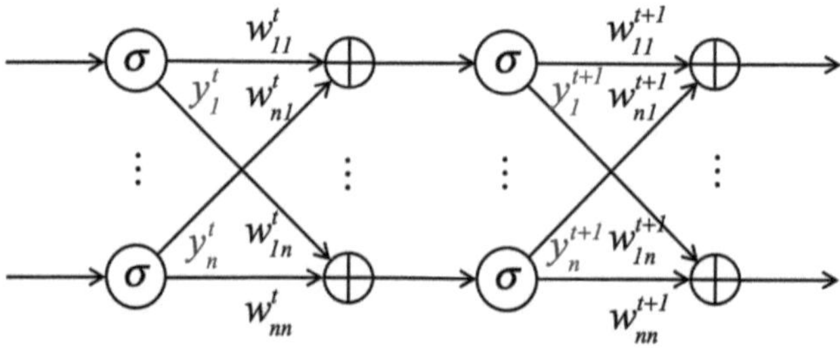

Fig. 1. The architecture of a DNN.

An AN has unfortunately little resemblance to a biological spiking neuron, from which the thresholding idea originated, as its output is definitely not spiking. However, as shown later, ANNs are in fact very closely related to non-spiking neural networks, if one assumes that all outgoing synapses of a neuron have the same activation function (with same parameters). Let us look first at ResNets [10].

3   Residual Neural Networks

DNNs with a very large number of hidden layers, were found to suffer from the so called degradation problem, which persisted even if the so called vanishing gradient problem was curated [10,11]. In short, the problem resulted from the inability of the DNNs to accurately learn the identity function. This manifested in the fact that a neural network with more layers, had poorer results, both in training and validation, than one with less layers. However, the first could have been obtained by simply extending the second with identity layers.

Since identities were hard to learn, they were just added to the DNNs in form of skip connections. The resulting architecture, as shown in Fig. 2, was called a

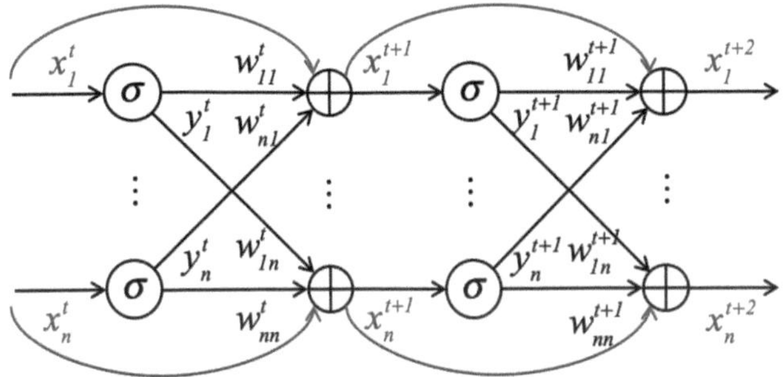

Fig. 2. The architecture of a ResNet.

residual neural network (ResNet) [10,11][1]. In ResNets, the outputs x_i^t of the sums are distinguished from the outputs y_i^t of the sigmoids. Mathematically:

$$x_i^{t+1} = x_i^t + \sum_{j=1}^{n} w_{ji}^t \, y_j^t \qquad y_j^t = \sigma(x_j^t, \mu_j^t) \tag{2}$$

This distinction is very important from a biophysical point of view, although this was not explicitly recognized this way. The main idea is that neurons may just play the role of summation, and the sigmoidal transformation is a transformation happening in the synapses. In fact, one can put the weights in the synaptic transformation, too, which leads to the equivalent equations:

$$x_i^{t+1} = x_i^t + \sum_{j=1}^{n} y_{ji}^t \qquad y_{ji}^t = w_{ji}^t \, \sigma(x_j^t, \mu_j^t) \tag{3}$$

Note that in Eq. 3, all outgoing synapses of pre-synaptic neuron j are assumed to have the same dynamic $\sigma(x_j^t, \mu_j^t)$. This might, or might not, be the case in nature. The weight w_{ji} can be thought of as a maximum conductance of the input dependent synaptic transformation. These transformations are indeed in nature graded, that is nonspiking. As we will see later on, this observation is important to making the connection to biophysical models of nonspiking neurons. Before we further investigate this new idea, let us first look at NeuralODEs [3].

4   Neural Ordinary Differential Equations

Equation 3, can be simply understood as the Euler discretization of a differential equation, where the time step is just taken to be one [3,7]. Mathematically:

$$\dot{x}_i(t) = \sum_{j=1}^{n} y_{ji}(t) \qquad y_{ji}(t) = w_{ji}(t) \, \sigma(x_j(t), \mu_j(t)) \tag{4}$$

---

[1] In [11], x_i^t skips the first sum and it is added directly to x_i^{t+2}. Hence, the architecture shown in Fig. 2, can be regarded as ResNets with finest skip granularity.

In this equation, not only $x_i(t)$ and $y_i(t)$ change continuously in time, but so do the synaptic weights $w_{ji}(t)$. This might be not such a big problem, as during the discrete integration and optimization, one obtains ResNets anyway, and their weights can be learned. One just has to unfold the differential equation $N = T/dt$ times, where T is the time horizon and dt the integration step.

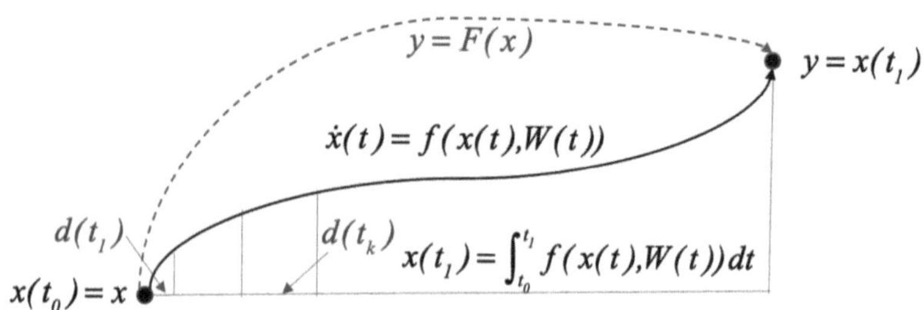

Fig. 3. Function approximation with ODEs.

At this point it is useful to make an excursion in function approximation with ODEs [7]. Suppose we are given a set of data points (x_i, y_i) for $i = 1:S$, and that we would like to learn the function $y = F(x)$ which best approximates the data, and generalizes to data points not seen yet. ANNs with a single hidden layer directly learn F with regression. They are universal approximators [5].

However, if the approximation is learned with a ResNet in its differential form of Eq. 4, then one actually learns the derivative f of F, such that $\dot{x}(t) = f(x(t), W(t))$, $x(t_0) = x$ and $y = x(t_1)$. To approximate F we thus need to compute the integral of f from t_0 to t_1, as shown in Fig. 3. In general, the derivative is simpler, and therefore learning DNNs is simpler, too.

ODE approximation can also be understood as a controllability problem [7]: For any given position x, is it possible to find an input $W(t)$ for $f(x(t), W(t))$, that steers x to y, within a finite horizon $T = t_1 - t_0$? The answer is affirmative, as ANNs with one layer are universal approximators, which implies that so are DNNs, and therefore ResNets, and consequently their ODE version, too.

Now suppose we make $W(t)$ time invariant, that is, $W(t) = W$. Are we still going to have a universal ODE approximator? The answer is yes, as we will show in next section. In this case, the differential equations are as follows:

$$\dot{x}_i(t) = \sum_{j=1}^{n} y_{ji}(t) \qquad y_{ji}(t) = w_{ji}\, \sigma(x_j(t), \mu_j) \tag{5}$$

This is the form of Neural Ordinary Differential Equations (NeuralODEs) [3, 7]. The N times unfolding is still a ResNet but dramatically more succinct: it requires only W parameters instead of $W \cdot N$! Could this work? This does indeed seem to be the case, with some slight extensions as discussed below. One intuitive explanation, is that modern numerical integrators would, for stability and efficiency reasons, use an adaptive time stepping d(t), as shown in Fig. 3:

$$x_i(t + d(t)) = x_i(t) + \sum_{j=1}^{n} y_{ji}(t) \qquad y_{ji}(t) = w_{ji}\, d(t)\, \sigma(x_j(t), \mu_j) \qquad (6)$$

Now the actual weights used are $W(t) = W d(t)$, which are time dependent, and can vary depending on the stiffness of the network [7].

NeuralODEs as approximators have the same dimensionality for the input, hidden state, and output. This is problematic, as ODEs do not allow trajectories to cross each other [6]. However, adding sufficient dimensions, one can always avoid crossings. This can be achieved with the embedding of the input in the internal state, and a projection at the end of the state to the output, as follows:

$$x(t_0) = [x, 0]^t \qquad y = \pi_S(x(t_1)) \qquad (7)$$

where the input is extended with zeroes as necessary, and the output is projected to the required subset. By applying such transformations, one arrives to what is called augmented NeuralODEs [6]. Unfortunately, like NeuralODEs, augmented NeuralODEs are harder to train than ResNets, which is no wonder, as their size and parametrization is considerably more succinct. However, for learning, one can use the adjoint equation, and employ efficient numerical solvers [3,7].

5   Continuous-Time Recurrent Neural Networks

One way to improve learning, is to add constraints to the approximation problem, or equivalently, to the control problem. Such constraints, can be understood as regularization constraints. Continuous-time recurrent neural networks (CT-RNN), introduce a stability regularization [8]. Their form is as follows:

$$\dot{x}_i(t) = -w_i x_i(t) + \sum_{j=1}^{n} y_{ji}(t) \qquad y_{ji}(t) = w_{ji}\, \sigma(x_j(t), \mu_j) \qquad (8)$$

The leading term $-w_i x_i(t)$ has the role to bring the system back to the equilibrium state, when no input is available. Hence, a small perturbation is forgotten, that is, the system is stable. This can be understood as a regularization constraint for the weights W that have to be learned. Note that, except for this term, CT-RNNs are equivalent to NeuralODEs.

CT-RNNs have been proven to be universal approximators, and the leading term does not play an important role in this proof [8]. Hence, NeuralODEs and augmented NeuralODEs are universal approximators, too.

6   Neural Regulatory Networks

Neural regulatory networks (NRNs), are a biophysical model for the neural system of small species, such as C.elegans [14,15], and of the retina of large species.

Their neuron model is called an electrical equivalent circuit in neuroscience [13]. Due to the small dimension of C.elegans, less than 1mm, the neural transmission happens passively, in the analog domain, without considerable attenuation. Hence, the neurons do not need to spike for an accurate signal transmission. The thickness of the retina is very small, too, only a half of a millimeter.

In large species, spiking happens at the beginning of the axon of a neuron. Its role is to transform the amplitude of the analog signal, computed by the body of the neuron, in frequency, that is, in a train of spikes. The axon has a myeline stealth (an insulator for passive transmission), interrupted every millimeter by Nodes of Ranvier, which reinforce the signal. When the signal arrives at the synapse, it is converted back into a neurotransmitter concentration. Hence, the synapses work in an analog domain, too. Except for the digital transmission part, neurons and their synapses are therefore analog computation units.

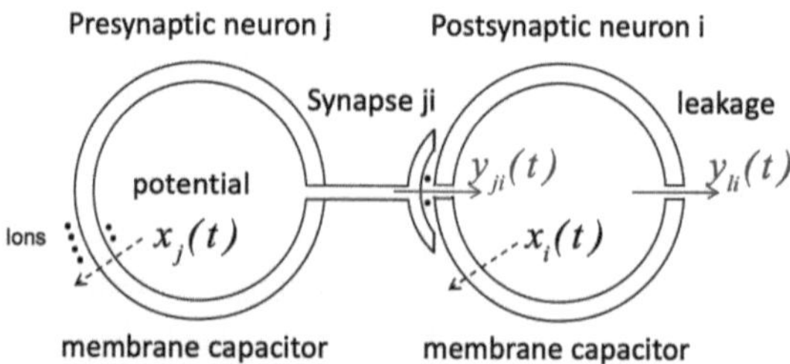

Fig. 4. The electric representation of a nonspiking neuron.

As shown in Fig. 4, the membrane of a neuron is an insulator, with ions both on its outside and inside. Electrically, it plays the role of a capacitor. The difference of the inside and outside ionic concentrations defines the membrane potential x. The rate of change of x depends on the currents y passing through the membrane, which are either external currents (ignored here), a leakage current, and synaptic currents. For simplicity, we consider only chemical synapses. The capacitor equation, with capacitance $C = 1$, is then as follows:

$$\dot{x}_i(t) = w_{li}\left(E_{li} - x_i(t)\right) + \sum_{j=1}^{n} y_{ji}(t) \quad y_{ji}(t) = w_{ji}\,\sigma(x_j(t), \mu_{ji})\left(E_{ji} - x_i(t)\right) \quad (9)$$

where $E_{li} = -70\,\mathrm{mV}$ is the resting potential of the neuron, w_{li} its leaking conductance, and E_{ji} the synaptic potentials, $0\,\mathrm{mV}$ in case of excitatory synapses (potential x_j is negative so the current is positive), or $-90\,\mathrm{mV}$ for inhibitory synapsess (this is smaller than any potential so the current is negative). For a detailed description of the biological parameters in Eq. 9, and their associated value in C.elegans, the interested reader is referred to [15].

This equation is very similar to the equation of CT-RNNs. First, it has a leaking current, which ensures the stability of the network. Second, it has a presynaptic-neuron controlled conductance $\sigma(x_j(t), \mu_{ji})$ for for each synapse s_{ji} from pre-synaptic neuron n_j to the post-synaptic neuron n_i. This synapse has a maximum conductance w_{ji}. However, in this case, each individual synapse s_{ji} may have a different mean μ_{ji}, and if desired, a different variance. Moreover, this conductance is multiplied with a difference of potential $(E_{ji} - x_i(t))$, to get a current. This biophysical constraint, makes it very different from CT-RNNs.

So what is the significance of this last term for learning? One way to look at it, is that for some fixed paramterization W, the actual weight of the synapse from neuron j to neuron i is $w_{ji}(t) = w_{ji}\, d(t)\, x_i(t)$. Hence, it not only varies in time with the integration step $d(t)$, but also with the actual input x provided to the network, and propagated as $x_i(t)$. In other words, an NRN generates for every particular input, a particular ResNet through unfolding, even though, there might be no single ResNet that would be appropriate for all inputs.

This flexibility makes NRNs M times more succinct than NeuralODEs and CT-RNNs, where M is proportional to the size of the training set. As the latter are N times more succinct than ResNets, NRNs are in total $M \cdot N$ times more succinct than ResNets. Hence, for a given approximation task, the learned NRNs are very small and explainable in terms of the architectural motifs, such as the binary motifs activation, inhibition, sequentialization, mutual exclusion, and synchronization, or more powerful higher order motifs.

Another way of looking at $w_{ji}(t) = w_{ji}\, d(t)\, x_i(t)$, is that the importance of a synaptic weight $w_{ji}(t) = w_{ji}\, d(t)$ between neuron j and neuron i, is weighted by the actual value $x_i(t)$ of neuron i. If the value $x_i(t)$ is low, the synapse is regarded as less important, and if the value $x_i(t)$ is high, as more important. This weighting mechanism of $w_{ji}(t)$ by $x_i(t)$, is also called gating in recurrent neural networks such as LSTMs, GRUs, and MGUs [4,12,16]. It shown there to play a decisive role in the ability of learning long-term dependencies

Finally, the importance of the multiplicative factor x can be explained in terms of interpretability [1]. Starting from a linear regression $f(x) = \theta^t x$, which is interpretable as a weighted sum of features in x, the authors suggest the generalization $f(x) = \theta(x)^t h(x)$, where parameters vector $\theta(x)$ is allowed to depend on x, and $h(x)$ is a vector of higher-order features. In case of DNNs, $h(x)$ can be taken as identity, as each successive layer introduces higher level features. Moreover, the input dependent parametrization $\theta(x)$ becomes $w^t \sigma(x, \mu)$. But $f(x) = \sum_j w_{ji}\, \sigma(x_j, \mu_j)\, x_i$ are precisely the synaptic currents of $\dot{x}_i$ in NRNs. They ensure robustness, too: small changes of x lead to small changes of $f(x)$. This is not the case for DNNs, ResNets, NeuralODEs, and CT-RNNs.

Once one realizes that neurons are just additive elements, and that chemical synapses are responsible for the sigmoidal transformations, one can proceed and add electric (or Ohmic) synapses to a network as well. Such synapses are encountered in nature, and they have a corresponding current y_{ji}^e as below:

$$y_{ji}^e(t) = w_{ji}\left(x_j(t) - x_i(t)\right) \tag{10}$$

In this case, the leaking current itself, can be understood as an electric synapse where $x_j(t) = E_{li}$, and it can be added to the structure of the network.

7   Conclusions

We have shown that ResNets, (augmented) Neural ODEs, and CT-RNNs are all a particular simplification of NRNs, a biphysical model for nonspiking neurons and their synapses. Such neurons are found for example in small species, such as the C.elegans nematode, or in the retina of large species.

We have also shown that NRNs are $M \cdot N$ times more succinct than ResNets, where M is the size of the quotient structure induced by the output on the input space, and N is the number of integration steps required for computing the approximation. For a given approximation task, the learned NRN is thus very small, and explainable in terms of its architectural motifs. This gives considerable hope towards the stated goal of explainable machine learning.

Acknowledgements. The author would like to thank his PhD students Sophie Grünbacher, Ramin Hasani, Mathias Lechner, and Zahra Babaiee, and his colleagues Daniela Rus, Scott Smolka, Jacek Cyranka, Flavio Fenton, Tom Henzinger, and Manuel Zimmer, for their feedback and fruitful discussions.

References

1. Alvarez-Melis, D., Jaakkola, T.: Towards robust interpretability with self-explaining neural networks. In: Proceedings of NIPS'18, the 32nd Conference on Neural Information Processing Systems, Montreal, Canada (2018)
2. Bishop, C.: Neural Networks for Pattern Recognition. Claredon Press, Oxford (1995)
3. Chen, T.Q., Rubanova, Y., Bettencourt, J., Duvenaud, D.K.: Neural ordinary differential equations. In: Advances in Neural Information Processing Systems, pp. 6571–6583 (2018)
4. Chung, J., Gulcehre, C., Cho, K., Bengio, Y.: Empirical evaluation of gated recurrent neural networks on sequence modeling. arXiv preprint arXiv:1412.3555 (2014)
5. Cybenko, G.: Approximation by superpositions of a sigmoidal function. Math. Control Signals Syst. **2**(4), 303–314 (1989)
6. Dupont, E., Doucet, A., Teh, Y.: Augmented neural odes (2019). arXiv, eprint 1904.01681
7. Weinan, E.: A proposal on machine learning via dynamical systems. Commun. Math. Stat. **5**, 1–11 (2017)
8. Funahashi, K., Nakamura, Y.: Approximation of dynamical systems by continuous time recurrent neural networks. Neural Netw. **6**(6), 801–806 (1993)
9. Goodfellow, I., Bengio, Y., Courville, A.: Deep Learning. MIT Press (2016)
10. He, K., Zhang, X., Ren, S., Sun, J.: Deep residual learning for image recognition. CoRR, abs/1512.03385 (2015)
11. He, K., Zhang, X., Ren, S., Sun, J.: Identity mappings in deep residual networks. CoRR, abs/1603.05027 (2016)
12. Hochreiter, S., Schmidhuber, J.: Long short-term memory. Neural Comput. **9**(8), 1735–1780 (1997)
13. Kandel, E., Schwartz, J., Jessell, T., Siegelbaum, S., Hudspeth, A.: Principles of Neural Science. McGraw-Hill Education and Medical (2012)

14. Lechner, M., Hasani, R., Zimmer, M., Henzinger, T., Grosu, R.: Designing worm-inspired neural networks for interpretable robotic control. In: Proceedings of the 2019 International Conference on Robotics and Automation (ICRA), Montreal, Canada (2019)
15. Wicks, S.R., Roehrig, C.J., Rankin, C.H.: A dynamic network simulation of the nematode tap withdrawal circuit: predictions concerning synaptic function using behavioral criteria. J. Neurosci. **16**(12), 4017–4031 (1996)
16. Zhou, G.-B., Wu, J., Zhang, C.-L., Zhou, Z.-H.: Minimal gated unit for recurrent neural networks. Int. J. Autom. Comput. **13**(3), 226–234 (2016). https://doi.org/10.1007/s11633-016-1006-2

Dominant Strategies for Hyperproperties

Bernd Finkbeiner$^{(\boxtimes)}$ and Noemi Passing

CISPA Helmholtz Center for Information Security, Saarbrücken, Germany
`finkbeiner@cispa.de`

Abstract. We introduce dominant strategies for hyperproperties, such as observational determinism and other notions related to information flow, knowledge, and robustness. The approach is inspired by joint work with Werner Damm on the synthesis of distributed systems with dominant strategies, i.e., with strategies that perform at least as well as the best alternative strategy. If the specification admits a dominant strategy, the distributed system can be built compositionally, considering one component at a time. This paper extends the theory of dominant strategies from trace properties, which refer to individual traces, to k-hyperproperties, which relate several traces. We present a method for checking whether a given finite-state strategy is dominant, establish the compositionality of dominance for k-hypersafety properties and fully-informed architectures, and show how to derive weakest environment assumptions from dominant strategies.

1 Introduction

The reactive synthesis problem, also known as Church's problem, is to construct a reactive system from a formal specification in such a way that the system is guaranteed to satisfy the specification. One of the most intriguing open questions about this problem is how to make the construction process *compositional*, i.e., how to build complex systems from small building blocks, so that the correctness of the system follows directly from the correctness of the individual components. Compositionality is widely considered to be the key to making synthesis scale to large systems; however, it is still rarely achieved in practice. The challenge is that, on its own, a component typically cannot guarantee that its specification is satisfied under all circumstances; rather, it relies on the cooperative behavior of the other components. As a result, it is difficult to synthesize individual components without immediately considering the entire system.

Some years ago, Werner Damm and the first author developed a compositional approach that exploits the availability of *dominant strategies* [5,7]. We view synthesis as a game where the component plays against its environment; the component wins the game if the specification is satisfied. While individual components usually do not have a winning strategy, they often have *dominant strategies*, which are strategies that perform *at least as well* as the best alternative strategy. In situations where no strategy wins the game, the dominant

M. Fränzle et al. (Eds.): Werner Damm Festschrift, LNCS 15471, pp. 298–311, 2026.
https://doi.org/10.1007/978-3-031-97537-0_18

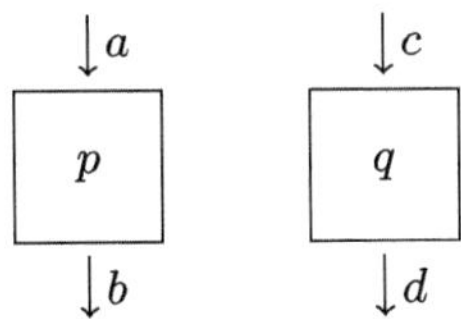

Fig. 1. Example architecture where component p has input a and output b, and component q has input c and output d. Both inputs are controlled by the environment.

strategy is not expected to win the game either. Applying this idea in a synthesis algorithm, we do not attempt to build a component that guarantees the specification under all possible inputs from its environment, but instead search for a solution that satisfies the specification for all inputs where it is, in principle, possible to satisfy the specification. When dominant strategies exist, they radically simplify the construction of distributed systems because the component strategies can be built in isolation, independently of the remainder of the system. A dominant strategy furthermore uniquely identifies a weakest condition on the rest of the system that, when satisfied, guarantees that the component can satisfy the specification. This condition is called the *weakest environment assumption*, where *environment* refers to the entire environment of the component, including both the external environment and the other system components.

The initial focus of our work on dominant strategies was on safety properties [7]. More recently, the notion has been extended to liveness properties [11,15]. In this paper, we propose a further, substantial extension of the theory of dominant strategies to hyperproperties. *Hyperproperties* [4] generalize trace properties, which are sets of traces, to sets of sets of traces. A system satisfies a trace property if its set of traces is contained in the trace property. A system satisfies the hyperproperty if its set of traces is an *element* of the hyperproperty. While trace properties express properties of individual traces, a hyperproperty can express relations between different traces. As result, hyperproperties capture many properties of interest that cannot be expressed as trace properties, including many notions related to information flow, knowledge, and robustness.

A classic example of a hyperproperty is *observational determinism* [18], which expresses that the system appears deterministic to an outside observer, even though the observer cannot see certain secret inputs. In other words, the behavior is completely determined by its publicly visible inputs, and, hence, independent of its secrets. Formally, observational determinism checks if all pairs of traces that agree on their input (in the first position of the trace) also have identical output (throughout the trace).

Consider now a system that consists of two components, p and q, where component p has input a and output b, and component q has input c and output d. This architecture is shown in Fig. 1. Observational determinism would require that all traces that agree on the inital values of a and c must also agree on all valuations of b and d. Obviously, component p cannot, by itself, guarantee observational nondeterminsim. If q, for example, copies the values of its input c to its output d, then component p has no chance to guarantee that observational determinism holds. Nevertheless, the best possible strategy for p is to act deterministically in the initial value of a. If q correspondingly acts determinstically in the initial value of c, then the overall system satisfies observational determinism. If q acts nondeterministically in the initial value of c, then no strategy of p could establish observational determinism. Acting deterministically in the initial value of a is, hence, a *dominant strategy* for component p. This strategy induces a *weakest assumption* on q, which is to act deterministically in the initial value of c.

In the remainder of the paper, we first give an overview on hyperproperties (Sect. 2), then formally introduce dominant strategies for hyperproperties (Sect. 3), discuss the compositionality of dominant strategies (Sect. 4), and finally describe how to derive weakest environment assumptions (Sect. 5). We conclude with an outlook on future work (Sect. 6).

2   Hyperproperties

Hyperproperties [4] generalize trace properties, which are sets of traces, to sets of sets of traces. In this section, we give some basic definitions and provide a brief introduction to the specification of hyperproperties with temporal logic. For a more comprehensive introduction, we refer the reader to a recent overview paper [8].

2.1   Trace Properties vs. Hyperproperties

Let AP be a finite set of atomic propositions. A *trace* over AP is a map $t\colon \mathbb{N} \to 2^{AP}$, denoted by $t(0)t(1)t(2)\cdots$. Let $(2^{AP})^{\omega}$ denote the set of all traces over AP. A *trace property* is a subset of $(2^{AP})^{\omega}$, a *hyperproperty* a set of subsets of $(2^{AP})^{\omega}$. In this paper, we focus on k-*hyperproperties*, which can be expressed in terms of a relation $R \subseteq ((2^{AP})^{\omega})^k$ on k traces for some natural number k. Such a relation defines the hyperproperty H, where a set T of traces is an element of H iff for all $t(1), t(2), \ldots t(k) \in T$ it holds that $(t(1), t(2), \ldots t(k)) \in R$.

Of particular interest are k-hyperproperties that are also hypersafety. Such k-hyperproperties are called k-hypersafety. For trace properties, safety is characterized as the properties where each violation is caused after a finite time [1]. For a finite trace $u \in \Sigma^*$ and an infinite trace $t \in \Sigma^{\omega}$, let $u \prec t$ denote that u is a prefix of t. A trace property P is *safety* if it holds that for every trace $t \notin P$, there exists a prefix $u \prec t$ such that for every infinite trace t' with $u \prec t'$, we have $t' \notin P$. Liveness properties characterize that something good happens eventually, i.e., a trace property P is *liveness* if for every finite trace $u \in \Sigma^*$, there exists an infinite trace $t \in \Sigma^{\omega}$ with $u \prec t$ and $t \in P$. Analogously, hyperproperties are characterized as hypersafety and hyperliveness [3]. We lift the prefix relation $\prec$ to sets of traces: a set $U \subseteq \Sigma^*$ of finite traces is a prefix of a set $T \subseteq \Sigma^{\omega}$ (written $U \prec T$) if, for every $u \in U$, there exists a $t \in T$ such that $u \prec t$.

A hyperproperty H is *hypersafety* if for every $T \subseteq \Sigma^\omega$ with $T \notin H$, there exists a finite set $U \subseteq \Sigma^*$ with $U \prec T$ such that, for every $T' \subseteq \Sigma^\omega$ with $U \prec T'$, we have $T' \notin H$. A property H is *hyperliveness* if for every finite set $U \subseteq \Sigma^*$, there exists $T \subseteq \Sigma^\omega$ with $U \prec T$ and $T \in H$.

2.2   Logical Specifications of Hyperproperties

In the following, we first briefly review the syntax and semantics of linear-time temporal logic (LTL) and then introduce HyperLTL and HyperQPTL, two extensions of LTL for hyperproperties.

LTL. The formulas of linear-time temporal logic (LTL) [16] are generated by the following grammar:

$$\varphi ::= a \mid \neg\varphi \mid \varphi \wedge \varphi \mid \mathbf{X}\,\varphi \mid \varphi\,\mathbf{U}\,\varphi$$

where $a \in AP$ is an *atomic proposition*, the Boolean connectives $\neg$ and $\wedge$ have the usual meaning, $\mathbf{X}$ is the temporal *next* operator, and $\mathbf{U}$ is the temporal *until* operator. We also consider the usual derived Boolean connectives, such as $\vee$, $\rightarrow$, and $\leftrightarrow$, and the derived temporal operators *eventually* $\mathbf{F}\,\varphi \equiv true\,\mathbf{U}\,\varphi$, *globally* $\mathbf{G}\,\varphi \equiv \neg\mathbf{F}\,\neg\varphi$, *weak until* $\varphi\,\mathbf{W}\,\psi \equiv (\varphi\,\mathbf{U}\,\psi) \vee \mathbf{G}\,\varphi$, and *release* $\varphi\,\mathbf{R}\,\psi \equiv \neg(\neg\varphi\,\mathbf{U}\,\neg\psi)$. The satisfaction of an LTL formula φ over a trace t at a position $i \in \mathbb{N}$, denoted by $t, i \models \varphi$, is defined as follows:

$$
\begin{aligned}
t, i &\models a &&\text{iff}\quad a \in t(i), \\
t, i &\models \neg\varphi &&\text{iff}\quad t, i \not\models \varphi, \\
t, i &\models \varphi_1 \wedge \varphi_2 &&\text{iff}\quad t, i \models \varphi_1 \text{ and } t, i \models \varphi_2, \\
t, i &\models \mathbf{X}\,\varphi &&\text{iff}\quad t, i+1 \models \varphi, \\
t, i &\models \varphi_1\mathbf{U}\varphi_2 &&\text{iff}\quad \exists k \geq i : t, k \models \varphi_2 \,\wedge\, \forall i \leq j < k : t, j \models \varphi_1.
\end{aligned}
$$

We say that a trace t satisfies an LTL formula φ, denoted by $t \models \varphi$, if $t, 0 \models \varphi$. For example, the LTL formula $\mathbf{G}\,(a \rightarrow \mathbf{F}\,b)$ specifies that every position in which a is true must eventually be followed by a position where b is true.

HyperLTL. In this paper, we will mainly use HyperLTL to specify hyperproperties. The formulas of HyperLTL [2] are generated by the grammar

$$\phi ::= \exists\pi.\ \phi \mid \forall\pi.\ \phi \mid \psi$$

$$\psi ::= a_\pi \mid \neg\psi \mid \psi \wedge \psi \mid \mathbf{X}\,\psi \mid \psi\,\mathbf{U}\,\psi$$

where a is an atomic proposition from a set AP and π is a trace variable from a set $\mathcal{V}$. Further Boolean connectives and the temporal operators $\mathbf{F}$, $\mathbf{G}$, $\mathbf{W}$, and $\mathbf{R}$ are derived as for LTL.

The semantics of HyperLTL is defined with respect to a trace assignment, a partial mapping $\Pi : \mathcal{V} \rightarrow (2^{AP})^\omega$. The assignment with empty domain is denoted by $\Pi_\emptyset$. Given a trace assignment Π, a trace variable π, and a trace t, we denote by $\Pi[\pi \to t]$ the assignment that coincides with Π everywhere but at π, which is mapped to t. The satisfaction of a HyperLTL formula ϕ over a trace assignment Π and a set of traces T at a position $i \in \mathbb{N}$, denoted by $T, \Pi, i \models \phi$, is defined as follows:

$$
\begin{aligned}
T, \Pi, i &\models a_\pi & &\text{iff} & &a \in \Pi(\pi)(i), \\
T, \Pi, i &\models \neg\psi & &\text{iff} & &T, \Pi, i \nvDash \psi, \\
T, \Pi, i &\models \psi_1 \wedge \psi_2 & &\text{iff} & &T, \Pi, i \models \psi_1 \text{ and } T, \Pi, i \models \psi_2, \\
T, \Pi, i &\models \mathbf{X}\,\psi & &\text{iff} & &T, \Pi, i+1 \models \psi, \\
T, \Pi, i &\models \psi_1 \,\mathbf{U}\, \psi_2 & &\text{iff} & &\exists k \geq i.\ T, \Pi, k \models \psi_2 \\
& & & & &\wedge \forall i \leq j < k.\ T, \Pi, j \models \psi_1, \\
T, \Pi, i &\models \exists\pi.\ \phi & &\text{iff} & &\exists t \in T.\ T, \Pi[\pi \mapsto t], i \models \phi, \\
T, \Pi, i &\models \forall\pi.\ \phi & &\text{iff} & &\forall t \in T.\ T, \Pi[\pi \mapsto t], i \models \phi.
\end{aligned}
$$

We say that a set T of traces satisfies a formula ϕ, denoted by $T \models \phi$, if $T, \Pi_\emptyset, 0 \models \phi$. In the following, we focus on k-hyperproperties, which are expressed as formulas with k universal quantifiers and no existential quantifiers. An example of a commonly used 2-hyperproperty is observational determinism. A system satisfies *observational determinism* [18] if every pair of traces with the same low-security input in the initial position remains indistinguishable for an observer throughout the execution; i.e., the program appears to be deterministic to low-security users. For a system with low-security input l and output o, observational determinism can be expressed in HyperLTL as follows:

$$
\forall\pi.\forall\pi'.\ (l_\pi \leftrightarrow l_{\pi'}) \;\to\; \mathbf{G}\,(o_\pi \leftrightarrow o_{\pi'})
$$

HyperQPTL. While we will mainly use HyperLTL as our specification language, the extension to HyperQPTL [17], which captures the *ω-regular hyperproperties* [9], will be useful as well. HyperQPTL extends HyperLTL with quantification over atomic propositions. To easily distinguish quantification over traces and propositions, we use $\exists\pi, \forall\pi$ for quantification over traces and $\exists p, \forall p$ for quantification over propositions. The formulas of HyperQPTL are generated by the following grammar:

$$
\begin{aligned}
\phi &::= \exists\pi.\ \phi \mid \forall\pi.\ \phi \mid \psi \mid \exists p.\ \phi \mid \forall p.\ \phi \\
\psi &::= a_\pi \mid p \mid \neg\psi \mid \psi \wedge \psi \mid \mathbf{X}\,\psi \mid \mathbf{F}\,\psi
\end{aligned}
$$

where $a, p \in AP$ and $\pi \in \mathcal{V}$. The semantics of HyperQPTL corresponds to the semantics of HyperLTL with additional rules for propositional quantification:

$$
\begin{aligned}
T, \Pi, i &\models \exists q.\ \phi & &\text{iff} & &\exists t \in (2^{\{q\}})^\omega.\ T, \Pi[\pi_q \mapsto t], i \models \phi, \\
T, \Pi, i &\models \forall q.\ \phi & &\text{iff} & &\forall t \in (2^{\{q\}})^\omega.\ T, \Pi[\pi_q \mapsto t], i \models \phi, \\
T, \Pi, i &\models q & &\text{iff} & &q \in \Pi(\pi_q)(i).
\end{aligned}
$$

HyperQPTL can express all properties expressible in HyperLTL and, additionally, ω-regular hyperproperties like "on even positions, proposition a has the same value on all traces." In the following HyperQPTL formula, even positions are identified by the existentially quantified proposition *even*:

$$\exists even. \forall \pi. \forall \pi'. \ even \wedge \mathbf{G}\left((even \leftrightarrow \mathbf{X} \neg even) \wedge (even \rightarrow (a_\pi \leftrightarrow a_{\pi'}))\right)$$

2.3   Synthesis of Distributed Systems

In the synthesis of distributed systems, we are interested in finding an implementation that corresponds to a given system architecture and that satisfies a given specification. To simplify the discussion, we focus in this paper on distributed systems with only two components. An *architecture* with two components p and q is given as a tuple $(I_p, I_q, O_p, O_q, O_e)$, where I_p, I_q, O_p, O_q, and O_e are all subsets of AP. O_p and O_q are the *output variables* of p and q. O_e are the output variables of the uncontrollable environment. The three sets O_p, O_q and O_e form a partition of AP. I_p and I_q are the *input variables* of components p and q, respectively. For each component, the inputs and outputs are disjoint, i.e., $I_p \cap O_p = \emptyset$ and $I_q \cap O_q = \emptyset$. The inputs I_p and I_q of the components are all either outputs of the environment or outputs of the other component, i.e., $I_p \subseteq O_q \cup O_e$ and $I_q \subseteq O_p \cup O_e$. We call architectures where both components can read the outputs of the environment, i.e., where $O_e \subseteq I_p$ and $O_e \subseteq I_q$, a *fully-informed* architecture.

An *implementation* of an architecture $(I_p, I_q, O_p, O_q, O_e)$ is a pair (s_p, s_q), consisting of a strategy for each of the two components. A *strategy* for a component p is a function $s_p : (2^{I_p})^* \rightarrow (2^{O_p})$ that maps finite sequences of valuations of p's input variables (i.e., *histories* of inputs) to a valuation of p's output variables. A strategy s is *finite-state* iff it can be represented as a finite-state Mealy automaton. A (finite-state) *Mealy automaton* $\mathcal{M} = (Q, \Sigma, \Omega, q_0, \delta, \lambda)$ consists of a (finite) set of states Q, an input alphabet Σ, an output alphabet Ω, a transition function $\delta : Q \times \Sigma \rightarrow Q$ and an output function $\lambda : Q \times \Sigma \rightarrow \Omega$. A run of $\mathcal{M}$ on a finite input word $w = w(0)w(1)\ldots w(k) \in \Sigma^*$ is a finite sequence $r = q(0)q(1)\ldots q(k+1) \in Q^*$ with $q(0) = q_0$ and $q(i+1) = \delta(q(i), w(i))$ for all $0 \leq i \leq k$. The output of $\mathcal{M}$ on input w is $\lambda(q(k), w(k))$. A Mealy automaton with input alphabet $\Sigma = 2^{I_p}$ and output alphabet $\Omega = 2^{O_p}$ represents a strategy s_p iff, for all $w \in (2^{I_p})^*$, the output of $\mathcal{M}$ on w is $s_p(w)$.

The *composition* $s_p \| s_q$ of the strategies of the components is the strategy $s : (2^{O_e})^* \rightarrow (2^{O_p \cup O_q})$ that maps finite sequences of valuations of the environment's output variables to valuations of all output variables. We define $s_p \| s_q(u) = s_p(f_p(u)) \cup s_q(f_q(u))$ for $u \in (2^{O_e})^*$, where f_p and f_q map sequences of environment outputs to sequences of component inputs: $f_p(\epsilon) = \epsilon, f_p(u \cdot x) = f_p(u) \cdot ((x \cup s_q(f_q(u))) \cap I_p)$ and $f_q(\epsilon) = \epsilon, f_q(u \cdot x) = f_q(u) \cdot ((x \cup s_p(f_p(u))) \cap I_q)$ for $u \in (2^{O_e})^*$ and $x \in 2^{O_e}$.

For each strategy $s : (2^I)^* \rightarrow 2^O$ for some $I, O \subseteq AP$, we define a function $trace_s : (2^{AP \setminus O})^\omega \rightarrow (2^{AP})^\omega$ that maps sequences of valuations (of all atomic propositions *except* for the outputs of the strategy) to traces:

$$trace_s(i_0 i_1 \ldots) = (s(\epsilon) \cup i_0)\,(s(i_0 \cap I) \cup i_1)\,\ldots$$

For a set of sequences T, $traces_s(T)$ is the set $\{trace_s(t) \mid t \in T\}$. We abbreviate $traces(s_p, s_q) = traces_{s_p \| s_q}((2^{O_e})^\omega)$, and say that the implementation *satisfies* the specification if this set of traces satisfies the specification, i.e., $traces(s_p, s_q) \models \varphi$.

3   Dominant Strategies

Strategic dominance describes a situation where one strategy is at least as good as any other strategy, no matter how the opponent plays. For trace properties, we can define *dominance* as follows [6]: let $I, O \subseteq AP$ and let $s, t : (2^I)^* \to 2^O$ be two strategies. Strategy t *is dominated by* strategy s, denoted by $t \preceq s$, iff, for every sequence $\gamma \in (2^I)^\omega$ for which the trace $trace(t, \gamma)$ resulting from strategy t satisfies the trace property T, the trace $trace(s, \gamma)$ resulting from strategy s also satisfies T. A strategy s is *dominant* iff, for all strategies $t : (2^I)^* \to 2^O$, $t \preceq s$. An implementation (s_p, s_q) is dominant iff $s_p \| s_q$ is dominant.

For a k-hyperproperty H, we generalize the notion of dominance $t \preceq s$ as follows. We compare s and t on all sets $T \subseteq (2^I)^\omega$ of at most k sequences. If, whenever the trace set $traces_t(T)$ resulting from t satisfies the hyperproperty H, the traces $traces_s(T)$ resulting from s also satisfies H, then s dominates t: $t \preceq s$. As before, a strategy s is *dominant* iff, for all strategies t, $t \preceq s$.

Example 1. As a first example, consider the architecture $(I_p, I_q, O_p, O_q, O_e)$ from Fig. 1, where component p has input $I_p = \{a\}$ and output $O_p = \{b\}$, component q has input $I_q = \{c\}$ and output $O_q = \{d\}$, and both inputs are controlled by the environment: $O_e = \{a, c\}$. The HyperLTL specification

$$\varphi_1 = \forall \pi, \pi'.\; \mathbf{G}\left((b_\pi \leftrightarrow b_{\pi'}) \wedge (d_\pi \leftrightarrow d_{\pi'})\right)$$

states that the outputs are independent from the inputs. Any strategy s_p for component p that produces an output b that is independent of the input a is a dominant strategy: since φ_1 is a 2-hyperproperty, we consider trace sets of size 2; if the two sequences differ in the values of d, then φ_1 is violated for all possible strategies of component p; if, on the other hand, the two sequences agree on the values of d, then φ_1 is satisfied, because s_p is independent of a.

Example 2. As a second example, consider the same architecture, and the HyperLTL specification

$$\varphi_2 = \forall \pi, \pi'.\; (a_\pi \leftrightarrow a_{\pi'}) \to \mathbf{X}\,(b_\pi \leftrightarrow b_{\pi'}).$$

The formula φ_2 states that the information in a is passed on to b: if a has different values during the first step, then b must have different values during the second step. The strategy s_p that immediately copies the value of a to b is, obviously, a dominant strategy; so is the alternative strategy which sets b to the negation of a.

Example 3. Now change the specification to

$$\varphi_3 = \forall \pi, \pi'. \ (c_\pi \leftrightarrow c_{\pi'}) \rightarrow \mathbf{X} \ (b_\pi \leftrightarrow b_{\pi'}),$$

which states that the information in c (which is not visible to component p) is passed on to b. Suppose strategy s_p sets b to the constant value *true*. Is s_p dominant? At first glance, it might appear that this is not the case, because the trace set $\{t, t'\}$ where $c \in t(0), b \in t(1), c \notin t'(0), b \in t'(1)$ results in a violation of φ_3, while the trace set $\{t, \hat{t}'\}$, where $b \notin \hat{t}'(1)$ and $\hat{t}$ is otherwise identical to t', satisfies φ_3. The fault in this reasoning is, however, that t and $\hat{t}'$ *cannot* be traces from the same strategy, because the input c is not visible to component p. In fact, φ_3 is violated on all pairs of traces that differ in the value of c, for *all* possible strategies. Strategy s_p is, therefore, indeed dominant.

Checking whether a given finite-state strategy is dominant is decidable for HyperLTL specifications. We show this via a reduction to the HyperQPTL model checking problem. We use the quantification over propositions available in Hyper-QPTL to quantify over the decisions of an alternative strategy. For a component with inputs $I = \{i_1, i_2, \ldots, i_m\}$ and outputs $O = \{o_1, o_2, \ldots, o_n\}$, and a universal HyperLTL specification $\varphi = \forall \pi_1, \pi_2, \ldots, \pi_k. \ \phi$, where ϕ is a quantifier-free formula over $I \cup O$, we consider the HyperQPTL formula

$$\psi = \forall \pi_1, \pi_2, \ldots, \pi_k. \ \forall_{x \in \{1,\ldots,n\}, y \in \{1,\ldots,k\}} \hat{o}_{x,y}.$$
$$(consistent \wedge \phi[o_{x,\pi_y} \mapsto \hat{o}_{x,y}]_{x \in \{1,\ldots,n\}, y \in \{1,\ldots,k\}}) \rightarrow \phi.$$

In the definition of ψ, $\phi[o_{x,\pi_y} \mapsto \hat{o}_{x,y}]_{x \in \{1,\ldots,n\}, y \in \{1,\ldots,k\}}$ denotes the formula that is obtained from ϕ by substituting for each occurrence of some indexed output variable o_{x,π_y} the corresponding quantified proposition $\hat{o}_{x,y}$. The formula *consistent* expresses the requirement discussed in the example that the assignment to the quantified variables $\hat{o}_{i,j}$ actually corresponds to some strategy. I.e., for any two traces, the inputs must differ *before* there can be a difference in the outputs:

$$\bigwedge_{y \in \{1,\ldots,k\}, y' \in \{1,\ldots,k\}} \left(\left(\bigvee_{x \in \{1,\ldots,m\}} i_{x,y} \leftrightarrow i_{x,y'} \right) \mathbf{R} \bigwedge_{x \in \{1,\ldots,n\}} \hat{o}_{x,y} \leftrightarrow \hat{o}_{x,y'} \right)$$

The decidability of dominance checking then follows directly from the decidability of HyperQPTL model checking [17].

Theorem 1. *Checking whether a given finite-state strategy is dominant for a k-hyperproperty given as a universal HyperLTL formula is decidable in NLOG-SPACE in the size of the strategy.*

4 Compositionality

We are interested in constructing individual implementations for the components that then compose into correct systems. This approach is sound if the notion of

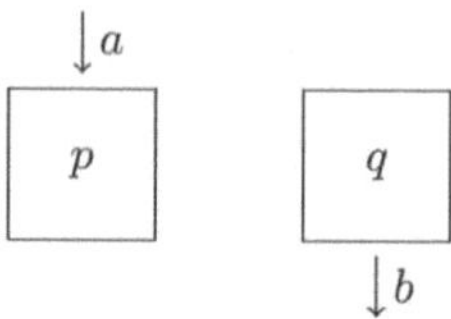

Fig. 2. Architecture from Example 4. Component p has input a and no output, and component q has no input and output b.

dominance is compositional, i.e., if the composition of two dominant strategies is again dominant. For trace properties, dominance is compositional for safety, but not for liveness [7]: consider, for example, the specification $(\mathbf{F}\,a) \wedge (\mathbf{F}\,b)$ where a is the output of component p and b is the output of component q. A strategy s_p of component p that waits for b to become *true* before setting, in the next step, a to *true*, is clearly dominant: the specification can only be satisfied if b is eventually set to *true*; in this case, however, a is set to *true* as well. Analogously, strategy s_q of component q, which waits for a to become *true* before setting b to true, is dominant as well. In the composition $s_p \| s_q$, neither a nor b ever become *true*. The composition thus violates the specification, even though there clearly exists a strategy that satisfies $(\mathbf{F}\,a) \wedge (\mathbf{F}\,b)$. The composition $s_p \| s_q$, hence, is not dominant, despite the fact that both s_p and s_q are dominant. Since liveness is a special case of hyperliveness, we therefore cannot hope for compositionality for hyperliveness. What about hypersafety?

Example 4. Consider the architecture shown in Fig. 2, where component p has input a and no output, and component q has no input and output b. The specification states that any pair of traces that disagree on the initial value of a must also disagree on the next value of b:

$$\varphi_4 : \forall \pi, \pi'.\ (a_\pi \not\leftrightarrow a_{\pi'}) \rightarrow \mathbf{X}\,(b_\pi \not\leftrightarrow b_{\pi'})$$

Since component p has no output, there is only one strategy $s_p : (2^{\{a\}})^* \rightarrow \{\emptyset\}$. Since it is the only strategy, s_p is dominant. For component q, let s_q be a strategy that sets b to some arbitrary constant value. This strategy violates φ_4 on any pair of traces π, π' where $a_\pi \not\leftrightarrow a_{\pi'}$. However, since component q has no inputs, we have that $\mathbf{X}\,(b_p \leftrightarrow b_{\pi'})$ on any other strategy as well. Strategy s_q is, hence, dominant as well. However, the composition $s_p \| s_q$ is not dominant. Like s_q, $s_p \| s_q$ violates φ_4 on traces that differ in the initial value of a. There are, however, strategies that satisfy φ_4: for example, the strategy that simply copies the value of a to b.

The example suggests that the issue is that the composition $s_p \| s_q$ is compared to strategies that have access to the full input from the environment, while s_p and s_q only see the individual inputs of components p and q. And indeed, we obtain compositionality if we restrict to fully-informed architectures. The proof of the following theorem is a straightforward adaptation of the proof for safety trace properties [7]:

Theorem 2. *For a k-hypersafety property φ, an architecture $(I_p, I_q, O_p, O_q, O_e)$ with $O_e \subseteq I_p$ and $O_e \subseteq I_q$, and an implementation (s_p, s_q), it holds that if s_p and s_q are dominant strategies, then $s_p \| s_q$ is dominant as well.*

Proof. Suppose, by way of contradiction, that s_p and s_q are dominant strategies and that $s_p \| s_q$ is not dominant. Since $s_p \| s_q$ is not dominant, there exists a set $E \subseteq (2^{O_e})^\omega$ of at most k sequences and a strategy $t : (2^{O_e})^* \rightarrow 2^{O_p \cup O_q}$ such that φ is satisfied by $traces_t(E)$ but not by $traces_{s_p \| s_q}(E)$. Since φ is hypersafety, there furthermore exists a set U of at most k finite traces such that $U \prec traces_{s_p \| s_q}(E)$, and for every $T' \subseteq \Sigma^\omega$ with $U \prec T'$, we have that T' violates φ. Among all the sets $U \prec traces_{s_p \| s_q}(E)$ where the finite traces have equal length, pick the one with the shortest length such that every $T' \subseteq \Sigma^\omega$ with $U \prec T'$ violates φ. This length cannot be zero, because φ is satisfied by $traces_t(E)$. The last position of the traces in U contains decisions of s_p and s_q. We distinguish the following cases:

- Case 1: there is a modification U' of U where the last decisions of strategy s_p are replaced with some other outputs from O_p, and an infinite extension $T'' \subseteq \Sigma^\omega$ with $U' \prec T''$ that satisfies φ. In other words, the violation of φ is the fault of strategy s_p. In this case, s_p is not dominant, because the input sequences to component p that result from restricting T'' to the input variables I_p causes s_p to violate φ, while an alternative strategy, producing the outputs of T'' would satisfy φ. This alternative strategy exists, because the architecture is fully informed, and component p can, hence, distinguish the prefixes in U.
- Case 2: there is no such modification U' of U, i.e., the violation of φ is (at least also) the fault of strategy s_q. Let U'' be a modification of U where the traces are the same except for possibly different outputs of components p and q in the last position, such that there exists an infinite extension $T'' \subseteq \Sigma^\omega$ with $U'' \prec T''$ that satisfies φ. Such a modification exists, because U was chosen to be the trace set with the shortest length such that no extension exists that satisfies φ. Then, s_q is not dominant, because the input sequences to component q that result from restricting T'' to the input variables I_q causes s_q to violate φ, while an alternative strategy, producing the outputs of T'' would satisfy φ. Analogously to the first case, the alternative strategy exists, because the architecture is fully informed, and component q can, hence, distinguish the prefixes in U.

Both cases contradict the assumption that s_p and s_q are dominant. $\qquad\square$

5   Synthesis of Environment Assumptions

A dominant strategy satisfies the specification for all inputs for which it is *possible* to satisfy the specification. A side effect of this definition is that the dominant strategy identifies the inputs that make it *impossible* to satisfy the specification, i.e., the inputs that must be excluded by the environment. This observation can

$$\{\langle \sigma, \sigma' \rangle \mid d \in \sigma \leftrightarrow d \in \sigma'\}$$

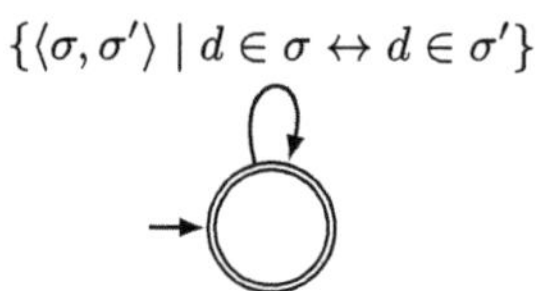

Fig. 3. Büchi automaton over alphabet $(2^{\{a,b,c,d\}})^2$. The automaton expresses the hyperproperty $\psi = \forall \pi, \pi'.\ \mathbf{G}\,(d_\pi \leftrightarrow d_{\pi'})$, which is the environment assumption for the architecture and specification from Example 1.

be used to derive a *weakest assumption* on the environment behavior; i.e., a condition that is both necessary and sufficient to guarantee that the system can satisfy the specification. For trace properties, the assumption can be expressed as the language of an automaton [7]. In this section, we generalize this result to k-hyperproperties.

We represent k-hyperproperties as sets of infinite sequences over the tuple alphabet $(2^{AP})^k$: for a given k-hyperproperty $R \subseteq (\Sigma^\omega)^k$, let

$$zip(R) = \{\langle \sigma_{1,0}, \sigma_{2,0}, \ldots, \sigma_{k,0}\rangle \langle \sigma_{1,1}, \sigma_{2,1}, \ldots, \sigma_{k,1}\rangle, \ldots$$
$$\mid \langle \sigma_{1,0}\sigma_{1,1}\ldots,\ \sigma_{2,0}\sigma_{2,1}\ldots,\ \ldots,\ \sigma_{k,0}\sigma_{k,1}\ldots\rangle \in R\}.$$

Using this representation, we can recognize a k-hyperproperties as the language of a Büchi automaton. A *Büchi automaton* $\mathcal{B} = (Q, \Sigma, q_0, F, \delta)$ is an automaton over infinite words. A run of $\mathcal{B}$ on an infinite word $w = w(0)w(1)\cdots \in \Sigma^\omega$ is an infinite sequence $r = q(0)q(1)\cdots \in Q^\omega$ with $q(0) = q_0$ and $q(i+1) \in \delta(q(i), w(i))$ for all $i \in \mathbb{N}$. A run r is accepting if there exist infinitely many $i \in \mathbb{N}$ such that $q(i) \in F$. We use a Büchi automaton $\mathcal{A}$ over the alphabet $(2^{AP})^k$ to represent the k-hyperproperty $R' \subseteq ((2^{AP})^k)^\omega$ with $\mathcal{L}(\mathcal{A}) = R'$.

Consider again Example 1, with $I_p = \{a\}$, $O_p = \{b\}$, $I_q = \{c\}$, $O_q = \{d\}$, $O_e = \{a, c\}$, and HyperLTL specification

$$\varphi_1 = \forall \pi, \pi'.\ \mathbf{G}\,((b_\pi \leftrightarrow b_{\pi'}) \wedge (d_\pi \leftrightarrow d_{\pi'})).$$

Let s_p be the dominant strategy for component p that constantly sets $b = true$. Then the environment assumption is the specification

$$\psi = \forall \pi, \pi'.\ \mathbf{G}\,(d_\pi \leftrightarrow d_{\pi'})$$

which is expressed by the Büchi automaton in Fig. 3. The proof of the following theorem gives a general recipe for the construction of the environment assumption.

Theorem 3. *For a k-hyperproperty given as an HyperLTL formula*

$$\varphi = \forall \pi_1, \forall \pi_2 \ldots \forall \pi_k.\ \phi$$

where θ is a quantifier-free formula, an architecture $(I_p, I_q, O_p, O_q, O_e)$, and a dominant finite-state strategy $s_p : I_p^* \to O_p$, there exists a Büchi automaton $\mathcal{B}$ over alphabet $(2^{AP})^k$ such that for any implementation (s_p', s_q'), the following holds:

1. $traces(s'_p, s'_q) \models \varphi$ implies that $zip(traces(s'_p, s'_q)^k) \subseteq \mathcal{L}(\mathcal{B})$, and
2. $zip((traces_{s'_q}(2^{AP \setminus O_q}))^k) \subseteq \mathcal{L}(\mathcal{B})$ implies that (s_p, s'_q) satisfies φ.

Proof. Let s_p be given as a finite-state Mealy automaton $\mathcal{A}_s = (Q_s, 2^{I_p}, 2^{O_p}, q_{s,0}, \delta_s, \lambda_s)$. We first construct the k-fold self-composition $\mathcal{A}_s^k$ of $\mathcal{A}_s$: $\mathcal{A}_s^k = (Q_s^k, (2^{I_p})^k, (2^{O_p})^k, q_{s,0}^k, \delta_s^k, \lambda_s^k)$ where

$$\delta_s^k(\langle q_1, q_2, \ldots, q_k \rangle, \langle \sigma_1, \sigma_2, \ldots, \sigma_k \rangle) = \langle \delta_s(q_1, \sigma_1), \delta_s(q_2, \sigma_2), \ldots, \delta_s(q_k, \sigma_k) \rangle,$$
$$\lambda_s^k(\langle q_1, q_2, \ldots, q_k \rangle, \langle \sigma_1, \sigma_2, \ldots, \sigma_k \rangle) = \langle \lambda_s(q_1, \sigma_1), \lambda_s(q_2, \sigma_2), \ldots, \lambda_s(q_k, \sigma_k) \rangle.$$

We then translate ϕ into a Büchi automaton $\mathcal{A}_\phi$ over the alphabet $(2^{AP})^k$, such that a sequence of tuples $(\sigma_{1,0}, \sigma_{2,1}, \ldots, \sigma_{k,1})(\sigma_{2,0}, \sigma_{2,1}, \ldots, \sigma_{k,2}) \ldots$ is accepted by $\mathcal{A}_\phi$ iff the assignment

$$\pi_1 \mapsto \sigma_{1,0}\sigma_{1,1}, \ldots, \pi_2 \mapsto \sigma_{2,0}\sigma_{2,1}, \ldots, \ldots, \pi_k \mapsto \sigma_{k,0}\sigma_{k,1}, \ldots$$

satisfies ϕ. Finally, we obtain $\mathcal{B}$ by combining $\mathcal{A}_s^k$ and $\mathcal{A}_\phi$, so that a sequence of k-tuples is accepted iff it either is rejected by $\mathcal{A}_s^k$, i.e., it does not conform to s_p, or it is accepted by $\mathcal{A}_\phi$, i.e., it satisfies ϕ. $\mathcal{B}$ thus accepts those sequences where, if s_p is applied, ϕ holds. We now argue why the two claims are correct.

1. For the first claim, let (s'_p, s'_q) be an implementation that satisfies φ, i.e., $traces(s'_p, s'_q)$ satisfies φ, and, hence, $zip(traces(s'_p, s'_q)^k) \subseteq \mathcal{L}(\mathcal{A}_\phi)$. Suppose, by way of contradiction, that there is a tuple of k traces from $traces(s'_p, s'_q)$ such that the corresponding sequence c of tuples is not in the language of $\mathcal{B}$. This means that while s'_p satisfies φ on the valuations of O_e and I_p from c, strategy s_p does not. This contradicts the assumption that s_p is a dominant strategy.

2. For the second claim, suppose, by way of contradiction, that all tuples of k traces from $(traces_{s'_q}(2^{O_e}))^k$ are, when represented as a sequence of tuples, contained in $\mathcal{L}(\mathcal{B})$, but (s_p, s'_q) does not satisfy φ, i.e., there is a counterexample of k traces from $traces(s_p, s'_q)$ that violates ϕ. Let c be a sequence of tuples corresponding to such a counterexample. Since c is in $zip(traces(s'_p, s'_q)^k)$ it is also in $\mathcal{L}(\mathcal{B})$; this, however, is impossible, because $\mathcal{L}(\mathcal{B})$ only contains sequences that satisfy ϕ when combined with the outputs according to s_p.

$\square$

6   Conclusions and Future Work

We have introduced dominant strategies for hyperproperties. As shown in Theorem 2, dominance is compositional for k-hypersafety properties and fully-informed architectures. Theorem 3 additionally shows that an environment assumption can be derived from the dominant strategy. These results lay the foundation for a compositional synthesis component similar to the one for trace properties [5,7]. The synthesis of dominant strategies for hyperproperties is, however, harder than for trace properties. Theorem 1 only establishes that the *verification* of dominance is decidable for k-hyperproperties given in HyperLTL. *Synthesis* is already undecidable for the case of winning strategies for universal hyperproperties [10]. A compositional synthesis method for hyperproperties will therefore need to rely on approaches like *bounded synthesis* [12], which introduces a bound on the number of states.

An intriguing open question for future work is how to extend dominance beyond k-hyperproperties, such as to HyperLTL formulas with quantifier alternation. This would make the approach applicable to an even larger set of hyperproperties, including information flow policies like noninference [14] and generalized noninterference [13].

Acknowledgement. This work was supported by the European Research Council (ERC) Grant HYPER (No. 101055412).

References

1. Alpern, B., Schneider, F.B.: Defining liveness. Inf. Process. Lett. **21**(4) (1985). https://doi.org/10.1016/0020-0190(85)90056-0
2. Clarkson, M.R., Finkbeiner, B., Koleini, M., Micinski, K.K., Rabe, M.N., Sánchez, C.: Temporal logics for hyperproperties. In: Proceedings of the 3rd Conference on Principles of Security and Trust (POST), pp. 265–284 (2014)
3. Clarkson, M.R., Schneider, F.B.: Hyperproperties. In: IEEE Computer Security Foundations Symposium, CSF 2008. IEEE (2008). https://doi.org/10.1109/CSF.2008.7
4. Clarkson, M.R., Schneider, F.B.: Hyperproperties. J. Comput. Secur. **18**(6), 1157–1210 (2010). https://doi.org/10.3233/JCS-2009-0393
5. Damm, W., Finkbeiner, B.: Does it pay to extend the perimeter of a world model? In: Butler, M.J., Schulte, W. (eds.) FM 2011: Formal Methods - 17th International Symposium on Formal Methods, Limerick, Ireland, June 20-24, 2011. Proceedings. Lecture Notes in Computer Science, vol. 6664, pp. 12–26. Springer (2011). https://doi.org/10.1007/978-3-642-21437-0_4
6. Damm, W., Finkbeiner, B.: Does it pay to extend the perimeter of a world model? In: Butler, M., Schulte, W. (eds.) FM 2011. LNCS, vol. 6664, pp. 12–26. Springer, Heidelberg (2011). https://doi.org/10.1007/978-3-642-21437-0_4
7. Damm, W., Finkbeiner, B.: Automatic compositional synthesis of distributed systems. In: FM 2014: Formal Methods - 19th International Symposium, Singapore, May 12-16, 2014. Proceedings. Lecture Notes in Computer Science, vol. 8442, pp. 179–193. Springer (2014). https://doi.org/10.1007/978-3-319-06410-9_13
8. Finkbeiner, B.: Logics and algorithms for hyperproperties. ACM SIGLOG News **10**(2), 4–23 (2023). https://doi.org/10.1145/3610392.3610394
9. Finkbeiner, B., Hahn, C., Hofmann, J., Tentrup, L.: Realizing ω-regular hyperproperties. In: Computer Aided Verification, pp. 40–63. Springer International Publishing, Cham (2020)
10. Finkbeiner, B., Hahn, C., Lukert, P., Stenger, M., Tentrup, L.: Synthesis from hyperproperties. Acta Informatica **57**(1-2), 137–163 (2020). https://doi.org/10.1007/s00236-019-00358-2

11. Finkbeiner, B., Passing, N.: Synthesizing dominant strategies for liveness. In: Dawar, A., Guruswami, V. (eds.) 42nd IARCS Annual Conference on Foundations of Software Technology and Theoretical Computer Science, FSTTCS 2022, December 18-20, 2022, IIT Madras, Chennai, India. LIPIcs, vol. 250, pp. 37:1–37:19. Schloss Dagstuhl - Leibniz-Zentrum für Informatik (2022). https://doi.org/10.4230/LIPIcs.FSTTCS.2022.37
12. Finkbeiner, B., Schewe, S.: Bounded synthesis. Int. J. Softw. Tools Technol. Transf. **15**(5-6), 519–539 (2013). https://doi.org/10.1007/s10009-012-0228-z
13. McCullough, D.: Noninterference and the composability of security properties. In: Proceedings of IEEE Symposium on Security and Privacy, pp. 177–186 (1988)
14. McLean, J.: A general theory of composition for trace sets closed under selective interleaving functions. In: Proceedings of IEEE Symposium on Security and Privacy, pp. 79–93 (1994)
15. Passing, N.: Compositional Synthesis of Reactive Systems. Ph.D. thesis, Saarland University (2023)
16. Pnueli, A.: The temporal logic of programs. In: FOCS 1977, pp. 46–57 (1977)
17. Rabe, M.N.: A Temporal Logic Approach to Information-flow Control. Ph.D. thesis, Saarland University (2016)
18. Zdancewic, S., Myers, A.C.: Observational determinism for concurrent program security. In: Proceedings of CSFW'03 (2003)

Author Index